AF352833

Application of Nanomedicine in Immunotherapy: Recent Advances and Prospects

Application of Nanomedicine in Immunotherapy: Recent Advances and Prospects

Editors

João Paulo Figueiró Longo
Lúis Alexandre Muehlmann

MDPI • Basel • Beijing • Wuhan • Barcelona • Belgrade • Manchester • Tokyo • Cluj • Tianjin

Editors
João Paulo Figueiró Longo
University of Brasília
Brasília, Brazil

Lúis Alexandre Muehlmann
University of Brasília
Brasília, Brazil

Editorial Office
MDPI
St. Alban-Anlage 66
4052 Basel, Switzerland

This is a reprint of articles from the Special Issue published online in the open access journal *Pharmaceutics* (ISSN 1999-4923) (available at: https://www.mdpi.com/journal/pharmaceutics/special_issues/nanomedicine_in_immunotherapy).

For citation purposes, cite each article independently as indicated on the article page online and as indicated below:

LastName, A.A.; LastName, B.B.; LastName, C.C. Article Title. *Journal Name* **Year**, *Volume Number*, Page Range.

ISBN 978-3-0365-8240-5 (Hbk)
ISBN 978-3-0365-8241-2 (PDF)

About the Editors

João Paulo Figueiró Longo

João Paulo Figueiró Longo holds a Master's degree in Molecular Pathology and a PhD in Animal Biology from the University of Brasília. Currently, they are an associate professor at the same university, working in the fields of Nanobiotechnology and Nanomedicine. Their primary research projects are related to the investigation of the effects of nanomaterials on biological systems at different levels of complexity. As a manager, they have served as the Coordinator of the Postgraduate Program, Master's and Doctorate, in Nanoscience and Nanobiotechnology. They are also the founder of the startup Glia Innovation, a company focused on the development and production of intelligent raw materials for life sciences industries. They have collaborated with the company on the development of some products.

Luis Alexandre Muahlmann

Muehlmann has been an Assistant Professor of Biochemistry at the University of Brasília, Brazil, since 2014. He graduated as a biochemist from the Federal University of Parana, Brazil, (2006), and holds a Master's degree in Cell Biology (2008) from the same university, as well as a PhD in Nanobiotechnology (2011) from the University of Brasília.

pharmaceutics

Editorial

Application of Nanomedicine in Immunotherapy: Recent Advances and Prospects

João Paulo Figueiró Longo [1,*,†] **and Luis Alexandre Muehlmann** [2,†]

1   Department of Genetics and Morphology, Institute of Biological Sciences, University of Brasília, Brasilia 70910-900, Brazil
2   Faculty of Ceilândia, University of Brasilia, Brasilia 72220-275, Brazil; luismuehlmann88@gmail.com
*   Correspondence: jplongo82@gmail.com; Tel.: +55-61-3107-3087; Fax: +55-61-81432755
†   These authors contributed equally to this work.

Citation: Longo, J.P.F.; Muehlmann, L.A. Application of Nanomedicine in Immunotherapy: Recent Advances and Prospects. *Pharmaceutics* **2023**, *15*, 1910. https://doi.org/10.3390/pharmaceutics15071910

Received: 10 June 2023
Accepted: 22 June 2023
Published: 9 July 2023

Nanomedicine is a special medical field focused on the application of nanotechnology to provide innovations for healthcare in different areas, including the treatment of a wide variety of diseases, including cancer [1,2], infections [3,4], and auto-immune disorders. The field emerged during the 1980s, aligning with the approval of the first regulatory-agency-approved nanomedical oncological drugs [2,5]. Additionally, nanotechnology has played a pivotal role in the development of mRNA vaccines utilized during the COVID-19 pandemic [6], further establishing its enduring significance in the domains of science and biomedical innovation.

The reasons for the use of nanotechnology in biomedicine can vary but are mostly the protection and/or delivery of bioactive molecules to target tissues. The general idea is to create nanoscopic platforms that can interact differently with biological systems, either through pharmacokinetic modifications or through the preferential activation of some biological pathway [5,7].

The intricate biomolecular interactions that underlie the functioning of the biological system take place within the nanoscale. So, nanoparticles can be designed to be similar in size to key biological structures like large biomolecules or small organelles, enabling them to interact with biological systems in unique ways, which can improve the effects of drugs used in the treatment of certain diseases. In other words, highly distinct and improved biological effects can emerge from such interactions [8,9].

For this Special Issue, we sought submissions specifically focusing on the utilization of nanomedicine in various types of immunotherapies. In terms of quantity, a total of ten papers were published, six of them original articles and four review articles. The contributions were made by a diverse group of sixty-nine authors hailing from eight countries: Austria, Brazil, China, France, Italy, Japan, Korea, and the United States of America. Currently, the published articles have been accessed by nearly twenty thousand readers.

Regarding the topics, Chen et al. (2023) [10], Sasaki et al. (2022) [11], and Lodeserto et al. (2022) [12] applied nanotechnology to deliver different types of active molecules to target. Chen et al. (2023) [10] reported the development of an innovative polypeptide hydrogel to deliver immune checkpoint inhibitors and doxorubicin to improve tumor immunotherapy. As the main results, the authors observed a significant reduction in tumor recurrence by using this immunotherapy strategy. Following the delivery strategy, Sasaki et al. (2022) [11] described the use of lipid nanoparticles to deliver mRNA for dendritic cells. With this strategy, the authors showed a therapeutic antitumor effect against a pre-clinical tumor model, and, Lodeserto et al. (2022) [12] reported the preparation of a spermidine nanocarrier to target specific cells. The authors showed that spermidine can activate immune cells, and this feature is related to immunosuppression reversion in the tumor environment.

Following an immune activation strategy, Punz et al. (2022) [13] proposed the utilization of silica nanoparticles as platforms for delivering immune activation molecules.

The authors demonstrated the potential of these platforms in modulating antigen uptake and promoting the maturation of antigen-presenting cells. Following a similar immune activation direction, Rodrigues et al. (2022) [14] proposed the use of phthalocyanine nanoemulsions to induce immunogenic cell death and exploit these cells as a platform for immune activation. Through this approach, the authors demonstrated that PDT-treated cancer cells used as a vaccine hinder tumor growth in pre-clinical models.

Following the opposite direction, Lasola et al. (2021) [15] published an intriguing paper assessing the impact of various nanoparticles' surface chemistry on mitigating undesired immune system activation. They aimed to observe an anti-inflammatory immune response. This approach holds potential for employing nanomaterials in the management of severe inflammatory diseases, such as sepsis.

Regarding the review articles included in this Special Issue, there were a total of four publications. Dias et al. (2022) [16] reported an interesting review collating information related to the use of iron oxide for immunotherapy for oncology applications. Rodrigues et al. (2022) [17] and Seong and Kim (2022) [18] published two reviews reporting the use of cell lysates to trigger the immune system. Rodrigues focused on the aspects related to immunogenic cell death, while Seong included more general cell lysates, including necrotic-induced cell death protocols. Lastly, Nigam et al. (2022) [19] reported an important review presenting an update on the use of nanotechnology in the immunotherapy of type 1 diabetes.

In summary, this Special Issue presents a compilation of groundbreaking research papers and comprehensive reviews that delve into various dimensions of immunotherapy and nanomedicine. The invaluable insights shared in this publication offer profound insights into the potential of nanotechnology to enhance the efficacy of diverse immunotherapeutic modalities. We are confident that the wealth of high-quality information disseminated in this Issue will captivate the attention of both researchers and individuals seeking alternative cancer treatments, particularly those intrigued by the prospects of nanotechnological approaches for immunotherapies.

Author Contributions: J.P.F.L. and L.A.M. collaborated in the creation of the article. J.P.F.L. focused on crafting the initial paragraphs, providing an overview of Nanomedicine and Immunotherapy. On the other hand, L.A.M. worked on developing the highlight paragraphs for the articles. All authors have read and agreed to the published version of the manuscript.

Funding: This article is part of two research projects supported by the Brazilian agency FAP-DF/Brazil (00193.00001053/2021-24)/(00193-00000734/2021-75), CNPq (403536/2021-9)/(303182/2019-9), and Coordenação de Aperfeiçoamento de Pessoal de Nível Superior—Brazil (CAPES)—Finance Code 001.

Conflicts of Interest: The authors declare no conflict of interest.

References

1. Faria, R.S.; de Lima, L.I.; Bonadio, R.S.; Longo, J.P.F.; Roque, M.C.; de Matos Neto, J.N.; Moya, S.E.; de Oliveira, M.C.; Azevedo, R.B. Liposomal paclitaxel induces apoptosis, cell death, inhibition of migration capacity and antitumoral activity in ovarian cancer. *Biomed. Pharmacother.* **2021**, *142*, 112000. [CrossRef] [PubMed]
2. Longo, J.P.F.; Muehlmann, L.A.; Calderón, M.; Stockmann, C.; Azevedo, R.B. Nanomedicine in Cancer Targeting and Therapy. *Front. Oncol.* **2021**, *11*, 788210. [CrossRef] [PubMed]
3. Mengarda, A.C.; Iles, B.; Longo, J.P.F.; de Moraes, J. Recent trends in praziquantel nanoformulations for helminthiasis treatment. *Expert Opin. Drug Deliv.* **2022**, *19*, 383–393. [CrossRef] [PubMed]
4. Mengarda, A.C.; Iles, B.; Longo, J.P.F.; de Moraes, J. Recent approaches in nanocarrier-based therapies for neglected tropical diseases. *Wiley Interdiscip. Rev. Nanomed. Nanobiotechnol.* **2022**, *15*, e1852. [CrossRef] [PubMed]
5. Figueiro Longo, J.P.; Muehlmann, L.A. Nanomedicine beyond tumor passive targeting: What next? *Nanomedicine* **2020**, *15*, 1819–1822. [CrossRef] [PubMed]
6. Figueiró Longo, J.P.; Muehlmann, L.A. How has nanomedical innovation contributed to the COVID-19 vaccine development? *Nanomedicine* **2021**, *16*, 1179–1181. [CrossRef] [PubMed]
7. Anselmo Joanitti, G.; Ganassin, R.; Correa Rodrigues, M.; Paulo Figueiro Longo, J.; Jiang, C.-S.; Gu, J.; Milena Leal Pinto, S.; de Fatima, M.A.d.S.; Bentes de Azevedo, R.; Alexandre Muehlmann, L. Nanostructured Systems for the Organelle-specific Delivery of Anticancer Drugs. *Mini Rev. Med. Chem.* **2017**, *17*, 224–236. [CrossRef] [PubMed]

8. Coelho, J.M.; Camargo, N.S.; Ganassin, R.; Rocha, M.C.O.; Merker, C.; Böttner, J.; Estrela-Lopis, I.; Py-Daniel, K.R.; Jardim, K.V.; Sousa, M.H. Oily core/amphiphilic polymer shell nanocapsules change the intracellular fate of doxorubicin in breast cancer cells. *J. Mater. Chem. B* **2019**, *7*, 6390–6398. [CrossRef] [PubMed]

9. Ganassin, R.; Horst, F.H.; Camargo, N.S.; Chaves, S.B.; Morais, P.C.; Mosiniewicz-Szablewska, E.; Suchocki, P.; Figueiro Longo, J.P.; Azevedo, R.B.; Muehlmann, L.A. Selol nanocapsules with a poly (methyl vinyl ether-co-maleic anhydride) shell conjugated to doxorubicin for combinatorial chemotherapy against murine breast adenocarcinoma in vivo. *Artif. Cells Nanomed. Biotechnol.* **2018**, *46* (Suppl. 2), 1046–1052. [CrossRef] [PubMed]

10. Chen, Z.; Rong, Y.; Ding, J.; Cheng, X.; Chen, X.; He, C. Injectable Polypeptide Hydrogel Depots Containing Dual Immune Checkpoint Inhibitors and Doxorubicin for Improved Tumor Immunotherapy and Post-Surgical Tumor Treatment. *Pharmaceutics* **2023**, *15*, 428. [CrossRef] [PubMed]

11. Sasaki, K.; Sato, Y.; Okuda, K.; Iwakawa, K.; Harashima, H. mRNA-Loaded Lipid Nanoparticles Targeting Dendritic Cells for Cancer Immunotherapy. *Pharmaceutics* **2022**, *14*, 1572. [CrossRef] [PubMed]

12. Lodeserto, P.; Rossi, M.; Blasi, P.; Farruggia, G.; Orienti, I. Nanospermidine in Combination with Nanofenretinide Induces Cell Death in Neuroblastoma Cell Lines. *Pharmaceutics* **2022**, *14*, 1215. [CrossRef] [PubMed]

13. Punz, B.; Johnson, L.; Geppert, M.; Dang, H.-H.; Horejs-Hoeck, J.; Duschl, A.; Himly, M. Surface Functionalization of Silica Nanoparticles: Strategies to Optimize the Immune-Activating Profile of Carrier Platforms. *Pharmaceutics* **2022**, *14*, 1103. [CrossRef] [PubMed]

14. Rodrigues, M.C.; de Sousa Júnior, W.T.; Mundim, T.; Vale, C.L.C.; de Oliveira, J.V.; Ganassin, R.; Pacheco, T.J.A.; Vasconcelos Morais, J.A.; Longo, J.P.F.; Azevedo, R.B. Induction of Immunogenic Cell Death by Photodynamic Therapy Mediated by Aluminum-Phthalocyanine in Nanoemulsion. *Pharmaceutics* **2022**, *14*, 196. [CrossRef] [PubMed]

15. Lasola, J.J.M.; Cottingham, A.L.; Scotland, B.L.; Truong, N.; Hong, C.C.; Shapiro, P.; Pearson, R.M. Immunomodulatory Nanoparticles Mitigate Macrophage Inflammation via Inhibition of PAMP Interactions and Lactate-Mediated Functional Reprogramming of NF-κB and p38 MAPK. *Pharmaceutics* **2021**, *13*, 1841. [CrossRef] [PubMed]

16. Dias, A.M.M.; Courteau, A.; Bellaye, P.-S.; Kohli, E.; Oudot, A.; Doulain, P.-E.; Petitot, C.; Walker, P.-M.; Decréau, R.; Collin, B. Superparamagnetic Iron Oxide Nanoparticles for Immunotherapy of Cancers through Macrophages and Magnetic Hyperthermia. *Pharmaceutics* **2022**, *14*, 2388. [CrossRef] [PubMed]

17. Rodrigues, M.C.; Morais, J.A.V.; Ganassin, R.; Oliveira, G.R.T.; Costa, F.C.; Morais, A.A.C.; Silveira, A.P.; Silva, V.C.M.; Longo, J.P.F.; Muehlmann, L.A. An Overview on Immunogenic Cell Death in Cancer Biology and Therapy. *Pharmaceutics* **2022**, *14*, 1564. [CrossRef]

18. Seong, J.; Kim, K. Activation of Cellular Players in Adaptive Immunity via Exogenous Delivery of Tumor Cell Lysates. *Pharmaceutics* **2022**, *14*, 1358. [CrossRef] [PubMed]

19. Nigam, S.; Bishop, J.O.; Hayat, H.; Quadri, T.; Hayat, H.; Wang, P. Nanotechnology in Immunotherapy for Type 1 Diabetes: Promising Innovations and Future Advances. *Pharmaceutics* **2022**, *14*, 644. [CrossRef] [PubMed]

pharmaceutics

Article

Injectable Polypeptide Hydrogel Depots Containing Dual Immune Checkpoint Inhibitors and Doxorubicin for Improved Tumor Immunotherapy and Post-Surgical Tumor Treatment

Zhixiong Chen [†], Yan Rong [†], Junfeng Ding, Xueliang Cheng, Xuesi Chen and Chaoliang He *

CAS Key Laboratory of Polymer Ecomaterials, Changchun Institute of Applied Chemistry, Chinese Academy of Sciences, Changchun 130022, China
* Correspondence: clhe@ciac.ac.cn; Tel.: +86-431-8526-2901
† These authors contributed equally to this work.

Abstract: In this work, we developed a strategy for local chemo-immunotherapy through simultaneous incorporation of dual immune checkpoint blockade (ICB) antibodies, anti-cytotoxic T-lymphocyte-associated protein 4 (aCTLA-4) and anti-programmed cell death protein 1 (aPD-1), and a chemotherapy drug, doxorubicin (Dox), into a thermo-gelling polypeptide hydrogel. The hydrogel encapsulating Dox or IgG model antibody showed sustained release profiles for more than 12 days in vitro, and the drug release and hydrogel degradation were accelerated in the presence of enzymes. In comparison to free drug solutions or hydrogels containing Dox or antibodies only, the Dox/aCTLA-4/aPD-1 co-loaded hydrogel achieved improved tumor suppression efficiency, strengthened antitumor immune response, and prolonged animal survival time after peritumoral injection into mice bearing B16F10 melanoma. Additionally, after injection of Dox/aCTLA-4/aPD-1 co-loaded hydrogel into the surgical site following tumor resection, a significantly enhanced inhibition on tumor reoccurrence was demonstrated. Thus, the polypeptide hydrogel-based chemo-immunotherapy strategy has potential in anti-tumor therapy and the prevention of tumor reoccurrence.

Keywords: hydrogel; combination therapy; immunotherapy; polypeptide; ICB therapy

Citation: Chen, Z.; Rong, Y.; Ding, J.; Cheng, X.; Chen, X.; He, C. Injectable Polypeptide Hydrogel Depots Containing Dual Immune Checkpoint Inhibitors and Doxorubicin for Improved Tumor Immunotherapy and Post-Surgical Tumor Treatment. *Pharmaceutics* **2023**, *15*, 428. https://doi.org/10.3390/pharmaceutics15020428

Academic Editors: João Paulo Figueiró Longo and Lúis Alexandre Muehlmann

Received: 26 December 2022
Revised: 14 January 2023
Accepted: 19 January 2023
Published: 28 January 2023

1. Introduction

Malignant tumor is a global threat to the health and life of humans [1]. The commonly used clinical treatment approaches include surgical resection, chemotherapy, radiotherapy, and immunotherapy [2,3]. Immunotherapy is a rapidly developing cancer treatment technology recently, including immune checkpoint blocking agents, immune adjuvants, chimeric antigen receptor T (CAR-T) cells, and cancer vaccines [3]. Immune checkpoints are key regulators for immunological tolerance, which can protect tumors from the recognition and attack of the immune system. In the past decade, immune checkpoint blockade (ICB) therapy has witnessed great progress in cancer therapy [4–6]. Cytotoxic T-lymphocyte-associated protein 4 (CTLA-4) and programmed cell death protein 1 (PD-1) are typical immune checkpoint transmembrane proteins that play key roles in different immunosuppressive processes. CTLA-4, a transmembrane protein expressed on T cells, is able to inhibit antigen presentation from antigen-presenting cells (APCs) to T cells and, therefore, restrain successive activation of T cells, through competitively binding with B7 molecules on APCs [7]. In the effector stage, the tumor recognition and killing process by effector T cells can be inhibited by the binding of PD-1 on T cells to ligands (PD-L1 and PD-L2) on tumor cells. Thus, the ICB therapy based on antibodies that block the CTLA-4 and/or PD-1 pathways has achieved remarkable clinical outcomes recently in the treatment of several types of cancers [8].

It is worth mentioning that the anti-cancer efficiency of ICB therapy is still limited by some major shortcomings, especially the insufficient objective response rates (ORRs) and serious immune-related adverse events (irAEs) [8]. It has been found that the cooperative inhibition of the CTLA-4 and PD-1/PD-L1 pathways with anti-CTLA-4 (aCTLA-4) and anti-PD-1 (aPD-1) antibodies lead to significantly enhanced ORRs [9–19]. In addition, it has been shown that the anti-tumor immune response can be improved through the combination with other therapy approaches, such as chemotherapy and radiation therapy [20,21]. For instance, several chemotherapeutics such as anthracyclines, oxaliplatin and cyclophosphamide can induce immunogenic cell death (ICD) of tumor cells, leading to the generation of tumor-associated antigens (TAAs) and damage-associated molecular patterns (DAMPs) [22]. Subsequently, the release of TAAs and DAMPs can stimulate an antigen-specific immune response against tumors.

On the other hand, to reduce systemic irAEs of ICB therapy, localized delivery systems based on injectable hydrogels have attracted considerable attention in recent years [23–26]. Injectable, biodegradable hydrogels have shown several advantages as topical delivery systems, including facile encapsulation of both small-molecule drugs and biomacromolecules, minimally invasive drug administration, good biocompatibility, as well as prolonged drug release behavior at targeted sites [27,28]. After the injection of drug-loaded hydrogels to the tumor sites, the ICB antibodies and combined drugs can be released locally and sustainedly, which may lead to persistently elevated drug concentration at the tumor sites while reducing systemic side effects [29,30]. Thus, several hydrogel systems loaded with an ICB antibody and a chemotherapy drug have been developed recently for improved antitumor chemo-immunotherapy [31–36]. Nevertheless, a study on the localized co-delivery of dual or multiple ICB antibodies and chemotherapeutics with injectable hydrogels has not been reported yet.

In this work, an injectable drug depot based on a thermosensitive polypeptide hydrogel was developed for the topical co-delivery of aCTLA-4, aPD-1 and an anthracycline chemotherapy drug, doxorubicin (Dox), for antitumor chemo-immunotherapy (Scheme 1). As the hydrogel precursor, the methoxy poly(ethylene glycol)-*block*-poly (γ-ethyl-L-glutamate) (mPEG-PELG) aqueous solutions presented sol–gel phase transitions with the temperature rising from room temperature to 37 °C. This property facilitated the mixing of antitumor agents with the solutions at a low temperature, and spontaneous formation of drug-loaded hydrogels at the physiological temperature. The sustained release of Dox and a model antibody from the hydrogel was investigated in PBS with or without proteinase K. The capacity of the Dox-loaded hydrogel to cause ICD in B16F10 melanoma cells was revealed by testing the expression of calreticulin (CRT). The antitumor efficacy and systemic side effects of the aCTLA-4/aPD-1/Dox co-loaded hydrogel in vivo were evaluated by peritumoral injection of the hydrogel into C57BL/6 mice bearing B16F10 melanoma. The immune response of the treatment was investigated by analyzing immune cells and pro-inflammatory cytokines. Additionally, the inhibition of tumor reoccurrence after tumor resection surgery by the treatment of the hydrogel depot was further studied in a tumor resection model of melanoma-bearing mice.

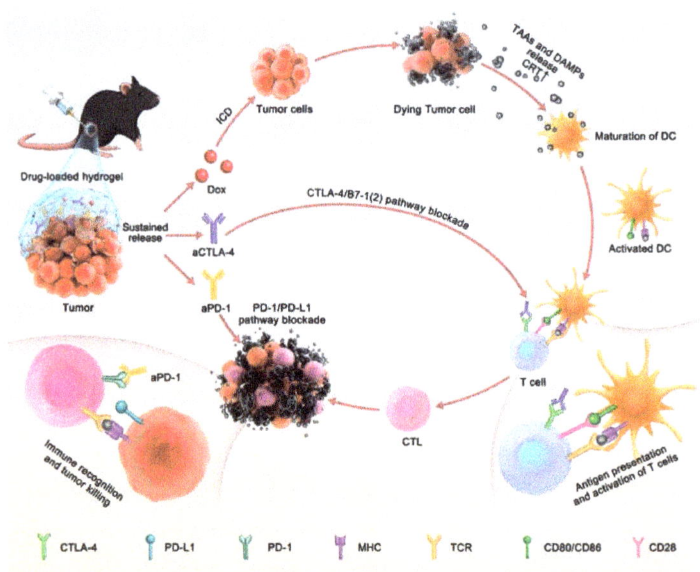

Scheme 1. Schematic illustration of sustained co-delivery of Dox, aCTLA-4 and aPD-1 by an injectable hydrogel for antitumor chemo-immunotherapy. After peritumoral injection of the multiple agent co-loaded hydrogel into tumor-bearing mice, the Dox release can induce ICD of tumor cells to release DAMPs and TAAs, promoting the maturation of dendritic cells (DCs). The subsequent antigen presentation by DCs to T cells can be enhanced by inhibiting the binding between CTLA-4 and CD80/86, with aCTLA-4. Additionally, the continuous release of aPD-1 can block the PD-1/PD-L1 pathway, thereby inhibiting the immune escape of tumor cells.

2. Materials and Methods

2.1. Materials

Amino-terminated monomethyl poly(ethylene glycol) (mPEG-NH$_2$, M_n = 2000) was bought from Jemkem Inc., China. Tetrahydrofuran (THF), N,N-dimethylformamide (DMF) and toluene were refluxed with CaH$_2$ and distilled under N$_2$ before use. γ-Ethyl-L-glutamate N-carboxyanhydride (ELG NCA) was prepared according to our previous method [37]. All the other chemical reagents were bought from Sinopharm Chemical Reagent Co., Ltd. (Shanghai, China), China and used as obtained.

InvivoMab anti-mouse PD-1 (CD279) and invivoMab anti-mouse CTLA-4 (CD152) were purchased from Bioxcell Inc. (West Lebanon, NH, USA). Dox was purchased from Beijing HVSF United Chemical Materials Co., Ltd. (Beijing, China). CD3-FITC, CD4-PE and CD8-APC antibodies were bought from Biolegend Inc (San Diego, CA, USA). Treg cell test kit was bought from Ebioscience Inc. (San Diego, CA, USA). ELISA kits for the detection of IL-2, TNF-α and IFN-γ were obtained from Shanghai Lengton Bioscience Co., Ltd. (Shanghai, China).

2.2. Characterization

^{1}H NMR spectra of mPEG-PELG solutions in deuterated trifluoroacetic acid (CF$_3$COOD) were recorded on a Bruker AV 500 NMR spectrometer (Bruker Optik GmbH, Ettlingen, Germany). Number-average molecular weight (M_n) and polydispersity index (PDI) were examined by gel permeation chromatography (GPC), which was equipped with 2 Styragel® HMW 6E columns (7.8 mm * 300 mm) and Waters 515 HPLC pump with a Waters 2414 refractive index detector (Waters, Milford, MA, USA). The eluant was DMF containing 0.05 M LiBr at a flow rate of 1.0 mL/min at 50 °C. Monodispersed poly(methyl methacrylate) standards were used to generate the calibration curve. The ellipticity of polymer aqueous solution (0.05 wt%) was obtained on a Chirascan CD spectrometer (Applied Photophysics, Leatherhead, UK) as a function of temperature in the range of 10–60 °C. The microstructure of freeze-dried hydrogel sample was observed by field emission scanning electron microscopy (SEM, Gemini 2, Carl Zeiss, Oberkochen, Germany).

2.3. Synthesis of mPEG-PELG

The mPEG-PELG block copolymer was synthesized via ring-opening polymerization (ROP) of γ-ethyl-L-glutamate N-carboxy anhydride (ELG NCA) using mPEG-NH$_2$ as the macroinitiator (Scheme S1) [37]. Briefly, mPEG-NH$_2$ (2 g, 1.0 mmol) was dissolved in toluene (150 mL) and underwent azeotropic distillation to remove the trace water. Anhydrous DMF (50 mL) and ELG NCA (3.0 g, 15.0 mmol) were then added, and the mixed solution was allowed to stir under N$_2$ atmosphere at room temperature for 3 days. After the reaction, the crude product was obtained by precipitation into cold diethyl ether and filtration. The product was then re-dissolved in DMF, purified by dialysis against deionized water for 3 days, and collected by lyophilization.

2.4. Phase Diagram

The sol–gel transition phase diagram of mPEG-PELG solutions was measured by the tube inversion method. The copolymer was dissolved using phosphate-buffered saline (PBS, 0.01 M, pH 7.4) at a given concentration in test tube with an inner diameter of 11 mm, and stirred at 0 °C for 72 h. The sol–gel status of the solution was observed with increasing the temperature at a rate of 2 °C per 10 min. The gelation temperature was determined when the solution kept no flow within 30 s after inverting the test tube.

2.5. Rheological Test

Rheological experiments were tested by placing 300 µL of copolymer solution between the plates with a diameter of 25 mm and a gap of 0.3 mm on the MCR 301 Rheometer (Anton Paar GmbH, Graz, Austria). A thin layer of silicone oil was used to seal the outer edge of the sample for preventing water evaporation. The data were tested with a fixed strain of 1% and a frequency of 1 Hz. The heating rate was set as 0.5 °C/min.

2.6. In Vitro Gel Degradation

The copolymer was dissolved in PBS at a concentration of 6 wt% and the solution was allowed to stir at 0 °C for 72 h. Then, 300 µL copolymer solution was placed into a glass vial of 11 mm in inner diameter and incubated at 37 °C for 10 min to obtain a hydrogel sample. Subsequently, 3 mL PBS with or without 5 U/mL proteinase K was added into the vials as the degradation medium and the hydrogels were incubated in an oscillating incubator (60 rpm) at 37 °C. At each preset time point, the solution was removed by gentle suction out from the vial with a pipette and the residual solution at the surface of the sample was further removed by a filter paper. The remaining mass of the hydrogel was weighed. Fresh medium was added to the vial. Three parallel samples were tested at each time point.

2.7. In Vitro Drug Release

The copolymer was dissolved in PBS with a concentration of 6 wt% and the solution was allowed to stir at 0 °C for 36 h. Dox or IgG was added to the copolymer solution, with a Dox or IgG concentration of 1 mg/mL. The solution was stirred for an additional 12 h. Then, 300 µL mixture was placed into a glass vial of 11 mm in inner diameter and incubated at 37 °C for 10 min to obtain a hydrogel sample. Following this, 3 mL PBS with or without proteinase K (5 U/mL) was added into each vial as the release medium, and all the vials were incubated in an oscillating incubator (60 rpm) at 37 °C. At a given time point, the release medium was collected to determine the amounts of released Dox and IgG, and an equal volume of fresh PBS with or without proteinase K was supplemented. The amount of Dox was quantified by a fluorophotometer with excitation and emission wavelengths of 480 nm and 592 nm, respectively. The procedure of the IgG content measurement experiment followed the instructions given by the IgG ELISA Kit (Invitrogen, Carlsbad, CA, USA). Three parallel samples were tested at each time point.

2.8. In Vivo Gel Degradation

Sprague Dawley rats were anesthetized and each rat was injected subcutaneously with 300 μL 6 wt% mPEG-PELG solution to form hydrogels in situ. The rats were sacrificed at predetermined intervals and dissected for observing the degradation grade of hydrogels.

2.9. ICD Assay

B16F10 cells were incubated in 6-well plates (3×10^5/well) for 12 h, with 3 mL complete medium in each well. Then, 300 μL mPEG-PELG hydrogel loaded with Dox (5 μg/mL) was added and incubated for another 4 h. The same dose of Dox solution was used as the positive control. After digestion with 1 mL trypsin, the cells were harvested and washed, and incubated with 100 μL PBS plus 2 μL Alexa Fluor® 488 calreticulin polyclonal antibody solution (Abcam, Cambridge, UK) for 1 h. Finally, the treated cells were detected and analyzed by flow cytometry (BD FASCCelesta™ Flow Cytometer, BD Biosciences, San Jose, CA, USA).

2.10. In Vivo Antitumor Tests

B16F10 cells (1×10^6) were injected subcutaneously into each C57BL/6 mouse to establish a B16F10-melanoma bearing mice model to evaluate the antitumor efficiency in vivo. The tumor-bearing mice were randomly divided into 7 groups one week after inoculation of melanoma cells ($n = 5$), as follows: PBS-treated group (denoted as PBS), free aCTLA-4 + aPD-1 mixed solution group (aCTLA-4 + aPD-1), aCTLA-4/aPD-1 co-loaded hydrogel group (aCTLA-4 + aPD-1)@gel, Dox solution group (Dox), Dox-loaded hydrogel group (Dox@gel), free Dox + aCTLA-4 + aPD-1 solution group (Dox + aCTLA-4 + aPD-1), and Dox/aCTLA-4/aPD-1 co-loaded hydrogel group (Dox + aCTLA-4 + aPD-1)@gel. The volume of paratumorally injected solution in each group was 50 μL. The dosages of Dox, aCTLA-4 and aPD-1 were controlled as 4 mg/kg, 1 mg/kg and 2 mg/kg, respectively, in the formulations containing each ingredient. The average melanoma volume and body weight were recorded and calculated in each group at predetermined time points. The melanoma volume was calculated according to the following equation: $V\ (\text{mm}^3) = L \times W^2/2$, where L (mm) means the length of melanoma and W (mm) represents the width of melanoma. The mice were euthanized when the volume of tumor was higher than 1500 mm^3, except those that died naturally. The survival time for all the treated groups was investigated in a separate study ($n = 7$).

2.11. Tumor Reoccurrence Inhibition after Tumor Resection

To evaluate the tumor reoccurrence inhibition efficacy after tumor resection, the B16F10 melanoma-bearing mice model was established in a similar procedure. After 12 days of melanoma incubation, tumor-bearing C57BL/6 mice with an average tumor volume of around 300–400 mm^3 were chosen for establishing the tumor-resection model. The mice were randomly divided into three groups with 12 mice in each group, including PBS group, free Dox + aCTLA-4 + aPD-1 mixed solution group (Dox + aCTLA-4 + aPD-1) and Dox/aCTLA-4/aPD-1 co-loaded hydrogel group (Dox + aPD-1 + aCTLA-4)@gel. After being anesthetized with 100 μL pentobarbital (1%), each melanoma-bearing mouse was excised ~90% of melanoma, surgically sutured and subsequently treated with subcutaneous injection of the given formulation at the surgical site. The volume of hydrogel and the dosages of drugs were the same as before. The volumes of melanoma and body weights of the animals were recorded at predetermined time points. At day 12, five mice in each group were randomly selected for analysis of immune cells, and the rest of the mice were continuously monitored for evaluation of survival time ($n = 7$).

2.12. Cytokines Assay

The serum from C57BL/6 mice of each group was collected, and the immune related cytokines including IL-2, TNF-α, IFN-γ were tested according to the specifications of the corresponding assay kit.

2.13. Immune Cell Assay

Spleens, lymph nodes and tumors were collected after sacrificing the animals and processed into single cell suspensions for analysis of CD4$^+$ cell, CD8$^+$ cell, Treg. The whole procedure of the mentioned immune cell processing was carried out according to the specifications of the test kit. Analysis of the stained immune cells was performed by flow cytometry.

2.14. Organ Damage Assay

To evaluate the safety of the treatments, the main organs, such as the lung, kidney, heart, liver and spleen, were collected after the mice were sacrificed. These organs were fixed by 4% paraformaldehyde for H&E staining analysis.

2.15. Animal Procedure

All animal procedures were conducted in accordance with the Guidelines for Care and Use of Laboratory Animals of Jilin University and approved by the Animal Ethics Committee of Changchun Institute of Applied Chemistry, CAS (No. 2020-010).

2.16. Statistical Analysis

All values were presented as mean $\pm$ standard deviation. Statistical analysis was performed by one-way ANOVA using the LSD or Tukey posttest for multiple comparison. Survival periods of animals were presented using a Kaplan$-$Meier curve, and analyzed by the log-rank test. Statistical significance is indicated as * $p < 0.05$, ** $p < 0.01$, and *** $p < 0.001$.

3. Results and Discussion

According to our previously reported method [37], mPEG-PELG was prepared via the ROP of ELG-NCA using mPEG-NH$_2$ as the macroinitiator. The chemical structure and molecule weight of the obtained block copolymer were identified by ^{1}H NMR and GPC, respectively (Figure S1). The degree of polymerization (DP) of mPEG-PELG was estimated as approximately 18 by comparing the integration of the peak at 0.82 ppm ascribed to the pendant methyl group of ELG residues with that at 3.09 ppm assigned to the terminal methyl group of mPEG in the ^{1}H NMR spectrum. Additionally, the monomodal distribution of the resulting copolymer in the GPC trace confirmed the successful synthesis of the diblock copolymer. The M_n was determined as 4100 Da with a PDI of 1.2 according to GPC data.

It was found that the mPEG-PELG aqueous solutions showed a sol–gel phase transition with increasing temperature, which was dependent on the polymer concentration. To investigate the gelation temperature range of the mPEG-PELG aqueous solutions, the thermo-induced sol–gel phase transitions of the polymer solutions were tested. As shown in Figure 1A, the mPEG-PELG solutions in PBS with polymer concentrations of 4~8 wt% exhibited sol–gel phase transitions with the increase in temperature, and the gelation temperature declined from 38 °C to 24 °C as the polymer concentration increased from 4 wt% to 8 wt% (Figures 1B and S2). The thermo-induced gelation of the mPEG-PELG solutions may be owing to the cooperative effects of partial dehydration of mPEG blocks and the maintained β-sheet conformation at elevated temperatures (Figure S3) [38,39]. The partial dehydration of mPEG segments may result in enhanced chain entanglement and aggregation of the mPEG-PELG micelles in aqueous solution, and the β-sheet conformation of the polypeptide blocks facilitates the formation of intermolecular hydrogen bonding. These effects promote the formation of a physical crosslinking network and hydrogel.

Considering the suitable gelation temperature (~30 °C), the 6 wt% mPEG-PELG solution in PBS was chosen for fabrication of the injectable hydrogel. Based on the rheological test, the 6 wt% mPEG-PELG solution showed an abrupt increase in the storage modulus (G′) with increasing temperature over 20 °C, indicating the hydrogel formation (Figure 1C). Moreover, the mixing of Dox and a model antibody, IgG, with the mPEG-PELG solution showed no obvious influence on the variations of G′ and G″ with the temperature increase, indicating that the Dox and antibody encapsulation did not markedly affect the thermo-induced gelation behavior of the polypeptide solution. Additionally, the freeze-dried hydrogel sample indicated a porous structure in the SEM image (Figure 1D), which may facilitate the encapsulation and transportation of the drugs and bioactive agents [27,28].

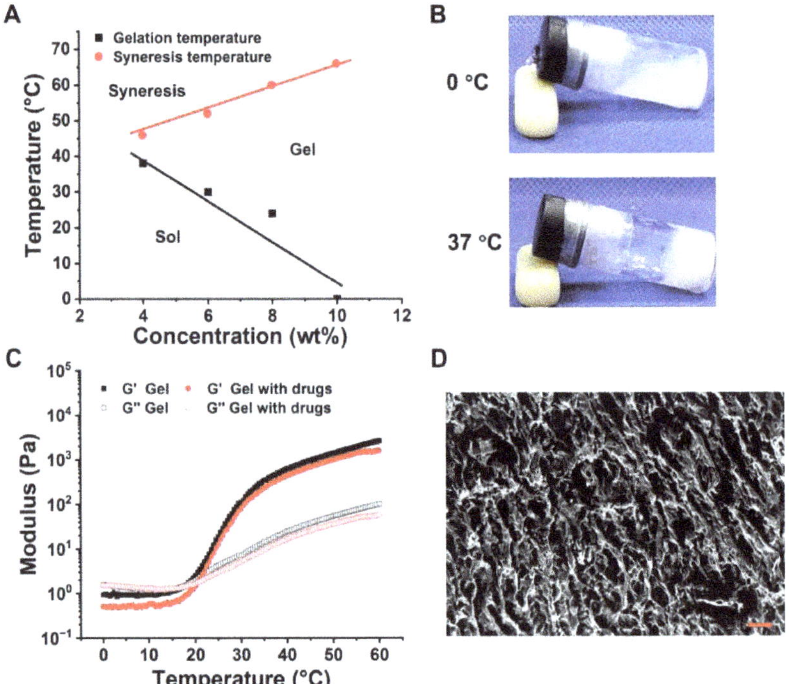

Figure 1. (**A**) Temperature-dependent sol–gel phase diagram of mPEG-PELG hydrogel. (**B**) Photographs of 6 wt% mPEG-PELG solution in PBS at 0 °C and the hydrogel formation at 37 °C. (**C**) Temperature-dependent variations in the storage modulus (G′) and loss modulus (G″) of the 6 wt% blank mPEG-PELG hydrogel or Dox/IgG-loaded hydrogel (Dox: 1 mg/mL; IgG: 1 mg/mL). (**D**) SEM image of lyophilized mPEG-PELG hydrogel (6 wt%). Scale bar = 20 μm.

The degradation of hydrogels plays an important role in affecting the release behaviors of encapsulated agents. The degradation behavior of the mPEG-PELG hydrogels was evaluated in vitro and in vivo, respectively. As shown in Figure 2A, the hydrogel showed ~30% mass loss in 15 days in PBS, which may be mainly due to the surface erosion of the hydrogel. When proteinase K was added, the hydrogel degradation was obviously accelerated with ~60% mass loss within 15 days. This is likely due to the enzymatic cleavage of the polypeptide segments in addition to the hydrogel erosion [40,41]. Moreover, after subcutaneous injection into rats, the 6 wt% hydrogels degraded continuously in the subcutaneous layer of rats and completely disappeared within 5 weeks in vivo (Figure S4). It is known that there are different enzymes, such as cathepsin B, cathepsin C and elastase, in the subcutaneous layer of mammals [42]. Therefore, the degradation of the hydrogel in the subcutaneous layer of rats may be accelerated by enzymatic hydrolysis in vivo. These results indicated that the hydrogels underwent continuous degradation in vitro and in vivo.

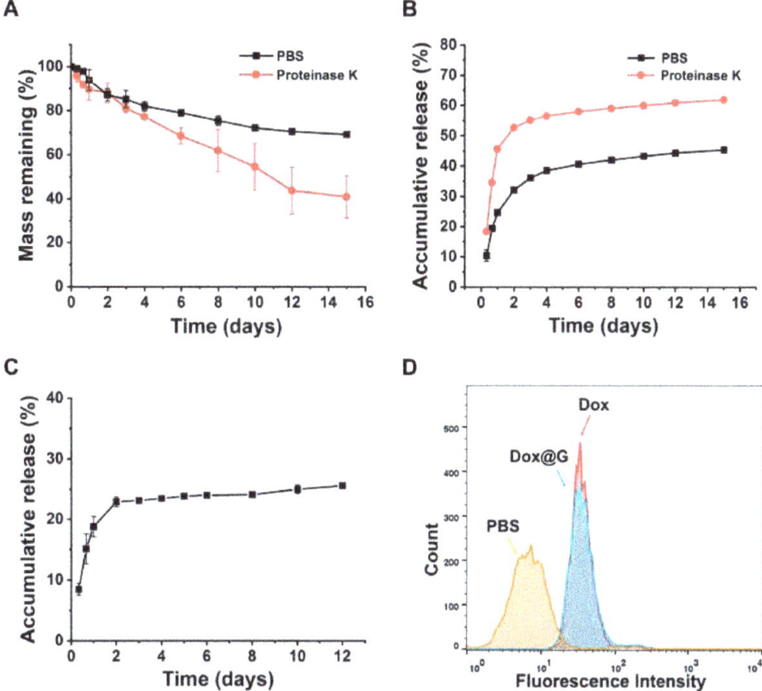

Figure 2. (**A**) In vitro degradation behaviors of 6 wt% hydrogel in PBS or PBS containing proteinase K (5 U/mL) (*n* = 3). (**B**) Accumulative Dox release profiles from 6 wt% hydrogels in PBS or PBS containing proteinase K (5 U/mL) (*n* = 3). (**C**) Accumulative IgG release from 6 wt% hydrogel in PBS (*n* = 3). (**D**) Fluorescence intensity of CRT expressed on B16F10 cells after different treatments.

Further, the release behaviors of chemotherapeutic drug and antibodies from the mPEG-PELG hydrogel were measured in vitro using Dox and IgG as the model drug and antibody, respectively. It was found that the Dox-loaded hydrogel showed a burst Dox release in the first 2 days, but exhibited more continuous and slower Dox release rates during the subsequent 13 days (Figure 2B). The two-stage release profile of Dox may be due to the rapid diffusion of Dox near the hydrogel surface at the first stage, and subsequent combined mechanisms of retarded Dox diffusion and slow hydrogel degradation [35]. In contrast, the Dox release was obviously accelerated with the addition of proteinase K, likely attributed to increased hydrogel disintegration in the presence of enzyme. Additionally, the release of IgG from the hydrogel was slower than Dox. This may be due to the slower diffusion rate of the antibody from the hydrogel than the small molecule (Figure 2C). The in vitro drug release data in the first 6 days were fitted to different kinetic models, including the zero-order model, first-order model, Korsmeyer–Peppas model and Higuchi model [43–45] (Figure S5). Through comparing the R-squared value, the Dox release profile from the hydrogel in PBS showed a relatively better degree of fitting to the Korsmeyer–Peppas model. It is noteworthy that the R-squared value was reduced for Dox release in the presence of proteinase K. Additionally, the IgG release from the hydrogel did not fit well with these kinetic models, which may be attributed to the fact that the IgG release behavior was influenced by complicated interactions between the protein and hydrogel network [46]. Therefore, the drug release tests in vitro suggested that the drug-loaded hydrogels exhibited continuous release profiles of both Dox and antibodies, which could be adjusted by the hydrogel degradation.

Calreticulin (CRT) is a typical marker on cells undergoing immunogenic cell death (ICD), which can promote antigen processing and presentation for adaptive immune response [47]. To reveal the ability of the drug-loaded hydrogel to induce ICD of tumor cells, the CRT expressions on the B16F10 melanoma cells after different treatments were tested (Figure 2D). It was found that incubation of tumor cells with either free Dox or Dox-loaded hydrogel caused enhanced CRT expression of B16F10 cells, indicating the occurrence of ICD in tumor cells.

To investigate the combination effects of Dox-mediated ICD of tumor cells and ICB therapy, Dox, aCTLA-4 and aPD-1 were co-loaded into the mPEG-PELG hydrogel to construct a depot for sustained antitumor chemo-immunotherapy. Different drug-containing systems were then peritumorally injected into C57BL/6 mice bearing B16F10 melanoma for evaluating the antitumor efficacy in vivo. The tumor-bearing mice were assigned into seven groups at random including PBS group, aCTLA-4 + aPD-1 group, (aCTLA-4 + aPD-1)@gel group, Dox solution group, Dox@gel group, Dox + aCTLA-4 + aPD-1 mixed solution group, and (Dox + aCTLA-4 + aPD-1)@gel group. It was observed that all the formulations containing Dox at a dose of 4 mg/kg exhibited an inhibition effect on tumor growth (Figure 3B). Moreover, the Dox/aCTLA-4/aPD-1 co-loaded hydrogel exhibited significantly improved tumor suppression efficiency compared to the formulations (hydrogels or solutions) containing either Dox or antibodies. Additionally, the Dox/aCTLA-4/aPD-1 co-loaded hydrogel led to a significantly extended animal survival time compared to single or multiple antitumor agent solutions, or hydrogels loaded with Dox or antibodies (Figure 3D). The results suggested that the sustained co-delivery of Dox, aCTLA-4 and aPD-1 using the hydrogel showed the best antitumor efficacy in vivo.

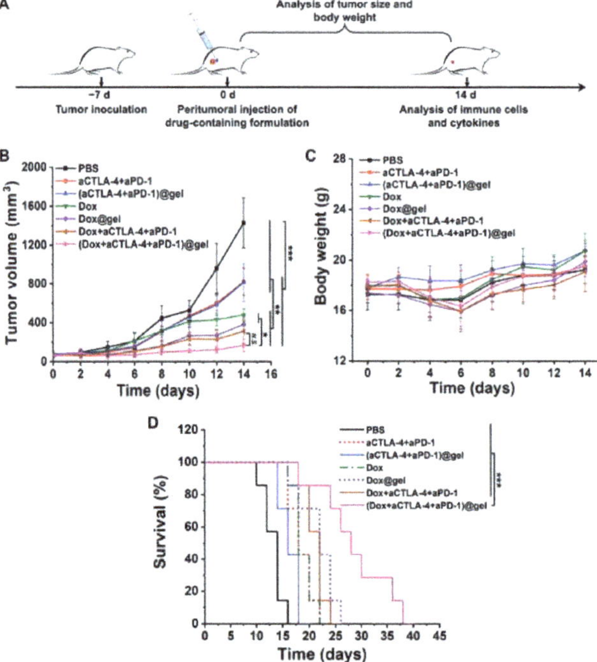

Figure 3. (**A**) Schematic illustration of localized treatment of melanoma-bearing mice with peritumoral injection of drug-containing systems. (**B**) Tumor volume curves during therapy ($n = 5$). (**C**) Changes of the body weight of mice during therapy ($n = 5$). (**D**) Survival periods of various experiment groups ($n = 7$). (* $p < 0.05$, ** $p < 0.01$, and *** $p < 0.001$).

Additionally, it was found that all the local treatments resulted in no obvious effect on the body weight in the mice (Figure 3C), indicating a reduced systemic side effect. After treatment for 14 days, the main organs of the mice, such as the heart, liver, spleen, lung and kidney, were taken and stained with H&E staining. Observation and pathological analysis were carried out under a microscope. As shown in Figure 4, no obvious pathological changes were found in any of the groups. This confirmed that the treatments with localized injection of drug-containing formulations showed no obvious toxic side effects to normal organs. Compared to systemic administration, the local and sustained release of Dox and ICB antibodies at tumor sites can markedly reduce the blood drug concentration and drug distribution in normal tissues [29]. Thus, low systemic toxicity was observed for the local treatments.

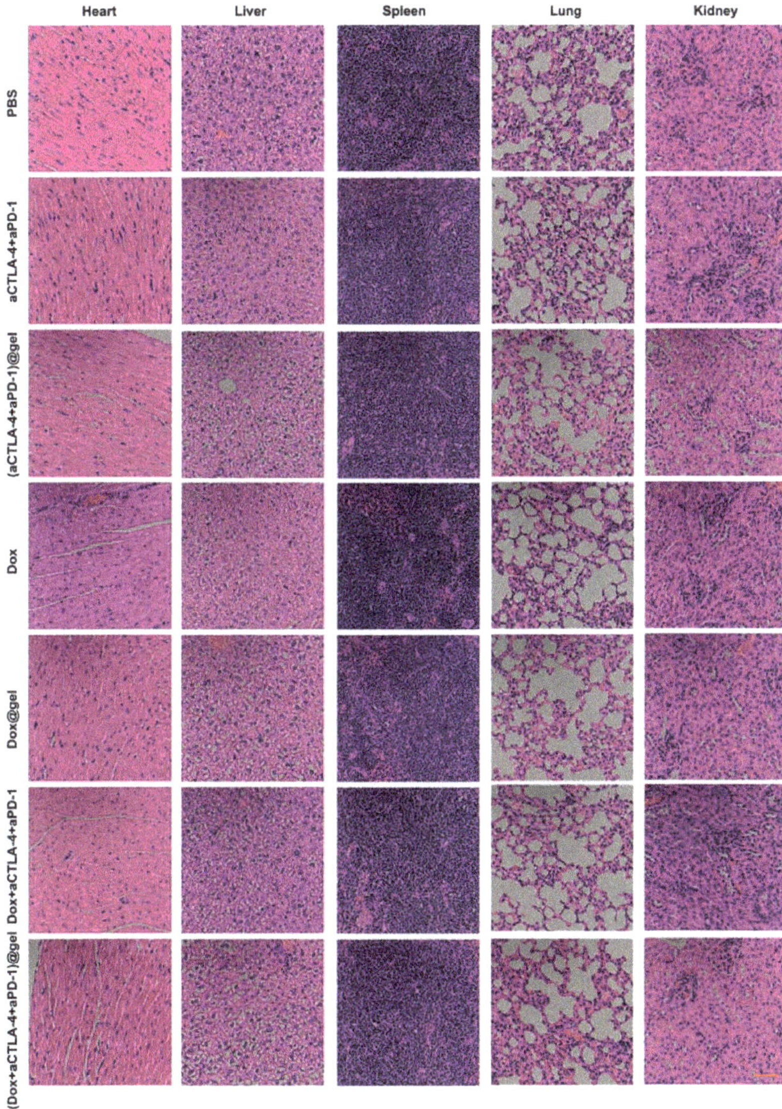

Figure 4. H&E staining images of the main organs of mice, including the heart, liver, spleen, lung and kidney, after 14 days of treatment with various therapeutic methods. Scale bar = 50 μm.

To analyze the immune response of the mice following treatments, typical immune cells and pro-inflammatory cytokines were evaluated. It was observed that the ratios of CD8$^+$ T cells in the spleen, lymph nodes and tumors were enhanced after the therapy with Dox/aCTLA-4/aPD-1 co-loaded hydrogel in Figure 5A–C. In addition, Dox/aCTLA-4/aPD-1 co-loaded hydrogel treatment reduced the ratio of immunosuppressive regulatory T cells (Tregs) in CD4$^+$ cells in tumors (Figure 5D). This indicated that the continuous, simultaneous release of Dox, aCTLA-4 and aPD-1 promoted the generation and tumor accumulation of tumor-killing CD8$^+$ T cells, and inhibited the accumulation of Tregs at tumor sites.

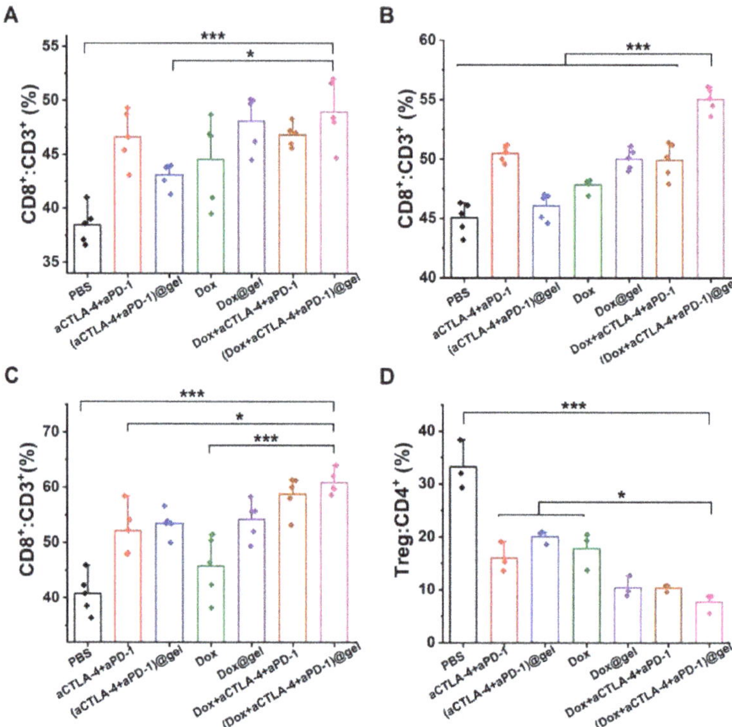

Figure 5. Immune cell analysis results of spleens, lymph nodes and tumors of the melanoma-bearing mice model after receiving various treatments, obtained by flow cytometry. (**A**) Ratio of CD8$^+$:CD3$^+$ in spleen (n = 5). (**B**) Ratio of CD8$^+$:CD3$^+$ in lymph (n = 5). (**C**) Ratio of CD8$^+$:CD3$^+$ in the tumor (n = 5). (**D**) Ratio of Treg:CD4$^+$ in the tumor (n = 3). (* p < 0.05 and *** p < 0.001).

It has been established that CTLA-4 acts on immunosuppressive functions in several aspects [7,48]. First, it inhibits antigen presentation from DCs to T cells through competitively binding CD80/CD86. Second, CTLA-4 is responsive for the functions of Tregs. Thus, the blockade of CTLA-4 by the sustained release of aCTLA-4 near the tumor site promoted the presentation of TAAs, which was generated by Dox-mediated ICD of tumor cells [22], resulting in enhanced T cell activation (Scheme 1). Moreover, the ratio of Tregs at the tumor site was also reduced. In addition, PD-1, another immune checkpoint receptor usually expressed on activated T cells, also acts as a "brake" during the tumor recognition and killing, through binding to its ligand (PD-L1) expressed on tumor cells, resulting in the immune escape of tumor cells [49]. Therefore, the inhibition of the PD-1/PD-L1 pathway by a localized release of aPD-1 near the tumor site led to enhanced antigen recognition and tumor killing by CD8$^+$ cytotoxic T lymphocytes (CTLs).

Moreover, the pro-inflammatory cytokines, including IL-2, TNF-α and IFN-γ, in the blood were examined. It was observed that the concentrations of pro-inflammatory cytokines were significantly enhanced after the treatment with the hydrogel loading Dox, aCTLA-4 and aPD-1 (Figure 6). IL-2 is a cytokine that undertakes immunoregulatory functions on T cells, NK cells and NK-T cells [50]. IL-2 plays a role in promoting the proliferation of lymphocytes including T cells and NK cells. TNF-α is able to exert antitumor effects by direct showing cytotoxicity on tumor cells and inducing an antitumor immune response. IFN-γ is a crucial cytokine for cell-mediated immunity, which plays key roles in stimulating antigen presentation, cytokine secretion, as well as the activation of macrophages, NK cells and neutrophils. Overall, the localized co-delivery of Dox, aCTLA-4 and aPD-1 from the hydrogel depot resulted in a significantly enhanced antitumor immune response, which may contribute to the increased tumor inhibition efficacy in vivo.

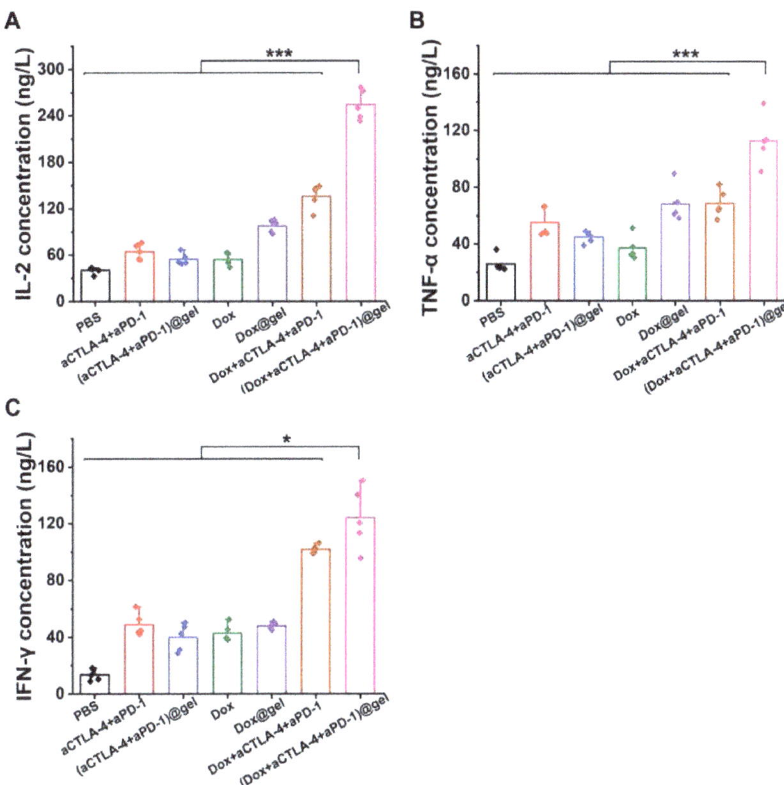

Figure 6. Pro-inflammatory cytokine concentration in the serum of various groups after treatment for 14 days. (**A**) The concentration of IL-2 ($n = 5$). (**B**) The concentration of TNF-α ($n = 5$). (**C**) The concentration of IFN-γ ($n = 5$). (* $p < 0.05$ and *** $p < 0.001$).

To further evaluate the ability of the Dox/aCTLA-4/aPD-1 co-loaded hydrogel to inhibit post-operative tumor reoccurrence, a tumor resection model was established using melanoma-bearing C57BL/6 mice. As shown in Figure 7, after ~90% of the tumor was resected, the Dox/aCTLA-4/aPD-1 co-loaded hydrogel was injected into the surgical site, with PBS or the mixed solution of the multiple agents as the controls. It was observed that the remaining tumors regrew rapidly after the tumor resection surgery with additional PBS treatment (Figure 7B). The local injection of Dox/aCTLA-4/aPD-1 mixed solution at the surgery site resulted in the effective inhibition of tumor reoccurrence. Notably, the local treatment of the Dox/aCTLA-4/aPD-1 co-loaded hydrogel showed a significantly stronger inhibition effect on the tumor reoccurrence, compared to either PBS or Dox/aCTLA-4/aPD-1 mixed solution. Moreover, the Dox/aCTLA-4/aPD-1 co-loaded hydrogel therapy led to a significantly extended survival time compared to other treatments (Figure 7D). The higher efficiency in tumor reoccurrence inhibition of the Dox/aCTLA-4/aPD-1 co-loaded hydrogel may be attributed to the sustained and prolonged release manners of the chemotherapy drug and ICB antibodies from the hydrogel depot [29,30].

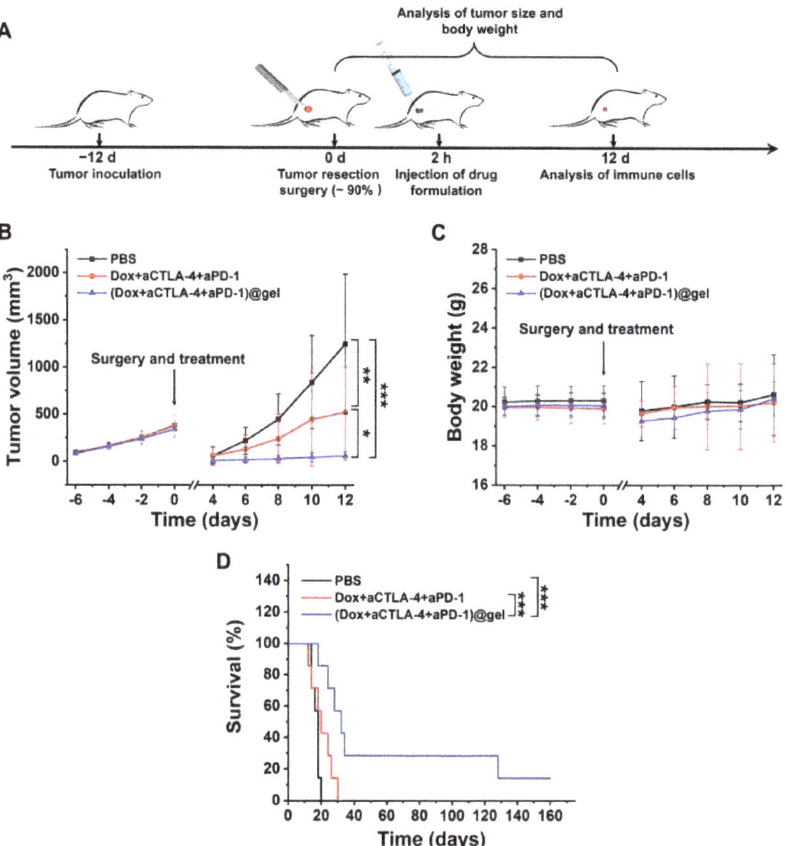

Figure 7. (**A**) Schematic illustration of post-surgical treatment by the injection of drug-containing systems to the surgical site after ~90% tumor resection. (**B**) Tumor volume curves during treatment (n = 12). (**C**) Changes of the body weight of mice during treatment (n = 12). (**D**) Survival periods of various groups after treatment (n = 7). (* $p < 0.05$, ** $p < 0.01$, and *** $p < 0.001$).

In addition, the immune cells in the spleens, lymph nodes and tumors of the treated mice were analyzed at day 12 post-surgery. It was observed that the Dox/aCTLA-4/aPD-1 co-loaded hydrogel treatment enhanced the ratio of CD8$^+$ T cells in the spleen and lymph node, and caused a decrease in the ratio of Treg cells in the tumor (Figure S6). Thus, the results suggested that the Dox/aCTLA-4/aPD-1 co-loaded hydrogel can effectively inhibit post-operative tumor reoccurrence and extend animal survival time, through strengthening the antitumor immune response.

In recent studies, hydrogel-based local delivery systems have been investigated for the co-delivery of chemotherapeutics and single ICB antibody, aPD-1 or anti-PD-L1, as a strategy for local chemo-immunotherapy [31–34,36]. It has been found that the chemotherapy-mediated ICD of tumor cells could promote the antitumor immune response when combining with ICB blockade. Moreover, some agents for modulating the immunosuppressive tumor micro-environment, e.g., indoleamine-(2,3)-dioxygenase (IDO) inhibitors, were further incorporated into the hydrogel depots for improving the antitumor efficacy [35,51]. In this study, a new formulation of a hydrogel depot co-loaded with dual ICB antibodies, aCTLA-4 and aPD-1, and a chemotherapeutics was developed for topical antitumor chemo-immunotherapy. The sustained release of aCTLA-4 at the tumor site can strengthen the antigen presentation of DCs to T cells, and the simultaneous delivery of aPD-1 is able to enhance the subsequent tumor recognition and killing. Based on the treatment of B16F10 melanoma-bearing mice and post-operative mice models, the Dox/aCTLA-4/aPD-1 co-loaded hydrogel showed a significantly increased antitumor immune response, enhanced tumor inhibition efficacy and prolonged animal survival time. Additionally, it was demonstrated that the local co-delivery of the ICB antibodies and chemotherapeutics caused no obvious systemic side effects, compared to the risk of severe irAEs by systemic administration of ICB antibodies [5].

4. Conclusions

In the present study, an antitumor chemo-immunotherapy system was prepared by the incorporation of Dox, aCTLA-4 and aPD-1 into thermosensitive mPEG-PELG hydrogel. The polypeptide hydrogel exhibited enzyme-accelerated degradation behavior in vitro for over 15 days, and gradually degraded in the subcutaneous layer of mice. The sustained release of Dox or IgG model antibody from the hydrogel persisted for over 12 days in vitro, which can be further accelerated by the addition of proteinase K. A combination antitumor strategy was developed by simultaneous, sustained delivery of Dox and the dual ICB antibodies through peritumoral injection of the multiple-agent co-loaded hydrogel into mice bearing B16F10 melanoma. The treatment with Dox/aCTLA-4/aPD-1 co-loaded hydrogel demonstrated markedly enhanced tumor inhibition efficacy, significantly extended animal survival time and strengthened antitumor immune response, compared to the mixed drug solutions or hydrogel containing Dox or antibodies. Moreover, the localized treatments with the hydrogel-based formulations showed no obvious systemic side effects. Additionally, the treatment with Dox/aCTLA-4/aPD-1 co-loaded hydrogel following tumor resection surgery achieved significantly increased efficiency in inhibiting tumor reoccurrence and extending animal survival time. In summary, the hydrogels encapsulating chemotherapeutics and dual ICB antibodies can serve as a promising candidate as depots for antitumor combination therapy and the prevention of post-operative tumor reoccurrence.

Supplementary Materials: The following supporting information can be downloaded at: https://www.mdpi.com/article/10.3390/pharmaceutics15020428/s1. Scheme S1. Synthesis route for mPEG-PELG. Figure S1. (A) ^{1}H NMR spectrum of mPEG-PELG in CF_3COOD. (B) GPC data of mPEG-PELG. Figure S2. Photographs of the mPEG-PELG (4–10 wt%) at different temperatures. The gelation temperatures and syneresis temperatures were marked specially. Figure S3. The ellipticity of polymer aqueous solution (0.05 wt%) with increasing temperature from 10 to 60 °C. Figure S4. Photographs of the hydrogels after injection into the subcutaneous layer of rats at different time points. Figure S5. Fitting of the in vitro drug release data to kinetic models of zero-order model (A), first-order model (B), Korsmeyer-Peppas model (C) and Higuchi model (D). (a) Dox release from Dox-loaded hydrogel in PBS; (b) Dox release from Dox-loaded hydrogel in PBS containing proteinase K; (c) IgG release from IgG-loaded hydrogel in PBS [43–45]. Figure S6. Immune cell analysis data of spleens, lymph nodes and tumors of the post-operative mice model after receiving various treatments, obtained by flow cytometry. (A) The ratio of $CD8^+:CD3^+$ in spleen ($n = 5$). (B) The ratio of $CD8^+:CD3^+$ in lymph nodes ($n = 5$). (C) The ratio of Tregs:$CD4^+$ in tumor ($n = 5$).

Author Contributions: Conceptualization, Z.C., X.C. (Xuesi Chen) and C.H.; Methodology, Z.C., Y.R., J.D., X.C. (Xuesi Chen), X.C. (Xueliang Cheng) and C.H.; Validation, C.H.; Investigation, Z.C. and Y.R.; Data curation, Z.C. and Y.R.; Writing—original draft preparation, Z.C. and Y.R.; Writing—review and editing, Y.R., J.D., X.C. (Xuesi Chen), X.C. (Xueliang Cheng) and C.H.; Supervision, X.C. (Xuesi Chen) and C.H.; Project administration, X.C. (Xuesi Chen) and C.H. All authors have read and agreed to the published version of the manuscript.

Funding: We thank the National Natural Science Foundation of China (No. 51973218, 51833010, 52173147, 52203202) and the Scientific and Technological Development Projects of Jilin Province (20210204136YY) for financial support.

Institutional Review Board Statement: All animal procedures were conducted in accordance with the Guidelines for Care and Use of Laboratory Animals of Jilin University and approved by the Animal Ethics Committee of Changchun Institute of Applied Chemistry, CAS (No. 2020-010).

Informed Consent Statement: Not applicable.

Data Availability Statement: The data supporting the study is available from the authors on request.

Conflicts of Interest: The authors declare no conflict of interest.

References

1. Sung, H.; Ferlay, J.; Siegel, R.L.; Laversanne, M.; Soerjomataram, I.; Jemal, A.; Bray, F. Global Cancer Statistics 2020: GLOBOCAN Estimates of Incidence and Mortality Worldwide for 36 Cancers in 185 Countries. *CA Cancer J. Clin.* **2021**, *71*, 209–249. [CrossRef] [PubMed]
2. Miller, A.B.; Hoogstraten, B.; Staquet, M.; Winkler, A. Reporting Results of Cancer Treatment. *Cancer* **1981**, *47*, 207–214. [CrossRef] [PubMed]
3. Dougan, M.; Dranoff, G. Immune Therapy for Cancer. *Annu. Rev. Immunol.* **2009**, *27*, 83–117. [CrossRef]
4. Wieder, T.; Eigentler, T.; Brenner, E.; Röcken, M. Immune Checkpoint Blockade Therapy. *J. Allergy Clin. Immunol.* **2018**, *142*, 1403–1414. [CrossRef] [PubMed]
5. Seidel, J.A.; Otsuka, A.; Kabashima, K. Anti-PD-1 and Anti-CTLA-4 Therapies in Cancer: Mechanisms of Action, Efficacy, and Limitations. *Front. Oncol.* **2018**, *8*, 86. [CrossRef]
6. Sharma, P.; Allison, J.P. The Future of Immune Checkpoint Therapy. *Science* **2015**, *348*, 56–61. [CrossRef] [PubMed]
7. Gaynor, N.; Crown, J.; Collins, D.M. Immune Checkpoint Inhibitors: Key Trials and an Emerging Role in Breast Cancer. *Semin. Cancer Biol.* **2022**, *79*, 44–57. [CrossRef] [PubMed]
8. Singh, S.; Hassan, D.; Aldawsari, H.M.; Molugulu, N.; Shukla, R.; Kesharwani, P. Immune Checkpoint Inhibitors: A Promising Anticancer Therapy. *Drug Discov. Today* **2020**, *25*, 223–229. [CrossRef]
9. Larkin, J.; Chiarion-Sileni, V.; Gonzalez, R.; Grob, J.J.; Cowey, C.L.; Lao, C.D.; Schadendorf, D.; Dummer, R.; Smylie, M.; Rutkowski, P.; et al. Combined Nivolumab and Ipilimumab or Monotherapy in Untreated Melanoma. *N. Engl. J. Med.* **2015**, *373*, 23–34. [CrossRef]
10. Buchbinder, E.I.; Desai, A. CTLA-4 and PD-1 Pathways: Similarities, Differences, and Implications of Their Inhibition. *Am. J. Clin. Oncol.* **2016**, *39*, 98–106. [CrossRef]
11. Zhang, B.; Zhou, Y.L.; Chen, X.; Wang, Z.; Wang, Q.; Ju, F.; Ren, S.; Xu, R.; Xue, Q.; Wu, Q. Efficacy and Safety of CTLA-4 Inhibitors Combined with PD-1 Inhibitors or Chemotherapy in Patients with Advanced Melanoma. *Int. Immunopharmacol.* **2019**, *68*, 131–136. [CrossRef] [PubMed]
12. Wei, S.C.; Anang, N.-A.A.S.; Sharma, R.; Andrews, M.C.; Reuben, A.; Levine, J.H.; Cogdill, A.P.; Mancuso, J.J.; Wargo, J.A.; Pe'er, D.; et al. Combination Anti–CTLA-4 plus Anti–PD-1 Checkpoint Blockade Utilizes Cellular Mechanisms Partially Distinct from Monotherapies. *Proc. Natl. Acad. Sci. USA* **2019**, *116*, 22699–22709. [CrossRef]
13. Callahan, M.K.; Postow, M.A.; Wolchok, J.D. CTLA-4 and PD-1 Pathway Blockade: Combinations in the Clinic. *Front. Oncol.* **2015**, *4*, 385. [CrossRef] [PubMed]
14. Mahoney, K.M.; Rennert, P.D.; Freeman, G.J. Combination Cancer Immunotherapy and New Immunomodulatory Targets. *Nat. Rev. Drug Discov.* **2015**, *14*, 561–584. [CrossRef] [PubMed]
15. Wolchok, J.D.; Kluger, H.; Callahan, M.K.; Postow, M.A.; Rizvi, N.A.; Lesokhin, A.M.; Segal, N.H.; Ariyan, C.E.; Gordon, R.-A.; Reed, K.; et al. Nivolumab plus Ipilimumab in Advanced Melanoma. *N. Engl. J. Med.* **2013**, *369*, 122–133. [CrossRef]
16. Khair, D.O.; Bax, H.J.; Mele, S.; Crescioli, S.; Pellizzari, G.; Khiabany, A.; Nakamura, M.; Harris, R.J.; French, E.; Hoffmann, R.M.; et al. Combining Immune Checkpoint Inhibitors: Established and Emerging Targets and Strategies to Improve Outcomes in Melanoma. *Front. Immunol.* **2019**, *10*, 453. [CrossRef]
17. Rotte, A. Combination of CTLA-4 and PD-1 Blockers for Treatment of Cancer. *J. Exp. Clin. Cancer Res.* **2019**, *38*, 255. [CrossRef]
18. Tarhini, A.; McDermott, D.; Ambavane, A.; Gupte-Singh, K.; Aponte-Ribero, V.; Ritchings, C.; Benedict, A.; Rao, S.; Regan, M.M.; Atkins, M. Clinical and Economic Outcomes Associated with Treatment Sequences in Patients with *BRAF*-Mutant Advanced Melanoma. *Immunotherapy* **2019**, *11*, 283–295. [CrossRef]
19. Curran, M.A.; Montalvo, W.; Yagita, H.; Allison, J.P. PD-1 and CTLA-4 Combination Blockade Expands Infiltrating T Cells and Reduces Regulatory T and Myeloid Cells within B16 Melanoma Tumors. *Proc. Natl. Acad. Sci. USA* **2010**, *107*, 4275–4280. [CrossRef]
20. Fan, W.; Yung, B.; Huang, P.; Chen, X. Nanotechnology for Multimodal Synergistic Cancer Therapy. *Chem. Rev.* **2017**, *117*, 13566–13638. [CrossRef]
21. Chen, Q.; Hu, Q.; Dukhovlinova, E.; Chen, G.; Ahn, S.; Wang, C.; Ogunnaike, E.A.; Ligler, F.S.; Dotti, G.; Gu, Z. Photothermal Therapy Promotes Tumor Infiltration and Antitumor Activity of CAR T Cells. *Adv. Mater.* **2019**, *31*, 1900192. [CrossRef] [PubMed]
22. Bezu, L.; Gomes-de-Silva, L.C.; Dewitte, H.; Breckpot, K.; Fucikova, J.; Spisek, R.; Galluzzi, L.; Kepp, O.; Kroemer, G. Combinatorial Strategies for the Induction of Immunogenic Cell Death. *Front. Immunol.* **2015**, *6*, 187. [CrossRef]
23. Chen, Q.; Chen, M.; Liu, Z. Local Biomaterials-Assisted Cancer Immunotherapy to Trigger Systemic Antitumor Responses. *Chem. Soc. Rev.* **2019**, *48*, 5506–5526. [CrossRef] [PubMed]

24. Li, J.; Mooney, D.J. Designing Hydrogels for Controlled Drug Delivery. *Nat. Rev. Mater.* **2016**, *1*, 16071. [PubMed]
25. Lei, K.; Tang, L. Surgery-Free Injectable Macroscale Biomaterials for Local Cancer Immunotherapy. *Biomater. Sci.* **2019**, *7*, 733–749.
26. Chen, W.-H.; Chen, Q.-W.; Chen, Q.; Cui, C.; Duan, S.; Kang, Y.; Liu, Y.; Liu, Y.; Muhammad, W.; Shao, S.; et al. Biomedical Polymers: Synthesis, Properties, and Applications. *Sci. China Chem.* **2022**, *65*, 1010–1075.
27. Yu, L.; Ding, J. Injectable Hydrogels as Unique Biomedical Materials. *Chem. Soc. Rev.* **2008**, *37*, 1473. [CrossRef]
28. He, C.; Kim, S.W.; Lee, D.S. In Situ Gelling Stimuli-Sensitive Block Copolymer Hydrogels for Drug Delivery. *J. Control. Release* **2008**, *127*, 189–207. [CrossRef]
29. Ma, H.; He, C.; Chen, X. Injectable Hydrogels as Local Depots at Tumor Sites for Antitumor Immunotherapy and Immune-Based Combination Therapy. *Macromol. Biosci.* **2021**, *21*, 2100039. [CrossRef]
30. Bai, Y.; Wang, T.; Zhang, S.; Chen, X.; He, C. Recent Advances in Organic and Polymeric Carriers for Local Tumor Chemo-Immunotherapy. *Sci. China Technol. Sci.* **2022**, *65*, 1011–1028. [CrossRef]
31. Wang, C.; Wang, J.; Zhang, X.; Yu, S.; Wen, D.; Hu, Q.; Ye, Y.; Bomba, H.; Hu, X.; Liu, Z.; et al. In Situ Formed Reactive Oxygen Species–Responsive Scaffold with Gemcitabine and Checkpoint Inhibitor for Combination Therapy. *Sci. Transl. Med.* **2018**, *10*, eaan3682. [CrossRef] [PubMed]
32. Wang, F.; Xu, D.; Su, H.; Zhang, W.; Sun, X.; Monroe, M.K.; Chakroun, R.W.; Wang, Z.; Dai, W.; Oh, R.; et al. Supramolecular Prodrug Hydrogelator as an Immune Booster for Checkpoint Blocker–Based Immunotherapy. *Sci. Adv.* **2020**, *6*, eaaz8985. [CrossRef] [PubMed]
33. Gong, Y.; Chen, M.; Tan, Y.; Shen, J.; Jin, Q.; Deng, W.; Sun, J.; Wang, C.; Liu, Z.; Chen, Q. Injectable Reactive Oxygen Species-Responsive SN38 Prodrug Scaffold with Checkpoint Inhibitors for Combined Chemoimmunotherapy. *ACS Appl. Mater. Interfaces* **2020**, *12*, 50248–50259. [CrossRef] [PubMed]
34. Shi, Y.; Li, D.; He, C.; Chen, X. Design of an Injectable Polypeptide Hydrogel Depot Containing the Immune Checkpoint Blocker Anti-PD-L1 and Doxorubicin to Enhance Antitumor Combination Therapy. *Macromol. Biosci.* **2021**, *21*, 2100049. [CrossRef] [PubMed]
35. Ding, J.; Wang, T.; Chen, Z.; Lin, Z.; Chen, X.; He, C. Enhanced Antitumor Chemo-immunotherapy by Local Co-delivery of Chemotherapeutics, Immune Checkpoint Blocking Antibody and IDO Inhibitor Using an Injectable Polypeptide Hydrogel. *J. Polym. Sci.* **2022**, *60*, 1595–1609. [CrossRef]
36. Wang, H.; Najibi, A.J.; Sobral, M.C.; Seo, B.R.; Lee, J.Y.; Wu, D.; Li, A.W.; Verbeke, C.S.; Mooney, D.J. Biomaterial-Based Scaffold for in Situ Chemo-Immunotherapy to Treat Poorly Immunogenic Tumors. *Nat. Commun.* **2020**, *11*, 5696. [CrossRef]
37. Cheng, Y.; He, C.; Xiao, C.; Ding, J.; Zhuang, X.; Huang, Y.; Chen, X. Decisive Role of Hydrophobic Side Groups of Polypeptides in Thermosensitive Gelation. *Biomacromolecules* **2012**, *13*, 2053–2059. [CrossRef]
38. Oh, H.J.; Joo, M.K.; Sohn, Y.S.; Jeong, B. Secondary Structure Effect of Polypeptide on Reverse Thermal Gelation and Degradation of L/DL-Poly(Alanine)–Poloxamer–L/DL-Poly(Alanine) Copolymers. *Macromolecules* **2008**, *41*, 8204–8209. [CrossRef]
39. Li, D.; Zhao, D.; He, C.; Chen, X. Crucial Impact of Residue Chirality on the Gelation Process and Biodegradability of Thermoresponsive Polypeptide Hydrogels. *Biomacromolecules* **2021**, *22*, 3992–4003. [CrossRef]
40. Li, D.; Shi, S.; Zhao, D.; Rong, Y.; Zhou, Y.; Ding, J.; He, C.; Chen, X. Effect of Polymer Topology and Residue Chirality on Biodegradability of Polypeptide Hydrogels. *ACS Biomater. Sci. Eng.* **2022**, *8*, 626–637. [CrossRef]
41. Shi, S.; Wang, J.; Wang, T.; Ren, H.; Zhou, Y.; Li, G.; He, C.; Chen, X. Influence of Residual Chirality on the Conformation and Enzymatic Degradation of Glycopolypeptide Based Biomaterials. *Sci. China Technol. Sci.* **2021**, *64*, 641–650. [CrossRef]
42. Jeong, Y.; Joo, M.K.; Bahk, K.H.; Choi, Y.Y.; Kim, H.-T.; Kim, W.-K.; Jeong Lee, H.; Sohn, Y.S.; Jeong, B. Enzymatically Degradable Temperature-Sensitive Polypeptide as a New in-Situ Gelling Biomaterial. *J. Control. Release* **2009**, *137*, 25–30. [CrossRef] [PubMed]
43. Mahdi Eshaghi, M.; Pourmadadi, M.; Rahdar, A.; Díez-Pascual, A.M. Novel Carboxymethyl Cellulose-Based Hydrogel with Core–Shell Fe$_3$O$_4$@SiO$_2$ Nanoparticles for Quercetin Delivery. *Materials* **2022**, *15*, 8711. [CrossRef] [PubMed]
44. Dash, S.; Murthy, P.N.; Nath, L.; Chowdhury, P. Kinetic Modeling on Drug Release from Controlled Drug Delicery Systems. *Acta Pol. Pharm.* **2015**, *67*, 217–223.
45. Siepmann, J.; Peppas, N.A. Higuchi Equation: Derivation, Applications, Use and Misuse. *Int. J. Pharm.* **2011**, *418*, 6–12. [CrossRef] [PubMed]
46. Coco, J.C.; Tundisi, L.L.; dos Santos, É.M.; Fava, A.L.M.; Alves, T.F.; Ataide, J.A.; Chaud, M.V.; Mazzola, P.G. PVA-CO-AAM and Peg-Co-Aam Hydrogels as Bromelain Carriers. *J. Drug Deliv. Sci. Technol.* **2021**, *63*, 102483. [CrossRef]
47. Gold, L.I.; Eggleton, P.; Sweetwyne, M.T.; Van Duyn, L.B.; Greives, M.R.; Naylor, S.; Michalak, M.; Murphy-Ullrich, J.E. Calreticulin: Non-endoplasmic Reticulum Functions in Physiology and Disease. *FASEB J.* **2010**, *24*, 665–683. [CrossRef]
48. Rowshanravan, B.; Halliday, N.; Sansom, D.M. CTLA-4: A Moving Target in Immunotherapy. *Blood* **2018**, *131*, 58–67. [CrossRef]
49. Kyi, C.; Postow, M.A. Checkpoint Blocking Antibodies in Cancer Immunotherapy. *FEBS Lett.* **2014**, *588*, 368–376. [CrossRef]
50. Borish, L.C.; Steinke, J.W. 2. Cytokines and Chemokines. *J. Allergy Clin. Immunol.* **2003**, *111*, S460–S475. [CrossRef]
51. Yu, S.; Wang, C.; Yu, J.; Wang, J.; Lu, Y.; Zhang, Y.; Zhang, X.; Hu, Q.; Sun, W.; He, C.; et al. Injectable Bioresponsive Gel Depot for Enhanced Immune Checkpoint Blockade. *Adv. Mater.* **2018**, *30*, 1801527. [CrossRef] [PubMed]

Article

mRNA-Loaded Lipid Nanoparticles Targeting Dendritic Cells for Cancer Immunotherapy

Kosuke Sasaki [†], Yusuke Sato [*,†], Kento Okuda, Kazuki Iwakawa and Hideyoshi Harashima [*]

Laboratory for Molecular Design of Pharmaceutics, Faculty of Pharmaceutical Sciences, Hokkaido University, Kita-12, Nishi-6, Kita-Ku, Sapporo 060-0812, Japan; ksksssk.sw.77@gmail.com (K.S.); okuken@eis.hokudai.ac.jp (K.O.); kazukiiwak78@eis.hokudai.ac.jp (K.I.)

* Correspondence: y_sato@pharm.hokudai.ac.jp (Y.S.); harasima@pharm.hokudai.ac.jp (H.H.); Tel.: +81-11-706-3734 (Y.S.); +81-11-706-3919 (H.H.)

† These authors contributed equally to this work.

Abstract: Dendritic cells (DCs) are attractive antigen-presenting cells to be targeted for vaccinations. However, the systemic delivery of mRNA to DCs is hampered by technical challenges. We recently reported that it is possible to regulate the size of RNA-loaded lipid nanoparticles (LNPs) to over 200 nm with the addition of salt during their formation when a microfluidic device is used and that larger LNPs delivered RNA more efficiently and in greater numbers to splenic DCs compared to the smaller counterparts. In this study, we report on the in vivo optimization of mRNA-loaded LNPs for use in vaccines. The screening included a wide range of methods for controlling particle size in addition to the selection of an appropriate lipid type and its composition. The results showed a clear correlation between particle size, uptake and gene expression activity in splenic DCs and indicated that a size range from 200 to 500 nm is appropriate for use in targeting splenic DCs. It was also found that it was difficult to predict the transgene expression activity and the potency of mRNA vaccines in splenic DCs using the whole spleen. A-11-LNP, which was found to be the optimal formulation, induced better transgene expression activity and maturation in DCs and induced clear therapeutic antitumor effects in an E.G7-OVA tumor model compared to two clinically relevant LNP formulations.

Keywords: lipid nanoparticles; particle size; dendritic cells; mRNA; delivery; cancer immunotherapy

Citation: Sasaki, K.; Sato, Y.; Okuda, K.; Iwakawa, K.; Harashima, H. mRNA-Loaded Lipid Nanoparticles Targeting Dendritic Cells for Cancer Immunotherapy. *Pharmaceutics* **2022**, *14*, 1572. https://doi.org/10.3390/pharmaceutics14081572

Academic Editor: Jesus Perez-Gil

Received: 9 July 2022
Accepted: 26 July 2022
Published: 28 July 2022

Publisher's Note: MDPI stays neutral with regard to jurisdictional claims in published maps and institutional affiliations.

1. Introduction

Nucleic acid vaccines are being evaluated for use in a number of clinical applications, including cancer, allergy, and infectious diseases [1–5]. It is particularly noteworthy that mRNA was first commercialized as a vaccine against the coronavirus disease 2019 (COVID-19) for the treatment of infections caused by the severe acute respiratory syndrome coronavirus 2 (SARS-CoV-2) [6–8], and its application to a wide variety of diseases is now being investigated. The mRNA molecule, unlike DNA, has no risk of unintentional insertion into genomic DNA [9]. In addition, mRNA does not require nuclear delivery and can efficiently express the encoded gene(s) if it is properly introduced into the cytoplasm. Recent developments in mRNA engineering, such as capping technology, sequence design technology, and chemically modified bases, as well as relatively simple synthesis by in vitro transcription, have greatly contributed to the development of this technology [10]. In addition, mRNA can express any peptide and protein structures depending on the sequence design, making it possible to present a variety of antigen peptides to major histocompatibility complex (MHC) classes I and II [11]. In addition, since mRNAs exhibit immunostimulatory properties through recognition by pattern recognition receptors (PPRs), including Toll-like receptors (TLRs) and retinoic acid-inducible gene I (RIG-I)-like receptor (RLR) families [12], which recognize pathogen-associated molecular patterns (PAMPs), they can activate innate immunity efficiently without the need for additional adjuvants [2,13].

While the above-mentioned advantages of mRNA and the progress in mRNA engineering technologies are encouraging the realization of its clinical applications, properties of the mRNA, including instability in an in vivo environment and cell membrane impermeability, prevent it from reaching the cytoplasm efficiently, where it functions. Therefore, it is essential to develop a technology to efficiently deliver mRNA to the cytoplasm of target cells. mRNA delivery systems include lipoplex and lipid nanoparticles (LNPs) [2,14–18]. A Comirnaty (Pfizer (New York, NY, USA)/BionTech (Mainz, Germany)) and Spikevax (Moderna (Cambrige, MA, USA)) are examples of LNP-based mRNA vaccines against COVID-19 [8].

LNPs are composed of pH-sensitive cationic lipids (CLs), phospholipids, cholesterol (Chol), and polyethylene glycol (PEG)-modified lipids. Chol contributes to the stability of LNPs [19,20]. Phospholipids mainly contribute to the formation and stabilization of the lipid membrane [21]. PEG lipids contribute to the regulation of particle size, dispersion stability, and blood retention [22–24]. pH-sensitive cationic lipids contain tertiary amino group(s) and are positively charged in a weakly acidic environment [25,26]. This is important for efficient mRNA encapsulation through electrostatic interactions. It is also important for delivering mRNA to the cytoplasm by escape via membrane fusion from acidified endosomes after the LNPs have been internalized into cells through endocytosis [27,28]. Therefore, the lipid composition of the LNPs has significant effects on physicochemical properties as well as potency for drug delivery and, therefore, should be carefully optimized for each specific application and payload due to differences in the properties needed for the specific targeting of cell types or applications [29–31].

Dendritic cells (DCs) are able to initiate antigen-specific immune responses in lymphoid tissues after infectious pathogens are sensed and are among the most powerful antigen-presenting cells (APCs) [32,33]. Therefore, DCs are attractive APCs to be targeted for vaccination. However, the systemic delivery of mRNA to DCs faces a number of technical challenges. Although an ex vivo pulsed DC vaccine would have great potential because of its specificity, flexibility, and efficiency [34,35], it requires a complicated process that involves collecting a sample from the patient to re-administer to the patient. The direct targeting of DCs in vivo would be a good strategy for overcoming that drawback. Several studies have reported on DC targeting by modifying ligand molecules on the surface of the nanoparticles [36,37]. While this approach is widely used to achieve specific targeting, the formulation process is very complex and often poses challenges in terms of cost, reproducibility, and difficulty in characterization. On the other hand, Kranz, et al. reported on an RNA-lipoplex (LPX) that was prepared by mixing mRNA and ligand-unmodified liposomes composed of 1,2-di-*O*-octadecenyl-3-trimethylammonium propane (DOTMA) and 1,2-dioleoyl-sn-glycero-3-phosphoethanolamine (DOPE) [2]. Focusing on charge balance, specific gene expression in the spleen was achieved by designing a formulation with a negative charge ratio, which resulted in inducing a relatively selective gene expression in DCs. Antigen-specific antitumor immunity was induced in several cancer models by using an mRNA encoding a cancer antigen. Currently, BioNTech is conducting phase II clinical trials on an mRNA cancer vaccine formulation containing a fixed set of cancer-associated antigens (referred to as the FixVac platform) against advanced melanomas (ClinicalTrials.gov Identifier: NCT04526899) and human papillomavirus 16-positive head and neck squamous cell carcinomas (AHEAD-MERIT study, ClinicalTrials.gov Identifier: NCT04534205) as of the writing this manuscript. However, there are still few reports of LNPs that enable efficient mRNA delivery to DCs after systemic administration, and there is clearly substantial room for improvement regarding efficiency. In this study, we optimized mRNA-loaded LNPs targeting DCs using our originally developed CLs and the recently reported broad particle size control technology of LNPs synthesized using a microfluidic device [38,39] and compared the optimized LNPs with clinically-relevant LNP formulations.

2. Materials and Methods

2.1. Materials

The pH-sensitive cationic lipid, 7-(4-(dipropylamino)butyl)-7-hydroxytridecane-1,13-diyl dioleate (CL4H6), was synthesized as described previously [38]. 7-Hydroxy-7-(4-morpholinobutyl)tridecane-1,13-diyl dioleate (CL7H6) was synthesized as described in the Supplementary Method. Chol was purchased from SIGMA Aldrich (St. Louis, MO, USA). DOTMA, 1,2-distearoyl-*sn*-glycero-3-phosphocholine (DSPC), DOPE, 1,2-dimirystoyl-*rac*-glycero, methoxyethyleneglycol 2000 ether (PEG-DMG), and 1,2-distearoyl-*rac*-glycero, methoxyethyleneglycol 2000 ether (PEG-DSG) were obtained from the NOF Corporation (Tokyo, Japan). DLin-MC3-DMA was purchased from Selleck Biotech (Houston, TX, USA). 3,3′-Dioctadecyloxacarbocyanine perchlorate (DiO) and Ribogreen were purchased from ThermoFisher Scientific (Waltham, MA, USA). Purified anti-mouse CD16/32 antibody (clone 93), PE Anti-mouse CD40 antibody (clone FGK45), FITC anti-mouse CD80 antibody (clone 16-10A1), phycoerythrin (PE) anti-mouse CD86 antibody (clone A17199A), allophycocyanin (APC) anti-mouse CD69 antibody (clone H1.2F3), PE anti-mouse CD3 antibody (clone 17A2), peridinin-chlorophyll-protein (PerCP)/Cyanine 7 anti-mouse CD19 antibody (clone 6D5), FITC anti-mouse NK1.1 antibody (clone PK136), APC anti-mouse CD11c antibody (clone N418), PerCP/Cyanine 5.5 anti-mouse I-A/I-E antibody (clone M5/114.15.2), and propidium iodide (PI) were purchased from BioLegend (San Diego, CA, USA). Anti-human polo-like kinase 1 (hPLK1) siRNA (siPLK1, sense: 5′-AGA uCA CCC uCC UUA AAu AUU-3′; antisense; 5′-UAU UUA AGG AGG GUG AuC UUU-3′, 2′-OMe-modified nucleotides are in lower case) was purchased from Hokkaido System Science Co., Ltd. (Sapporo, Japan). The ovalbumin (OVA)-encoding mRNA (cat# L-7610) and enhanced green fluorescent protein (EGFP)-encoding mRNA (cat# L-7601) were purchased from Trilink BioTechnologies (San Diego, CA, USA). A microfluidic device, an invasive lipid nanoparticle production (iLiNP) device, was fabricated as described previously [40].

2.2. Mice and Cell Cultures

E.G7-OVA cells, generated by transducing the chicken OVA gene into the murine lymphoma cell line EL4, were obtained from the American Type Culture Collection (Manassas, VA, USA). E.G7-OVA cells were grown in Roswell Park Memorial Institute (RPMI) 1640 (Sigma Aldrich, St. Louis, MO, USA) supplemented with 50 μM of β-mercaptoethanol, 10% fetal calf serum, 10 mM 2-[4-(2-Hydroxyethyl)-1-piperazinyl]ethanesulfonic acid (HEPES), 1 mM sodium pyruvate and 100 units/mL of penicillin/streptomycin. Cells were cultured at 37 °C in a 5% CO_2 incubator.

C57BL/6N mice (female, 6 weeks of age) were purchased from Japan SLC (Shizuoka, Japan).

2.3. In Vitro Transcription of mRNA

Nanoluciferase (Nluc)-encoding mRNA was synthesized from linearized pDNA by in vitro transcription using an mMESSAGE mMACHINE T7 Transcription Kit (ThermoFisher Scientific Inc., Waltham, MA, USA) according to the manufacturer's protocol. The in vitro transcribed mRNA was purified using a MEGAclear Transcription Clean-Up Kit (ThermoFisher Scientific Inc., Waltham, MA, USA) according to the manufacturer's protocol. The purified mRNA was qualified by denatured agarose gel electrophoresis, was quantified by absorbance, and was then stored at −80 °C until used.

2.4. Preparation of RNA-Loaded LNPs

An ethanol solution containing a pH-sensitive cationic lipid, a phospholipid, chol, and PEG-lipid at the indicated molar ratios was prepared at a total lipid concentration of 16 mM. The RNA was dissolved in 25 mM acetate buffer (pH 4.0) containing NaCl (0 to 400 mM). For MC3-LNPs, a fixed lipid composition of MC3:DSPC:chol:PEG-DMG = 50:10:40:1.5 (molar ratio) and acetate buffer without NaCl were used. LNPs were prepared by mixing the lipids in ethanol and an aqueous solution of mRNA using an iLiNP device at a total flow rate (TFR) of 0.5 mL/min and RNA-to-lipid flow rate ratio (FRR) of 3. RNA-to-lipid ratio was adjusted to 26.6 µg of RNA/µmol total lipid. Syringe pumps (Harvard apparatus, MA, USA or YMC Co., Ltd., Kyoto, Japan) were used to control the flow rate. The resulting LNP solution was then dialyzed for 2 h or more at 4 °C against 20 mM MES buffer (pH 6.0), followed by phosphate-buffered saline without Ca^{2+} and Mg^{2+} (PBS(-)) using Spectra/Por 4 dialysis membranes (molecular weight cut-off 12,000–14,000 Da, Spectrum Laboratories, Rancho Dominguez, CA, USA) to remove ethanol and adjust pH to neutral. For screening A and B, 0.1 mol% of DiO was added to the lipid solution, and a mixture of siPLK1 and mNluc at 19:1 (weight ratio) was used.

2.5. Preparation of RNA-Lipoplexes (RNA-LPX)

A lipid film was formed by the evaporation of an ethanolic solution containing DOTMA and DOPE at a molar ratio of 1:1 (10 mM, 400 µL). PBS(-) (800 µL) was then added to the lipid film, followed by incubation for 7 min at room temperature to hydrate the lipids. To form empty liposomes, the lipid film was then sonicated for approximately 30 s in a bath-type sonicator. After dilution with PBS(-) 5.92 times (final lipid concentration of 0.845 mM), equal volumes of the liposome suspension and the mRNA solution (0.2 mg/mL in PBS(-)) were mixed to form RNA-LPX with a nitrogen per phosphate ratio of 1.3/2. The resulting sample was used in in vivo experiments without further purification.

2.6. Characterization of RNA-Loaded LNPs

The ζ-average size, polydispersity index (PdI), and ζ-potential of the LNPs were measured by means of a Zetasizer Nano ZS ZEN3600 instrument (Malvern Instruments, Worchestershire, UK). The encapsulation efficiency and total concentration of RNA were measured by a Ribogreen assay, as described previously [41].

2.7. In Vivo Screening of RNA-Loaded LNPs

C57BL/6N mice were intravenously injected with the RNA-loaded LNPs at a dose of 1 mg RNA/kg (0.05 mg Nluc mRNA/kg). At twenty-four hours after the injection, the liver and half of the spleen tissues were harvested, frozen in liquid nitrogen, and stored at $-80\,°C$ for use in Nluc assays. Splenocytes were dissociated from the remaining spleen tissues and were then passed through a cell strainer (40 µm pore, BD Falcon, CA, USA). The recovered cells were spun down ($400\times g$, 4 °C, 5 min) to remove the supernatant, resuspended in red blood cell (RBC) Lysis buffer (1 mL, BioLegend) and incubated for 5 min at room temperature. The resulting treated cells were washed with Hanks' balanced salt solution (HBSS(-)) by spinning ($400\times g$, 4 °C, 5 min). The concentration of cells was adjusted to 1×10^7 cells/mL with fluorescence-activated cell sorting (FACS) buffer, and the resulting cells were treated with a 10 µg/mL solution of an anti-mouse CD16/32 antibody followed by incubation at 4 °C for 10 min. The resulting sample was then treated with fluorophore-conjugated anti-mouse antibodies at 4 °C for 30 min. Cells were washed with FACS buffer twice, filtered via a nylon filter, stained with propidium iodide (PI) (BioLegend), and analyzed for cellular uptake and cell sorting (SONY SH800 cell sorter, SONY, Tokyo, Japan). Three thousand living B cells (defined as PI^-CD19^+ cells), macrophages (Mac, defined as $PI^-F4/80\,^+CD11c^+$ cells), and DCs (defined as $PI^-I\text{-}A/I\text{-}E^+CD19^-$ cells) were collected in a sample tube containing 20 µL of 2 × passive lysis buffer.

Nluc activity was measured using the Nano-Glo Luciferase Assay Kit (Promega Corporation, Madison, WI, USA) according to the manufacturer's protocol. Luminescence was measured using a luminometer (Luminescencer-PSN, ATTO, Tokyo, Japan). For tissues, Nluc activity was corrected for protein quantity using a Pierce BCA Protein Assay Kit (ThermoFisher, Waltham, MA, USA) and expressed as relative light units (RLU) per mg protein. Cellular Nluc activity was expressed as RLU per 3000 cells.

2.8. Comparison of Transgene Expression Activity

At the tissue level, C57BL/6N mice were intravenously injected with Nluc mRNA-loaded A-11-LNPs, MC3-LNPs or RNA-LPX at a dose of 0.5 mg mRNA/kg. Twenty-four hours after the injection, liver, spleen, and inguinal lymph node (LN) were harvested, frozen in liquid nitrogen, and stored at −80 °C for use in a Nluc assay, as described in Section 2.7.

For splenic DC level, C57BL/6N mice were intravenously injected with EGFP mRNA-loaded A-11-LNPs, MC3-LNPs or RNA-LPX at a dose of 0.5 mg mRNA/kg. Twenty-four hours after the injection, the spleen was harvested. Splenocytes were dissociated from the remaining spleen tissues and were passed through a cell strainer (40 μm pore). The recovered cells were spun down (400× *g*, 4 °C, 5 min), and the supernatant was discarded. The resulting cells were resuspended in RBC Lysis buffer (1 mL) and incubated for 5 min at room temperature. These treated cells were washed with HBSS(-) by spinning (400× *g*, 4 °C, 5 min). The concentration of cells was adjusted to 1×10^7 cells/mL with FACS buffer, and the resulting cells were treated with a 10 μg/mL solution of an anti-mouse CD16/32 antibody and then incubated at 4 °C for 10 min. The resulting material was then treated with an APC anti-mouse CD11c antibody and a PerCP/Cyanine 5.5 anti-mouse I-A/I-E antibody at 4 °C for 30 min. The resulting cells were washed with FACS buffer twice, filtered via a nylon filter, and stained with PI, and the DCs were then analyzed for EGFP and I-A/I-E expression using CytoFLEX (Beckman Coulter, Inc., Brea, CA, USA).

2.9. Measurement of Activation Markers in Splenocytes

C57BL/6N mice were intravenously injected with OVA mRNA-loaded A-11-LNPs, MC3-LNPs or RNA-LPX at a dose of 0.03 mg mRNA/kg. Spleen tissues were harvested, and the resulting dissociated splenocytes were passed through a cell strainer (40 μm pore) 24 h after the injection. The recovered cells were spun down (400× *g*, 4 °C, 5 min), the supernatant discarded, and the resulting cells were resuspended in RBC Lysis buffer (1 mL) and incubated for 5 min at room temperature. These treated cells were washed with HBSS(-) by spinning (400× *g*, 4 °C, 5 min). The concentration of cells was adjusted to 1×10^7 cells/mL with FACS buffer, and the resulting cells were treated with a 10 μg/mL solution of an anti-mouse CD16/32 antibody, followed by incubation at 4 °C for 10 min. This preparation was then treated with fluorophore-conjugated antibodies at 4 °C for 30 min. Cells were washed with FACS buffer twice, filtered via nylon filter, stained with PI, and applied for flowcytometric analysis (CytoFLEX). DCs, T cells, B cells, and NK cells were defined as PI⁻I-A/I-E⁺CD11c⁺ cells, PI⁻CD3⁺CD19⁻ cells, PI⁻CD3⁻CD19⁺ cells, and PI⁻CD3⁻CD19⁻NK1.1⁺ cells, respectively. Relative expression of CD40, CD80, CD86 in DCs and CD69 in B cells, T cells, and NK cells were measured.

2.10. Prophylactic and Therapeutic Antitumor Effect of mRNA-Loaded LNPs on Mice

For the prophylactic antitumor experiment, C57BL6/N mice (6 weeks old) were intravenously injected with OVA mRNA-loaded A-11-LNPs at the indicated doses on day −14 and −7. On day 0, E.G7-OVA cells (8×10^6 cells/50 μL/mouse) were inoculated to the immunized mice subcutaneously in the right flank under isoflurane anesthesia. Tumor volumes were calculated using the equation shown below (tumor volume = major axis × minor axis² × 0.52) from day 6 to 21.

For therapeutic antitumor experiments, C57BL6/N mice (6 weeks old) were anesthetized with isoflurane. E.G7-OVA cells (8×10^6 cells/40 μL/mouse) were inoculated subcutaneously in the right flank. The tumor-bearing mice were intravenously injected with EGFP mRNA or OVA mRNA-loaded A-11-LNPs, OVA mRNA-loaded MC3-LNPs or OVA mRNA-loaded RNA-LPX at doses of 0.03 mg mRNA/kg on days 8 and 11. Tumor volumes were calculated by the above method from day 6 to 21.

2.11. Toxicity Test

C57BL6/N mice (6 weeks old) were intravenously injected with OVA mRNA-loaded A-11-LNPs at two doses of 0.03 mg mRNA/kg on days 0 and 3. Blood was obtained 24 h after the last dose and processed into plasma using heparin. Alanine transferase (ALT), aspartate transferase (AST), total bilirubin (T-BIL), lactate dehydrogenase (LDH), blood urea nitrogen (BUN), and creatinine (CRE) levels in plasma were measured at Oriental yeast Co., Ltd. (Shiga, Japan).

2.12. Statistical Analyses

Statistical data obtained by the design of the experiment (DOE) were analyzed using the JMP 14 software (SAS, Cary, NC, USA). Statistical significance was defined as *p*-values less than 0.05. Two independent experiments were performed for the DOE. To identify significant factors for each physicochemical property of the LNPs in a relatively vast experimental design space, a $3^4 \times 2^2$ definitive screening design (DSD) was used for screening A. Effective design-based model selection for DSD or the forward stepwise regression method with Akaike's information criterion and finite correction (c-AIC) was applied to each response. The forward stepwise regression method with c-AIC was applied only in cases where the number of both statistically significant main factors and interactions between 2 factors were less than 3. Responses with ranges of several digits (i.e., gene expression, cellular uptake) were converted to logarithms in order to ensure linearity. For screening B, a 2^4 fractional factorial design (FFD) was used. A standard least squares linear regression model was applied to each response.

Results are expressed as the mean + standard deviation (SD) or mean ± SD of independent repeats. For comparisons between the means of two variables, we used unpaired Student's *t*-tests. For comparisons between multiple groups, we used one-way analysis of variance (ANOVA) with the Tukey–Kramer posthoc tests. These statistical analyses were done using the JMP 14 software.

3. Results

3.1. Optimization Strategy for Splenic DCs

Microfluidic technology has made it possible to reproducibly produce relatively small and uniform LNPs and has recently been adopted as a major production method [42]. The rapid and reproducible mixing of two liquids (an alcohol solution of lipids and a buffer solution containing RNAs) is achieved by introducing them at a high flow rate into a microfluidic device equipped with a micromixer. While most RNA-loaded LNPs that are formed by microfluidic technology are less than 100 nm in size [43–46], we recently reported that the addition of a salt (e.g., NaCl) to the RNA-containing buffer significantly contributed to the formation of LNPs with sizes in excess of 100 nm based on the Derjaguin–Landau–Verwey–Overbeek (DLVO) theory [39]. The addition of a salt reduces electrostatic repulsion and promotes fusion between the initially produced liposome-like cationic particles, resulting in the formation of larger LNPs in a salt concentration-dependent manner. Both cellular uptake and the functional delivery of siRNA and mRNA in splenic DCs were significantly higher when larger LNPs were used compared to the smaller counterparts with a consistent lipid composition. This can be explained by the fact that macropinocytosis, which is a unique pathway characterized by the nonspecific internalization of large amounts of extracellular fluid, is constitutively active in immature DCs and, therefore, relatively larger-sized particles (e.g., 200 nm or higher) would be beneficial for targeting DCs [47,48]. Although the finding suggests that an increase in LNP size is a significant factor for the efficient delivery of RNAs to splenic DCs, the optimal ranges of the size and other factors, including lipid composition, have not yet been clarified.

In this study, we first conducted 2 steps of DOE for screening and optimizing both synthetic conditions (NaCl concentration) and the lipid composition of LNPs for the delivery of mRNA to splenic DCs. The experimental scheme is represented in Figure 1. The physicochemical properties and cellular uptake of the LNPs, which would be mediator variables, were also measured in an attempt to understand how these factors contribute to the functional delivery (final output). Nluc mRNA was used to quantify the transgene expression level. The long intracellular half-life of the Nluc protein (>6 h) attenuates the potential effects of protein degradation during the lengthy (at least several hours) experimental process from sacrifice to cell sorting and would be suitable for the measurement of enzymatic activity in the sorted cells. The bright signal from Nluc was also suitable for quantitatively detecting transgene expression levels in the limited number of sorted cells. DiO-labeled LNPs were used to assess cellular uptake. Three types of APCs, including DCs, macrophages (Mφ), and B cells, were used in the analysis.

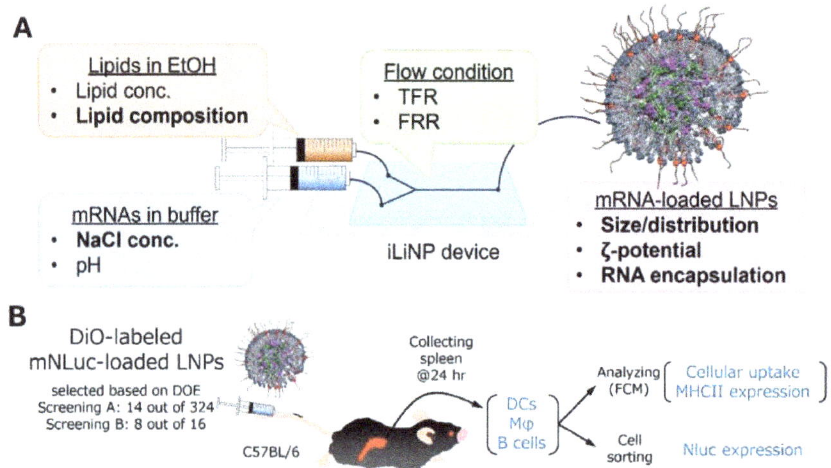

Figure 1. Schematic illustration of the experimental method for DOE. (**A**) Parameters and responses in manufacturing LNPs. Both parameters and responses examined in this study were expressed in bold. (**B**) Method for screening of LNPs in vivo. Cellular uptake, NHC class II expression, and Nluc expression in splenic DCs, Mφ, and B cells after intravenous injection of DiO-labeled mNLuc-loaded LNPs selected by DOE were measured to identify the optimal formulation.

Adjuvant effects associated with type I interferon (IFN) (IFN-I) stimulation have been suggested to lead to superior acquired immune responses, and pathways that activate IFN-I expression have been identified, including the TLRs, RLRs, and stimulator of interferon genes (STING) pathways [12]. Mouse TLR7, which recognizes single-stranded RNA (ssRNA) and activates the adaptor protein myeloid differentiation primary response 88 (MyD88), leads to the expression of a suite of inflammatory cytokines, including IFN-I [49]. It is generally thought that regular uridine-containing mRNA, which is an ssRNA, stimulates the innate immune system via TLR7. The cytoplasmic RNA sensor RIG-I binds to double-stranded RNA (dsRNA) and induces the expression of inflammatory cytokines such as IFNβ through the activation of the adapter protein mitochondrial antiviral signaling protein (MAVS) [50]. Trace amounts of dsRNA, a byproduct of the in vitro transcription of mRNA, contribute to IFNβ production via the RIG-I/MAVS pathway [51,52]. In this study, we used regular uridine-containing mRNAs, which were obtained from Trilink BioTechnologies, which are thought to induce IFN-I expression through stimulation of the innate immune system via the RIG-I/MAVS pathway and TLR7. IFN-I is known to drive a distinctive DC maturation program, including the continuous upregulation of MHC-II and antigen processing [53], and to be highly expressed by DCs (especially by plasmacytoid DCs that highly and constitutively express IFN response factor-7) [54]. Therefore, the efficient introduction of immunostimulatory mRNAs into DCs results in the efficient maturation of DCs and, for dendritic cells, the expression level of MHC class II (I-A/I-E), one of the maturation markers, was also quantified in the screening.

3.2. DOE-Based Optimization of LNPs

In the 1st DOE (screening A), 6 independent factors, including the molar percentage of CL (level: 40 to 60), PL (level: 10 to 40), PEG-lipid (level: 0.5 to 1.5), a type of CL (level: CL4H6 or CL7H6), PEG-lipid (level: PEG-DMG or PEG-DSG), and NaCl concentration (level: 0 to 400 mM), were systematically examined. Since TFR and FRR had only a limited impact on the size of LNPs in our previous study and have substantial issues (reduced reproducibility at lower levels and unintended dilution at higher levels, respectively), these parameters were fixed at 500 µL/min and 3, respectively. DSD was adopted in screening A to determine the significant contributing factor(s) for the responses (including physicochemical properties, cellular uptake, and gene expression) and to narrow down experimental space of the following 2nd DOE (screening B) (Table 1). The RNA-loaded LNPs were synthesized under 14 different formulation conditions (coded as A-1 to A-14) that were determined based on DSD. The diameter of the LNPs varied with a ζ-average from 88 to 754 nm (Table 1). The LNPs were typically uniform (PdI of 0.2 or less) except for A-4 and A-5. The encapsulation efficiency and ζ-potential of the LNPs were typically high (85% or over) and near neutral (within $\pm$ 5 mV), respectively, except for A-14 (72.4% encapsulation and -10.3 mV). The reproducibility of the synthesis of the LNPs between 2 technically independent experiments were confirmed (Figure S1). Statistical analysis revealed that an increase in NaCl concentration had the most significant effect on increasing the size of the LNP, an observation that is consistent with our previous study (Figure S2). A decrease in both %DOPE and %PEG also had a significant effect on the increase in size (Figure S2). This was also consistent with our previous study [39] and can be explained by the fact that the relatively bulky hydrophilic moieties of both DOPE and PEG-lipid resist the decrease in total surface area of the LNP with increasing particle size. Both CL4H6 and a higher %DOPE significantly contributed to a higher encapsulation efficiency (Figure S3). It is possible that an oxygen atom in the morpholino ring at the hydrophilic head of CL7H6 attracts electrons from the tertiary amine, thereby lowering the acid-dissociation constant and reducing the RNA encapsulation efficiency, especially under competitive conditions in the presence of NaCl. DOPE can facilitate encapsulation of RNA due to the fact that a primary amino group of DOPE, which is a proton donor, can form direct hydrogen bonds with phosphate groups of RNAs [30].

Table 1. Physicochemical properties of the LNPs examined in screening A.

Entry	%CL	%DOPE	%PEG	NaCl Conc. (mM)	CL	PEG	ζ-Average (nm)	PdI	ζ-Potential (mV)	%RNA Encapsulation
A-1	40	40	0.5	0	CL7H6	PEG-DMG	112	0.20	-3.2	90.3
A-2	50	10	0.5	0	CL4H6	PEG-DMG	121	0.19	-2.3	99.2
A-3	60	10	1	0	CL7H6	PEG-DSG	104	0.15	-2.3	91.0
A-4	60	40	1.5	0	CL7H6	PEG-DMG	128	0.41	-0.6	89.8
A-5	40	25	1.5	0	CL4H6	PEG-DSG	88	0.30	-1.5	99.2
A-6	60	40	0.5	200	CL4H6	PEG-DSG	155	0.13	1.9	98.8
A-7	50	25	1	200	CL4H6	PEG-DMG	138	0.12	-1.1	99.0
A-8	40	10	1.5	200	CL7H6	PEG-DMG	128	0.20	-2.6	85.0
A-9	50	25	1	200	CL7H6	PEG-DSG	171	0.15	-2.7	88.0
A-10	40	40	1	400	CL4H6	PEG-DMG	216	0.08	2.5	99.0
A-11	60	10	1.5	400	CL4H6	PEG-DSG	547	0.19	-1.4	89.2
A-12	50	40	1.5	400	CL7H6	PEG-DSG	108	0.07	-1.7	88.3
A-13	60	25	0.5	400	CL7H6	PEG-DMG	501	0.12	-1.2	91.2
A-14	40	10	0.5	400	CL4H6	PEG-DSG	754	0.17	-10.3	72.4

Cellular uptake and Nluc activity in 3 types of splenic APCs and whole spleens for the 14 types of LNPs are summarized in Table S1. The reproducibility of Nluc activity in the whole spleen, cellular uptake and Nluc activity in splenic DCs were confirmed (Figure S4). Scatter plots revealed that the cellular uptake in the 3 types of splenic APCs was well correlated (Figure 2A,B), suggesting that factors and their levels examined in screening A have moderate effects on cell specificity between the 3 types of splenic APCs. On the other hand, the amount of cellular uptake differed significantly (1 to 2 orders of magnitude) among the LNPs, and the order was Mφ > DCs >> B cells. Scatter plots of Nluc expression versus cellular uptake in DCs and Mφ showed a trend toward higher activity in DCs (Figure 2C). An analysis of covariance was performed to reveal statistical differences in the Nluc expression level between DCs and Mφ without the effect of cellular uptake level, and the findings revealed that Nluc expression in the DCs was significantly higher than that in Mφ, which is consistent with observations obtained in a previous report [2]. The slope of Nluc activity to cellular uptake was obviously higher than 1 for DCs, suggesting that more LNPs are taken up per cell or LNPs with characteristics that allow them to be more easily taken up lead to higher transgene expression in DCs. The maximum Nluc expression in B cells was less than 100 RLU/3000 cells, which was over 100-fold lower than that in both DCs and Mφ (Table S1). Scatter plots of Nluc expression in the whole spleen vs. splenic DCs showed a moderate correlation (R^2 = 0.3919) (Figure 2D). Although the transgene expression level in the whole spleen has been extensively measured in many studies on mRNA vaccine-oriented formulation development due to its simplicity [55–57], the data indicate that transgene expression level in the whole spleen would not be a good indicator of the corresponding process in DCs, even if the transgene expression level in DCs was high. This can be attributed to the fact that the population of splenic DCs is only ~1% of all splenocytes. Therefore, much lower Nluc signals derived from B cells (~50% of splenocytes) and other cell types that were not examined in this study would account for a non-negligible proportion.

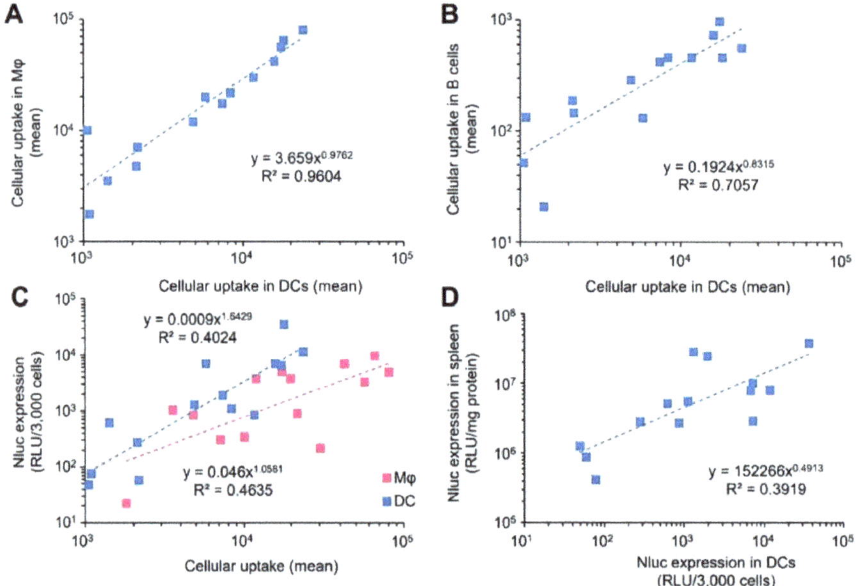

Figure 2. Correlations between outputs of in vivo experiments in screening A. (**A**) A dot-plot of cellular uptake in Mφ versus that in DCs. (**B**) A dot-plot of cellular uptake in B cells versus that in DCs. (**C**) A dot-plot of Nluc expression versus cellular uptake in both Mφ and DCs. (**D**) A dot-plot of Nluc expression in spleen versus in DCs.

Concerning Nluc activity in splenic DCs, a well-fitted regression model was obtained by the effective design-based model selection for DSD process (Figure 3A). The Nluc expression (RLU/3000 cells) was converted to logarithms before regression because the range of the Nluc activity spans 3 orders of magnitude. A total of 5 out of all 6 factors that were examined in screening A were found to significantly contribute to Nluc expression in splenic DCs (Figure 3B). CL showed the highest impact on the Nluc expression level, and the use of CL4H6 was preferable to CL7H6. A higher %CL was also important in terms of improving the Nluc expression level. In addition, lower %PEG and higher NaCl concentration were also significant in improving the Nluc activity. The cellular uptake and Nluc expression in splenic DCs were plotted against ζ-average because the two factors are also significant determinants of LNP size. The plot clearly indicated that cellular uptake sharply increases by the ζ-average for sizes of up to ~200 nm (Figure 3C), but the upward trend was less pronounced in the range above 200 nm. On the other hand, the cellular uptake of extremely large LNPs (>700 nm) by DCs tended to be decreased. The plot for Nluc expression also showed a similar trend to cellular uptake (Figure 3D). Particles that are too large have the potential to induce toxicity and cause the obstruction of pulmonary microvessels. Therefore, an optimal LNP diameter for splenic DCs would be in the range of 200 to 500 nm. These observations suggest that the ζ-average is an intermediate factor that links the examined factors and cellular uptake or transgene expression level. To confirm the validity of the above regression model, the worst 5 and the top 5 LNPs were selected from 14 LNPs that were tested in screening A, and the occupancy of each level in each factor was then visualized. Nluc activity, cellular uptake, and ζ-average between the 2 groups were significantly different (Figure 3E–G). For %PEG, NaCl concentration, and CL, the levels of occupancy were clearly reversed between the two groups (Figure 3H–J). For example, 80% CL7H6 and 80% CL4H6 were occupied in the worst 5 and the top 5, respectively, for CL (Figure 3J). These results are consistent with the regression model. For the remaining factors, including %CL, %PL, and PEG, differences between the 2 groups were not clear, while both %CL and PEG were significant in the regression model (Figure 3K–M). Given the above findings, PEG was again examined in the next DOE. The level of %CL was increased from 40–60 to 50–60%. The level of %PL was narrowed down to 10–20% to reduce its impact on the size of LNPs. The level of %PEG was narrowed down from 0.5–1.5 to 0.75–1.5% to avoid the production of extremely large LNPs. CL was fixed with CL4H6. NaCl concentration was determined to be 300 mM for producing LNPs with an appropriate size based on simulations using the regression model.

In screening B, 8 different LNPs were synthesized based on the FFD (Table S2) and evaluated in vivo (Table S3). The diameters of the LNPs were in the range of approximately 110 to 420 nm with a narrow distribution (PdI < 0.15) and high encapsulation (>80%). In vivo screening indicated positive correlations between LNP size and cellular uptake or transgene expression in splenic DCs (R^2 = 0.8186 or 0.9173, respectively) (Figure S5A,B), a trend similar to that observed in screening A. Substantial correlations were observed when data from screening A and B were combined (Figure S5C,D). On the other hand, correlations between LNP size and transgene expression in the whole spleen were clearly lower (R^2 = 0.1148) (Figure S6A). The correlation between transgene activity in splenic DCs and the whole spleen was also low when data from screening A and B were combined (R^2 = 0.3582) (Figure S6B). These results suggest that transgene activity in the whole spleen is not a good indicator of the same process in splenic DCs, and that evaluation at the cell type level is essential for selecting the optimal formulation.

Based on the screenings, the A-11-LNPs showed the highest transgene activity in splenic DCs (Tables S1 and S3). The A-11-LNPs showed the most enhanced I-A/I-E expression in splenic DCs among all of the LNPs that were tested (Figure S7). The size and RNA encapsulation of the A-11-LNPs remained constant at 4 °C for at least 24 days after their preparation, indicating a stable formulation (Figure S8). Considering the above data, we conclude that the A-11-LNPs were the optimal formulation for targeting splenic DCs as mRNA vaccines.

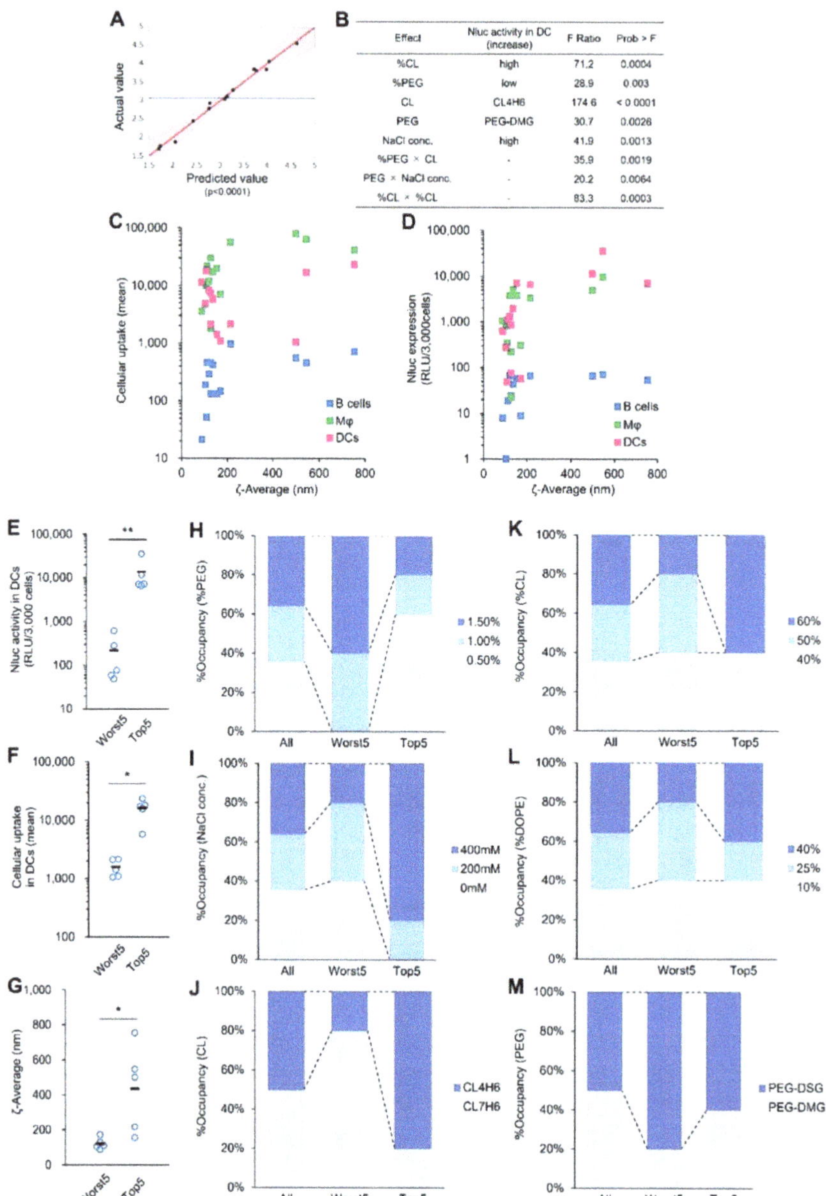

Figure 3. Analysis of significant factors for outputs in in vivo experiment in screening A. (**A**) A plot of predicted and actual values for Nluc expression in splenic DCs obtained by a regression model. (**B**) Statistically significant main factors and interactions for Nluc expression in splenic DCs. (**C,D**) Plots of cellular uptake (**C**) or Nluc expression (**D**) in splenic DCs versus ζ-average. Two independent experiments with one mouse in each preparation were performed. The plots are expressed as the average of the two experiments. (**E–G**) Comparison of Nluc expression (**E**), cellular uptake (**F**), and ζ-average (**G**) between top 5 and worst 5 LNPs on Nluc expression in splenic DCs. $n = 5$, * $p < 0.05$, ** $p < 0.01$. (**H–M**) Occupancy of each level of %PEG (**H**), NaCl conc. (**I**), CL (**J**), %CL (**K**), %DOPE (**L**), and PEG (**M**).

3.3. Comparison with Clinically Relevant Formulations

We next tested the A-11-LNPs against two clinically relevant RNA-loaded LNP formulations to determine their potential as an mRNA vaccine formulation. The first was the MC3-LNP, also known as Onpattro, a first-ever siRNA therapeutic developed by Alnylam pharmaceuticals for the treatment of transthyretin (TTR) amyloidosis [58,59]. The MC3-LNP was developed as a siRNA formulation targeting liver tissue. The MC3-LNP is considered to be the gold standard LNP formulation and is also widely used as a formulation for mRNA delivery [17,57,60,61]. For these reasons, it was used as a comparative formulation in this study. The second is RNA-LPX, which was reported by Kranz, et al. as a formulation that shows excellent cancer vaccine efficacy by selectively introducing mRNA into splenic DCs [2]. We selected this formulation because it is oriented for the same application as the A-11-LNP, and, because of this, we considered it to be the most suitable control. The MC3-LNP was 59 nm in diameter and showed a 97% mRNA encapsulation. The RNA-LPX was 319 nm in diameter and negatively charged (ζ-potential of −17 mV), consistent with previous reports [2]. The transgene expression level in secondary lymphoid tissues, the spleen and inguinal LN, was next evaluated using Nluc mRNA. The A-11-LNPs showed significantly higher transgene expression levels in both tissues over the two control formulations (Figure 4A). The RNA-LPX showed a superior transgene expression in the spleen compared to MC3-LNPs, which would be reasonable because of the fact that the RNA-LPX was reported as a spleen-selective mRNA delivery system. However, and interestingly, the MC3-LNPs showed a much higher transgene expression in inguinal LN over the RNA-LPX. The transgene expression level in splenic DCs was then measured using EGFP mRNA. The A-11-LNPs achieved approximately 9% of EGFP⁺ DCs, which was significantly higher than the two control formulations, indicating the superior transgene expression potency of the A-11-LNPs (Figure 4B). The highest I-A/I-E expression in splenic DCs was also observed for the A-11-LNPs (Figure 4C). These results suggest that the A-11-LNP would have a superior mRNA vaccine effect.

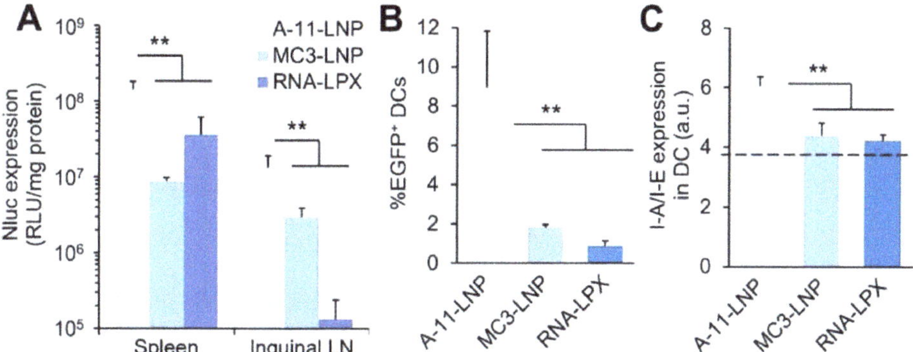

Figure 4. Comparison of the A-11-LNPs with clinically relevant formulations. (**A**) Nluc expression in spleen and inguinal LN 24 h after intravenous administration of Nluc-mRNA-loaded formulations at a dose of 0.5 mg mRNA/kg. (**B**) Percentage of EGFP⁺ splenic DCs 24 h after the intravenous administration of EGFP-mRNA-loaded formulations at a dose of 0.5 mg mRNA/kg. (**C**) I-A/I-E expression in splenic DCs 24 h after intravenous administration of EGFP-mRNA-loaded formulations at a dose of 0.5 mg mRNA/kg. *n* = 3. ** *p* < 0.01.

Motivated by the above results, a prophylactic antitumor experiment in E.G7-OVA tumor-bearing mouse model was conducted. Two sequential administrations of OVA mRNA-loaded A-11-LNPs completely rejected the tumor establishment at all doses tested (Figure 5A). We next examined the therapeutic antitumor effect of OVA mRNA-loaded A-11-LNPs in the same tumor-bearing mouse model. Since a clear shrinkage of tumor tissues was observed at doses of 0.015 mg mRNA/kg or higher in the dose-dependent study (Figure S9), the dose in subsequent experiments was set at 0.03 mg mRNA/kg. A comparative study of the therapeutic antitumor effects of the A-11-LNP and two control formulations (MC3-LNP and RNA-LPX) showed that only the A-11-LNP exhibited significant antitumor activity (Figure 5B). Increased serum IFNβ levels were observed only after the administration of the A-11-LNPs (Figure S10). The A-11-LNPs induced the maturation of splenic DCs, which resulted in the upregulation of the activation markers CD40, CD80, and CD86 (Figure 5C–E). The A-11-LNPs also activated splenic B, T, and natural killer cells, which upregulated the activation marker CD69 (Figure 5F–H). No antitumor activity was observed in the case of the EGFP mRNA-loaded A-11-LNP (Figure 5B), indicating that the antitumor effect depends on the development of an OVA antigen-specific immune response. In addition, B-8-LNPs, which showed significantly lower activity (2.1%) in splenic DCs but exhibited comparable activity (73%) in the whole spleen compared to the A-11-LNPs (Tables S1 and S3), showed no significant antitumor effect (Figure 5B), suggesting that mRNA was efficiently delivered to DCs rather than to the whole spleen, thus leading to antitumor activity. The limited induction of I-A/I-E expression in splenic DCs by the B-8-LNPs also supports this suggestion (Figure S7). Although clear antitumor effects were observed for two doses (0.03 mg mRNA/kg × 2) of the A-11-LNPs, the tumors tended to re-grow after day18 without having completely disappeared (Figure 5B). Based on this observation, the number of doses and dosages were increased (0.125 to 0.5 mg mRNA/kg × 3) to examine the therapeutic antitumor effect. Although there was one individual that experienced a complete response at doses of 0.25 and 0.5 mg mRNA/kg, the results showed a limited increase in the overall antitumor effect (Figure S11). This suggests that the immunosuppressive pathways are enhanced in association with the induction of antitumor immunity by the A-11-LNPs. In fact, E.G7-OVA cells express programmed cell death ligand-1 (PD-L1) [62], one of the major immune checkpoint molecules, and the combination of an adjuvant and an anti-PD-L1 antibody, an immune checkpoint inhibitor (ICI), appears to promote E.G7-OVA tumor shrinkage compared to the use of the adjuvant alone [63,64]. Therefore, the combination of the A-11-LNPs and ICIs would be a reasonable strategy for enhancing an antitumor effect in this tumor model. In addition, when the same tumor cells were re-challenged at 68 days after the initial tumor transplantation in the two individuals that completely responded, they completely rejected tumor engraftment (Figure S12), suggesting that treatment with the A-11-LNPs induced the formation of memory cells, which was also observed for the RNA-LPX in the previous study [2].

Finally, a hematological test was performed after two doses of the OVA mRNA-loaded LNPs at 0.03 mg mRNA/kg had been administered, and no significant alteration in any of the serum chemistry parameters tested was observed, suggesting that the A-11-LNPs are well tolerated under therapeutically relevant doses (Figure 6).

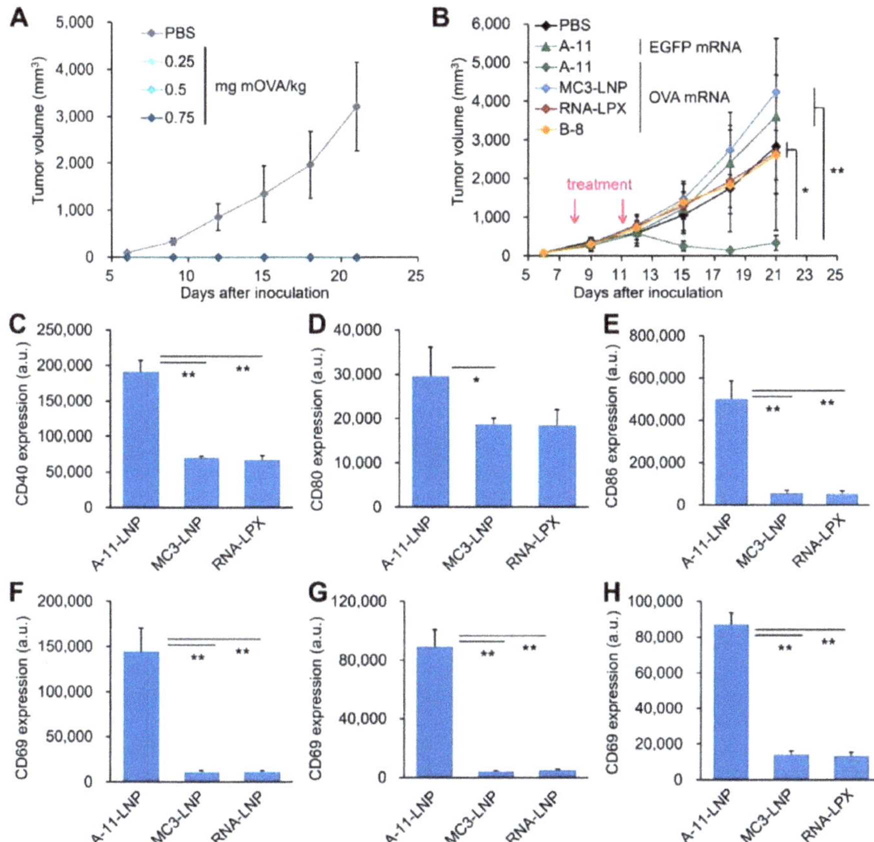

Figure 5. Prophylactic and therapeutic antitumor activity. (**A**) Prophylactic antitumor activity of the A-11-LNPs in E.G7-OVA tumor-bearing mice. The OVA mRNA-loaded A-11-LNPs were intravenously injected at the indicated doses twice on 14 and 7 days before tumor inoculation. $n = 3$–4. (**B**) The therapeutic antitumor activity of the A-11-LNPs, MC3-LNP, RNA-LPX, and B-8-LNPs. E.G7-OVA tumor-bearing mice were intravenously injected with OVA mRNA-loaded formulations at two doses of 0.03 mg mRNA/kg on day 8 and 11. $n = 5$. * $p < 0.05$, ** $p < 0.01$. (**C**–**E**) Expression of activation markers CD40 (**C**), CD80 (**D**), and CD86 (**E**) in splenic DCs 24 h after an intravenous injection of OVA mRNA-loaded formulations at a dose of 0.03 mg mRNA/kg. $n = 3$. * $p < 0.05$, ** $p < 0.01$. (**F**–**H**) Expression of an activation marker CD69 in splenic B cells (**F**), T cells (**G**), and NK cells (**H**) 24 h after an intravenous injection of OVA mRNA-loaded formulations at a dose of 0.03 mg mRNA/kg. $n = 3$. ** $p < 0.01$.

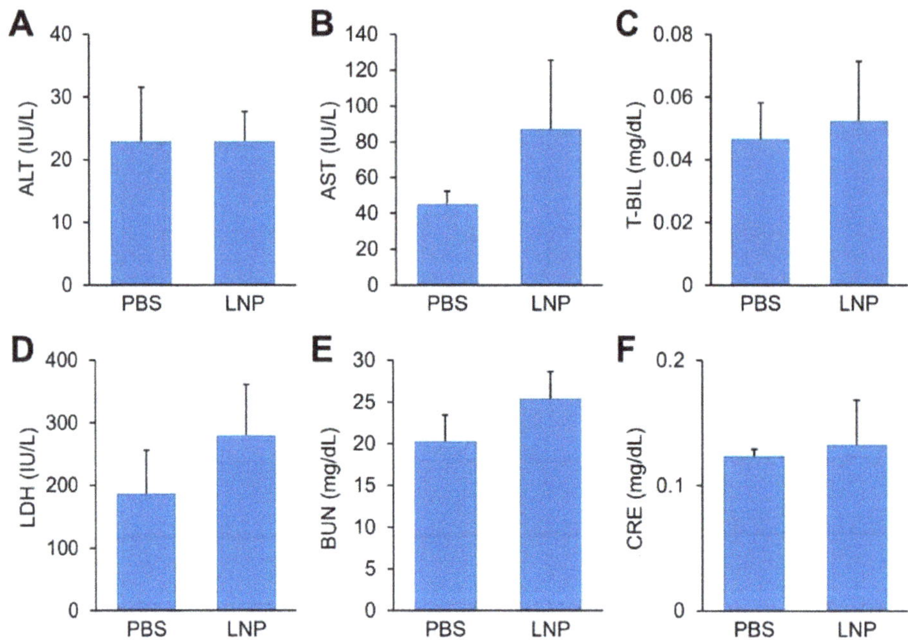

Figure 6. Safety of the A-11-LNPs. Serum chemistry parameters, ALT (**A**), AST (**B**), T-BIL (**C**), LDH (**D**), BUN (**E**), and CRE (**F**) were measured 24 h after the last injection of OVA mRNA-loaded A-11-LNPs at two doses of 0.03 mg mRNA/kg. $n = 3$.

4. Conclusions

In the present study, the DOE-based in vivo optimization of mRNA-loaded LNPs for mRNA vaccines was studied. The screening included a wide range of particle sizes that were controlled by the addition of NaCl in addition to the lipid type used and its composition. The results showed a clear correlation between particle size and uptake and gene expression activity in splenic DCs and indicated that a size range from 200 to 500 nm is appropriate for use in conjunction with splenic DCs. It was also found that transgene expression activity in the whole spleen, which can be easily evaluated, was not a valid predictor of transgene expression activity in splenic DCs and potency as mRNA vaccines. The A-11-LNP, which was found to have the optimal formulation, induced better transgene expression activity and maturation in DCs compared to two clinically relevant LNP formulations and induced clear therapeutic antitumor effects in an E.G7-OVA tumor model. The findings reported herein are expected to contribute to the development of future mRNA vaccines.

5. Patents

K. Sasaki, Y. Sato, and H. Harashima have filed intellectual property patents related to this publication.

Supplementary Materials: The following supporting information can be downloaded at: https://www.mdpi.com/article/10.3390/pharmaceutics14081572/s1, Synthesis of CL7H6, Figure S1: Reproducibility of physicochemical properties of the 14 LNPs (A-1 to A-14) examined in screening A, Figure S2: Statistical information for the effective design-based model selection for definitive screening design to predict the ζ-average of the LNPs (A-1 to A-14), Figure S3: Statistical information for the effective design-based model selection for definitive screening design to predict the percentage of RNA encapsulation of the LNPs (A-1 to A-14), Figure S4: Reproducibility of outputs of in vivo experiments in screening A, Figure S5: Correlation between ζ-average and cellular uptake or Nluc expression in splenic DCs, Figure S6: Correlation between ζ-average and Nluc expression, Figure S7: I-A/I-E expression in splenic DCs after intravenous injection of LNPs examined in screening A and B, Figure S8: Stability of the A-11-LNPs, Figure S9: Examination of dose of the OVA mRNA-loaded A-11-LNPs for therapeutic antitumor experiment, Figure S10: Serum IFNβ level after an intravenous injection of OVA mRNA-loaded formulations at a dose of 0.03 mg mRNA/kg, Figure S11: Therapeutic antitumor effect of the A-11-LNPs on high dose conditions, Figure S12: Examination of formation of memory cells, Table S1: Full data of outputs of in vivo experiment in screening A, Table S2: Formulation conditions and physicochemical properties of the LNPs examined in screening B, Table S3: Full data of outputs of in vivo experiment in screening B.

Author Contributions: Conceptualization, Y.S. and H.H.; methodology, Y.S. and H.H.; validation, Y.S. and H.H.; formal analysis, Y.S.; investigation, K.S., K.O. and K.I.; resources, Y.S.; data curation, K.S., Y.S., K.O. and K.I.; writing—original draft preparation, K.S., Y.S., K.O. and K.I.; writing—review and editing, K.S., Y.S., K.O., K.I. and H.H.; visualization, Y.S.; supervision, Y.S. and H.H.; project administration, Y.S. and H.H.; funding acquisition, Y.S. and H.H. All authors have read and agreed to the published version of the manuscript.

Funding: This work was supported, in part, by the Hokkaido University Support Program for Frontier Research, The Mochida Memorial Foundation for Medical and Pharmaceutical Research and Research Expenses from the Ministry of Education, Culture, Sports, Science and Technology.

Institutional Review Board Statement: The experimental protocols were reviewed and approved by the Hokkaido University Animal Care Committee in accordance with the guidelines for the care and use of laboratory animals (approval number: 16-0015, date 28 March 2016; 20-0176, date 1 March 2021).

Informed Consent Statement: Not applicable.

Data Availability Statement: Not applicable.

Acknowledgments: The authors wish to express their gratitude to both Manabu Tokeshi and Masatoshi Maeki (Division of Applied Chemistry, Faculty of Engineering, Hokkaido University) for providing a microfluidic device (iLiNP device). The authors also wish to thank Milton S. Feather for his helpful advice in preparing the English manuscript.

Conflicts of Interest: The authors declare no conflict of interest.

References

1. Moingeon, P.; Lombardi, V.; Saint-Lu, N.; Tourdot, S.; Bodo, V.; Mascarell, L. Adjuvants and Vector Systems for Allergy Vaccines. *Immunol. Allergy Clin. N. Am.* **2011**, *31*, 407–419. [CrossRef]
2. Kranz, L.M.; Diken, M.; Haas, H.; Kreiter, S.; Loquai, C.; Reuter, K.C.; Meng, M.; Fritz, D.; Vascotto, F.; Hefesha, H.; et al. Systemic RNA delivery to dendritic cells exploits antiviral defence for cancer immunotherapy. *Nature* **2016**, *534*, 396–401. [CrossRef]
3. Bogers, W.M.; Oostermeijer, H.; Mooij, P.; Koopman, G.; Verschoor, E.J.; Davis, D.; Ulmer, J.B.; Brito, L.A.; Cu, Y.; Banerjee, K.; et al. Potent Immune Responses in Rhesus Macaques Induced by Nonviral Delivery of a Self-amplifying RNA Vaccine Expressing HIV Type 1 Envelope with a Cationic Nanoemulsion. *J. Infect. Dis.* **2014**, *211*, 947–955. [CrossRef]
4. Pardi, N.; Hogan, M.J.; Pelc, R.S.; Muramatsu, H.; Andersen, H.; DeMaso, C.R.; Dowd, K.A.; Sutherland, L.L.; Scearce, R.M.; Parks, R.; et al. Zika virus protection by a single low-dose nucleoside-modified mRNA vaccination. *Nature* **2017**, *543*, 248–251. [CrossRef] [PubMed]
5. Richner, J.M.; Himansu, S.; Dowd, K.A.; Butler, S.L.; Salazar, V.; Fox, J.M.; Julander, J.G.; Tang, W.W.; Shresta, S.; Pierson, T.C.; et al. Modified mRNA Vaccines Protect against Zika Virus Infection. *Cell* **2017**, *168*, 1114–1125.e10. [CrossRef] [PubMed]
6. Lamb, Y.N. BNT162b2 mRNA COVID-19 Vaccine: First Approval. *Drugs* **2021**, *81*, 495–501. [CrossRef] [PubMed]
7. Topol, E.J. Messenger RNA vaccines against SARS-CoV-2. *Cell* **2021**, *184*, 1401. [CrossRef]
8. Schoenmaker, L.; Witzigmann, D.; Kulkarni, J.A.; Verbeke, R.; Kersten, G.; Jiskoot, W.; Crommelin, D.J.A. mRNA-lipid nanoparticle COVID-19 vaccines: Structure and stability. *Int. J. Pharm.* **2021**, *601*, 120586. [CrossRef] [PubMed]
9. Pardi, N.; Hogan, M.J.; Porter, F.W.; Weissman, D. mRNA vaccines—A new era in vaccinology. *Nat. Rev. Drug Discov.* **2018**, *17*, 261–279. [CrossRef] [PubMed]
10. Kwon, H.; Kim, M.; Seo, Y.; Moon, Y.S.; Lee, H.J.; Lee, K.; Lee, H. Emergence of synthetic mRNA: In vitro synthesis of mRNA and its applications in regenerative medicine. *Biomaterials* **2018**, *156*, 172–193. [CrossRef] [PubMed]
11. Pollard, C.; Rejman, J.; De Haes, W.; Verrier, B.; Van Gulck, E.; Naessens, T.; De Smedt, S.; Bogaert, P.; Grooten, J.; Vanham, G.; et al. Type I IFN Counteracts the Induction of Antigen-Specific Immune Responses by Lipid-Based Delivery of mRNA Vaccines. *Mol. Ther.* **2013**, *21*, 251–259. [CrossRef]

12. Li, K.; Qu, S.; Chen, X.; Wu, Q.; Shi, M. Promising Targets for Cancer Immunotherapy: TLRs, RLRs, and STING-Mediated Innate Immune Pathways. *Int. J. Mol. Sci.* **2017**, *18*, 404. [CrossRef] [PubMed]

13. Rettig, L.; Haen, S.P.; Bittermann, A.G.; von Boehmer, L.; Curioni, A.; Krämer, S.D.; Knuth, A.; Pascolo, S. Particle size and activation threshold: A new dimension of danger signaling. *Blood* **2010**, *115*, 4533–4541. [CrossRef] [PubMed]

14. Michel, T.; Luft, D.; Abraham, M.K.; Reinhardt, S.; Salinas Medina, M.L.; Kurz, J.; Schaller, M.; Avci-Adali, M.; Schlensak, C.; Peter, K.; et al. Cationic Nanoliposomes Meet mRNA: Efficient Delivery of Modified mRNA Using Hemocompatible and Stable Vectors for Therapeutic Applications. *Mol. Ther. Nucleic Acids* **2017**, *8*, 459–468. [CrossRef] [PubMed]

15. Habrant, D.; Peuziat, P.; Colombani, T.; Dallet, L.; Gehin, J.; Goudeau, E.; Evrard, B.; Lambert, O.; Haudebourg, T.; Pitard, B. Design of Ionizable Lipids To Overcome the Limiting Step of Endosomal Escape: Application in the Intracellular Delivery of mRNA, DNA, and siRNA. *J. Med. Chem.* **2016**, *59*, 3046–3062. [CrossRef]

16. Hajj, K.A.; Melamed, J.R.; Chaudhary, N.; Lamson, N.G.; Ball, R.L.; Yerneni, S.S.; Whitehead, K.A. A Potent Branched-Tail Lipid Nanoparticle Enables Multiplexed mRNA Delivery and Gene Editing In Vivo. *Nano Lett.* **2020**, *20*, 5167–5175. [CrossRef]

17. Miao, L.; Li, L.; Huang, Y.; Delcassian, D.; Chahal, J.; Han, J.; Shi, Y.; Sadtler, K.; Gao, W.; Lin, J.; et al. Delivery of mRNA vaccines with heterocyclic lipids increases anti-tumor efficacy by STING-mediated immune cell activation. *Nat. Biotechnol.* **2019**, *37*, 1174–1185. [CrossRef]

18. Cornebise, M.; Narayanan, E.; Xia, Y.; Acosta, E.; Ci, L.; Koch, H.; Milton, J.; Sabnis, S.; Salerno, T.; Benenato, K.E. Discovery of a Novel Amino Lipid That Improves Lipid Nanoparticle Performance through Specific Interactions with mRNA. *Adv. Funct. Mater.* **2022**, *32*, 2106727. [CrossRef]

19. Semple, S.C.; Chonn, A.; Cullis, P.R. Influence of cholesterol on the association of plasma proteins with liposomes. *Biochemistry* **1996**, *35*, 2521–2525. [CrossRef] [PubMed]

20. Senior, J.; Gregoriadis, G. Stability of small unilamellar liposomes in serum and clearance from the circulation: The effect of the phospholipid and cholesterol components. *Life Sci.* **1982**, *30*, 2123–2136. [CrossRef]

21. Schroeder, A.; Levins, C.G.; Cortez, C.; Langer, R.; Anderson, D.G. Lipid-based nanotherapeutics for siRNA delivery. *J. Intern. Med.* **2010**, *267*, 9–21. [CrossRef] [PubMed]

22. Mui, B.L.; Tam, Y.K.; Jayaraman, M.; Ansell, S.M.; Du, X.; Tam, Y.Y.; Lin, P.J.; Chen, S.; Narayanannair, J.K.; Rajeev, K.G.; et al. Influence of Polyethylene Glycol Lipid Desorption Rates on Pharmacokinetics and Pharmacodynamics of siRNA Lipid Nanoparticles. *Mol. Ther. Nucleic Acids* **2013**, *2*, e139. [CrossRef] [PubMed]

23. Sato, Y.; Note, Y.; Maeki, M.; Kaji, N.; Baba, Y.; Tokeshi, M.; Harashima, H. Elucidation of the physicochemical properties and potency of siRNA-loaded small-sized lipid nanoparticles for siRNA delivery. *J. Control. Release* **2016**, *229*, 48–57. [CrossRef] [PubMed]

24. Sato, Y. Development of Lipid Nanoparticles for the Delivery of Macromolecules Based on the Molecular Design of pH-Sensitive Cationic Lipids. *Chem. Pharm. Bull.* **2021**, *69*, 1141–1159. [CrossRef] [PubMed]

25. Semple, S.C.; Klimuk, S.K.; Harasym, T.O.; Dos Santos, N.; Ansell, S.M.; Wong, K.F.; Maurer, N.; Stark, H.; Cullis, P.R.; Hope, M.J.; et al. Efficient encapsulation of antisense oligonucleotides in lipid vesicles using ionizable aminolipids: Formation of novel small multilamellar vesicle structures. *Biochim. Biophys. Acta* **2001**, *1510*, 152–166. [CrossRef]

26. Sato, Y.; Hatakeyama, H.; Hyodo, M.; Harashima, H. Relationship Between the Physicochemical Properties of Lipid Nanoparticles and the Quality of siRNA Delivery to Liver Cells. *Mol. Ther.* **2016**, *24*, 788–795. [CrossRef]

27. Hafez, I.M.; Maurer, N.; Cullis, P.R. On the mechanism whereby cationic lipids promote intracellular delivery of polynucleic acids. *Gene Ther.* **2001**, *8*, 1188–1196. [CrossRef]

28. Sato, Y.; Hatakeyama, H.; Sakurai, Y.; Hyodo, M.; Akita, H.; Harashima, H. A pH-sensitive cationic lipid facilitates the delivery of liposomal siRNA and gene silencing activity in vitro and in vivo. *J. Control. Release* **2012**, *163*, 267–276. [CrossRef] [PubMed]

29. Hashiba, A.; Toyooka, M.; Sato, Y.; Maeki, M.; Tokeshi, M.; Harashima, H. The use of design of experiments with multiple responses to determine optimal formulations for in vivo hepatic mRNA delivery. *J. Control. Release* **2020**, *327*, 467–476. [CrossRef]

30. Suzuki, Y.; Onuma, H.; Sato, R.; Sato, Y.; Hashiba, A.; Maeki, M.; Tokeshi, M.; Kayesh, M.E.H.; Kohara, M.; Tsukiyama-Kohara, K.; et al. Lipid nanoparticles loaded with ribonucleoprotein-oligonucleotide complexes synthesized using a microfluidic device exhibit robust genome editing and hepatitis B virus inhibition. *J. Control. Release* **2021**, *330*, 61–71. [CrossRef] [PubMed]

31. Shobaki, N.; Sato, Y.; Harashima, H. Mixing lipids to manipulate the ionization status of lipid nanoparticles for specific tissue targeting. *Int. J. Nanomed.* **2018**, *13*, 8395–8410. [CrossRef] [PubMed]

32. Banchereau, J.; Steinman, R.M. Dendritic cells and the control of immunity. *Nature* **1998**, *392*, 245–252. [CrossRef] [PubMed]

33. Palucka, K.; Banchereau, J. Cancer immunotherapy via dendritic cells. *Nat. Rev. Cancer* **2012**, *12*, 265–277. [CrossRef]

34. Warashina, S.; Nakamura, T.; Sato, Y.; Fujiwara, Y.; Hyodo, M.; Hatakeyama, H.; Harashima, H. A lipid nanoparticle for the efficient delivery of siRNA to dendritic cells. *J. Control. Release* **2016**, *225*, 183–191. [CrossRef] [PubMed]

35. Gu, Y.Z.; Zhao, X.; Song, X.R. Ex vivo pulsed dendritic cell vaccination against cancer. *Acta Pharm. Sin.* **2020**, *41*, 959–969. [CrossRef] [PubMed]

36. Tacken, P.J.; de Vries, I.J.M.; Torensma, R.; Figdor, C.G. Dendritic-cell immunotherapy: From ex vivo loading to in vivo targeting. *Nat. Rev. Immunol.* **2007**, *7*, 790–802. [CrossRef]

37. Phua, K.K. Towards Targeted Delivery Systems: Ligand Conjugation Strategies for mRNA Nanoparticle Tumor Vaccines. *J. Immunol. Res.* **2015**, *2015*, 680620. [CrossRef]

38. Sato, Y.; Hashiba, K.; Sasaki, K.; Maeki, M.; Tokeshi, M.; Harashima, H. Understanding structure-activity relationships of pH-sensitive cationic lipids facilitates the rational identification of promising lipid nanoparticles for delivering siRNAs in vivo. *J. Control. Release* **2019**, *295*, 140–152. [CrossRef]
39. Okuda, K.; Sato, Y.; Iwakawa, K.; Sasaki, K.; Okabe, N.; Maeki, M.; Tokeshi, M.; Harashima, H. On the size-regulation of RNA-loaded lipid nanoparticles synthesized by microfluidic device. *J. Control. Release* **2022**, *348*, 648–659. [CrossRef]
40. Kimura, N.; Maeki, M.; Sato, Y.; Note, Y.; Ishida, A.; Tani, H.; Harashima, H.; Tokeshi, M. Development of the iLiNP Device: Fine Tuning the Lipid Nanoparticle Size within 10 nm for Drug Delivery. *ACS Omega* **2018**, *3*, 5044–5051. [CrossRef]
41. Sato, Y.; Kinami, Y.; Hashiba, K.; Harashima, H. Different kinetics for the hepatic uptake of lipid nanoparticles between the apolipoprotein E/low density lipoprotein receptor and the N-acetyl-d-galactosamine/asialoglycoprotein receptor pathway. *J. Control. Release* **2020**, *322*, 217–226. [CrossRef] [PubMed]
42. Maeki, M.; Kimura, N.; Sato, Y.; Harashima, H.; Tokeshi, M. Advances in microfluidics for lipid nanoparticles and extracellular vesicles and applications in drug delivery systems. *Adv. Drug Deliv. Rev.* **2018**, *128*, 84–100. [CrossRef] [PubMed]
43. Belliveau, N.M.; Huft, J.; Lin, P.J.; Chen, S.; Leung, A.K.; Leaver, T.J.; Wild, A.W.; Lee, J.B.; Taylor, R.J.; Tam, Y.K.; et al. Microfluidic Synthesis of Highly Potent Limit-size Lipid Nanoparticles for In Vivo Delivery of siRNA. *Mol. Ther. Nucleic Acids* **2012**, *1*, e37. [CrossRef] [PubMed]
44. Roces, C.B.; Lou, G.; Jain, N.; Abraham, S.; Thomas, A.; Halbert, G.W.; Perrie, Y. Manufacturing Considerations for the Development of Lipid Nanoparticles Using Microfluidics. *Pharmaceutics* **2020**, *12*, 1095. [CrossRef] [PubMed]
45. Terada, T.; Kulkarni, J.A.; Huynh, A.; Chen, S.; van der Meel, R.; Tam, Y.Y.C.; Cullis, P.R. Characterization of Lipid Nanoparticles Containing Ionizable Cationic Lipids Using Design-of-Experiments Approach. *Langmuir* **2021**, *37*, 1120–1128. [CrossRef] [PubMed]
46. Hassett, K.J.; Higgins, J.; Woods, A.; Levy, B.; Xia, Y.; Hsiao, C.J.; Acosta, E.; Almarsson, Ö.; Moore, M.J.; Brito, L.A. Impact of lipid nanoparticle size on mRNA vaccine immunogenicity. *J. Control. Release* **2021**, *335*, 237–246. [CrossRef] [PubMed]
47. Diken, M.; Kreiter, S.; Selmi, A.; Britten, C.M.; Huber, C.; Türeci, Ö.; Sahin, U. Selective uptake of naked vaccine RNA by dendritic cells is driven by macropinocytosis and abrogated upon DC maturation. *Gene Ther.* **2011**, *18*, 702–708. [CrossRef] [PubMed]
48. Lin, X.P.; Mintern, J.D.; Gleeson, P.A. Macropinocytosis in Different Cell Types: Similarities and Differences. *Membranes* **2020**, *10*, 177. [CrossRef]
49. Heil, F.; Hemmi, H.; Hochrein, H.; Ampenberger, F.; Kirschning, C.; Akira, S.; Lipford, G.; Wagner, H.; Bauer, S. Species-specific recognition of single-stranded RNA via toll-like receptor 7 and 8. *Science* **2004**, *303*, 1526–1529. [CrossRef] [PubMed]
50. Barral, P.M.; Sarkar, D.; Su, Z.Z.; Barber, G.N.; DeSalle, R.; Racaniello, V.R.; Fisher, P.B. Functions of the cytoplasmic RNA sensors RIG-I and MDA-5: Key regulators of innate immunity. *Pharmacol. Ther.* **2009**, *124*, 219–234. [CrossRef] [PubMed]
51. Nelson, J.; Sorensen, E.W.; Mintri, S.; Rabideau, A.E.; Zheng, W.; Besin, G.; Khatwani, N.; Su, S.V.; Miracco, E.J.; Issa, W.J.; et al. Impact of mRNA chemistry and manufacturing process on innate immune activation. *Sci. Adv.* **2020**, *6*, eaaz6893. [CrossRef] [PubMed]
52. Baiersdörfer, M.; Boros, G.; Muramatsu, H.; Mahiny, A.; Vlatkovic, I.; Sahin, U.; Karikó, K. A Facile Method for the Removal of dsRNA Contaminant from In Vitro-Transcribed mRNA. *Mol. Ther. Nucleic Acids* **2019**, *15*, 26–35. [CrossRef] [PubMed]
53. Simmons, D.P.; Wearsch, P.A.; Canaday, D.H.; Meyerson, H.J.; Liu, Y.C.; Wang, Y.; Boom, W.H.; Harding, C.V. Type I IFN drives a distinctive dendritic cell maturation phenotype that allows continued class II MHC synthesis and antigen processing. *J. Immunol.* **2012**, *188*, 3116–3126. [CrossRef]
54. Fitzgerald-Bocarsly, P.; Feng, D. The role of type I interferon production by dendritic cells in host defense. *Biochimie* **2007**, *89*, 843–855. [CrossRef] [PubMed]
55. Kowalski, P.S.; Capasso Palmiero, U.; Huang, Y.; Rudra, A.; Langer, R.; Anderson, D.G. Ionizable Amino-Polyesters Synthesized via Ring Opening Polymerization of Tertiary Amino-Alcohols for Tissue Selective mRNA Delivery. *Adv. Mater.* **2018**, *30*, e1801151. [CrossRef] [PubMed]
56. Cheng, Q.; Wei, T.; Farbiak, L.; Johnson, L.T.; Dilliard, S.A.; Siegwart, D.J. Selective organ targeting (SORT) nanoparticles for tissue-specific mRNA delivery and CRISPR-Cas gene editing. *Nat. Nanotechnol.* **2020**, *15*, 313–320. [CrossRef] [PubMed]
57. Liu, S.; Cheng, Q.; Wei, T.; Yu, X.; Johnson, L.T.; Farbiak, L.; Siegwart, D.J. Membrane-destabilizing ionizable phospholipids for organ-selective mRNA delivery and CRISPR-Cas gene editing. *Nat. Mater.* **2021**, *20*, 701–710. [CrossRef] [PubMed]
58. Akinc, A.; Maier, M.A.; Manoharan, M.; Fitzgerald, K.; Jayaraman, M.; Barros, S.; Ansell, S.; Du, X.; Hope, M.J.; Madden, T.D.; et al. The Onpattro story and the clinical translation of nanomedicines containing nucleic acid-based drugs. *Nat. Nanotechnol.* **2019**, *14*, 1084–1087. [CrossRef]
59. Jayaraman, M.; Ansell, S.M.; Mui, B.L.; Tam, Y.K.; Chen, J.; Du, X.; Butler, D.; Eltepu, L.; Matsuda, S.; Narayanannair, J.K.; et al. Maximizing the potency of siRNA lipid nanoparticles for hepatic gene silencing in vivo. *Angew. Chem. Int. Ed. Engl.* **2012**, *51*, 8529–8533. [CrossRef] [PubMed]
60. Davies, N.; Hovdal, D.; Edmunds, N.; Nordberg, P.; Dahlén, A.; Dabkowska, A.; Arteta, M.Y.; Radulescu, A.; Kjellman, T.; Höijer, A.; et al. Functionalized lipid nanoparticles for subcutaneous administration of mRNA to achieve systemic exposures of a therapeutic protein. *Mol. Ther. Nucleic Acids* **2021**, *24*, 369–384. [CrossRef]
61. Hassett, K.J.; Benenato, K.E.; Jacquinet, E.; Lee, A.; Woods, A.; Yuzhakov, O.; Himansu, S.; Deterling, J.; Geilich, B.M.; Ketova, T.; et al. Optimization of Lipid Nanoparticles for Intramuscular Administration of mRNA Vaccines. *Mol. Ther. Nucleic Acids* **2019**, *15*, 1–11. [CrossRef] [PubMed]

62. Shin, J.H.; Park, H.B.; Choi, K. Enhanced Anti-tumor Reactivity of Cytotoxic T Lymphocytes Expressing PD-1 Decoy. *Immune Netw.* **2016**, *16*, 134–139. [CrossRef] [PubMed]
63. Kataoka, K.; Shiraishi, Y.; Takeda, Y.; Sakata, S.; Matsumoto, M.; Nagano, S.; Maeda, T.; Nagata, Y.; Kitanaka, A.; Mizuno, S.; et al. Aberrant PD-L1 expression through 3′-UTR disruption in multiple cancers. *Nature* **2016**, *534*, 402–406. [CrossRef]
64. Nakamura, T.; Haloho, S.E.E.; Harashima, H. Intravenous liposomal vaccine enhances CTL generation, but not until antigen presentation. *J. Control. Release* **2022**, *343*, 1–12. [CrossRef] [PubMed]

 *pharmaceutics*

Article

Nanospermidine in Combination with Nanofenretinide Induces Cell Death in Neuroblastoma Cell Lines

Pietro Lodeserto [1,†], Martina Rossi [1,†], Paolo Blasi [1], Giovanna Farruggia [1,2] and Isabella Orienti [1,*]

[1] Department of Pharmacy and Biotechnology, University of Bologna, Via San Donato 19/2, 40127 Bologna, Italy; pietro.lodeserto@unibo.it (P.L.); martina.rossi12@unibo.it (M.R.); p.blasi@unibo.it (P.B.); giovanna.farruggia@unibo.it (G.F.)

[2] National Institute of Biostructures and Biosystems, Via delle Medaglie d'Oro 305, 00136 Rome, Italy

[*] Correspondence: isabella.orienti@unibo.it; Tel.: +39-345-439-8111

[†] These authors contributed equally to this work.

Abstract: A new strategy to cause cell death in tumors might be the increase of intracellular polyamines at concentrations above their physiological values to trigger the production of oxidation metabolites at levels exceeding cell tolerance. To test this hypothesis, we prepared nanospermidine as a carrier for spermidine penetration into the cells, able to escape the polyamine transport system that strictly regulates intracellular polyamine levels. Nanospermidine was prepared by spermidine encapsulation in nanomicelles and was characterized by size, zeta potential, loading, dimensional stability to dilution, and stability to spermidine leakage. Antitumor activity, ROS production, and cell penetration ability were evaluated in vitro in two neuroblastoma cell lines (NLF and BR6). Nanospermidine was tested as a single agent and in combination with nanofenretinide. Free spermidine was also tested as a comparison. The results indicated that the nanomicelles successfully transported spermidine into the cells inducing cell death in a concentration range (150–200 µM) tenfold lower than that required to provide similar cytotoxicity with free spermidine (1500–2000 µM). Nanofenretinide provided a cytostatic effect in combination with the lowest nanospermidine concentrations evaluated and slightly improved nanospermidine cytotoxicity at the highest concentrations. These data suggest that nanospermidine has the potential to become a new approach in cancer treatment. At the cellular level, in fact, it exploits polyamine catabolism by means of biocompatible doses of spermidine and, in vivo settings, it can exploit the selective accumulation of nanomedicines at the tumor site. Nanofenretinide combination further improves its efficacy. Furthermore, the proven ability of spermidine to activate macrophages and lymphocytes suggests that nanospermidine could inhibit immunosuppression in the tumor environment.

Keywords: intracellular polyamine levels; ROS increase; NLF; BR6; antitumor activity; cell motility; cell morphology; quantum phase imaging; spermidine immunomodulation

Citation: Lodeserto, P.; Rossi, M.; Blasi, P.; Farruggia, G.; Orienti, I. Nanospermidine in Combination with Nanofenretinide Induces Cell Death in Neuroblastoma Cell Lines. *Pharmaceutics* **2022**, *14*, 1215. https://doi.org/10.3390/pharmaceutics14061215

Academic Editors: James J. Moon and Carlo Irace

Received: 27 March 2022
Accepted: 6 June 2022
Published: 7 June 2022

1. Introduction

Over the last decades, polyamine research has continuously progressed, providing insight into the anabolic pathways and transport processes of polyamines. The effects of polyamines on cancer cells have also been explored in relation to different oncogenes and signaling pathways involved [1–3]. It has been demonstrated that oncogenes can affect the metabolism and function of polyamines by interfering with the expression and translation of key enzymes. Polyamines can also influence the expression of oncogenes in various ways, thus regulating the physiological function of cancer cells [4,5]. This research has prompted new ideas for cancer treatment. In particular, it has been demonstrated that the use of 2-difluoromethylornithine (DMFO), an inhibitor of ornithine decarboxylase (ODC), interferes with the polyamine biosynthesis and slows down the development of cancer in high-risk groups but DFMO has no significant effect on preventing cancer recurrence [6]. Moreover, the association of DFMO with inhibitors of polyamine transporters strongly improved the effect of single DFMO treatment in different tumor models by inducing significant levels of polyamine depletion in cells.

More recent attention has been given to polyamine analogs that upregulate polyamine catabolism and generate toxic compounds, such as H_2O_2, as a means to induce cancer cell death [7,8]. H_2O_2 is formed, with 3-acetoamidopropanal, by the acetylpolyamine oxidase (PAOX) catalyzed conversion of N-acetylspermine and N-acetylspermidine into spermidine and putrescine, respectively. It is also formed by the conversion of N-acetylspermine in N-acetylspermidine, catalyzed by the spermine oxidase (SMOX) (Figure 1). Polyamine analogs, such as N1,N11-diethylnorspermine (DENSpm), and N1-ethyl-N11-[(cyclopropyl) methyl]-4,8- diazaundecane (CPENSpm), have provided cytotoxic responses in several tumor types by their ability to upregulate SMOX and Spermidine/spermine N1-acetyltransferase (SSAT) with a consequent increase of H_2O_2 [9,10]. However, these molecules showed some drawbacks, such as poor specificity for cancer cells and induction of epithelial-mesenchymal transition-dedifferentiation in non-cancerous cells [11]. Overall, they showed poor positive outcomes in Phases I and II clinical trials [12].

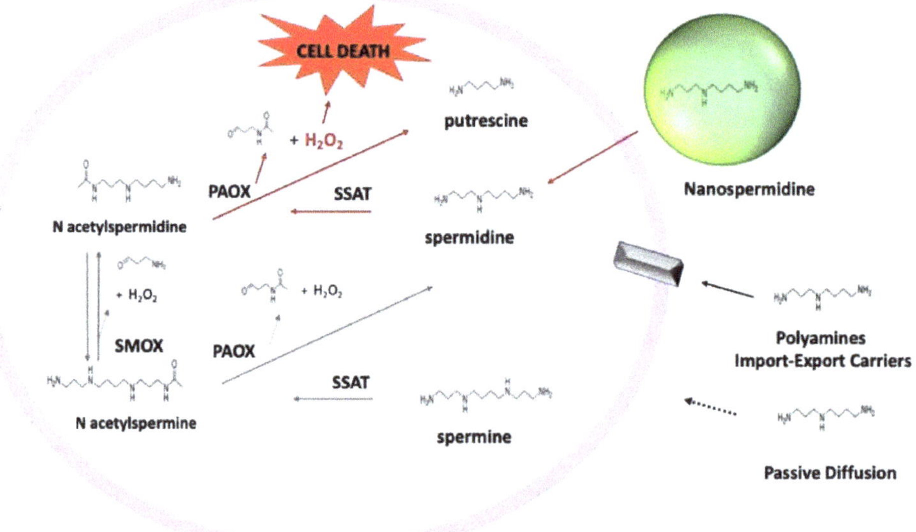

Figure 1. Schematic representation of the polyamine catabolic pathway and absorption mechanisms in cells. Spermidine/spermine N1-acetyltransferase (SSAT) catalyzes the acetyl-group transfer from acetyl-CoA to the aminopropyl end of spermidine or spermine, producing N1-acetylspermidine or N1-acetylspermine, respectively. These acetylated polyamines are either excreted from the cell or used as substrates for peroxisomal N1-acetylpolyamine oxidase (PAOX), producing H_2O_2, 3-acetoamidopropanal, and either putrescine or spermidine, depending on the starting substrate. Alternatively, N1-acetylspermine can be directly converted to spermidine by spermine oxidase (SMOX) while generating H_2O_2 and 3-aminopropanal. Polyamine absorption takes place by specific import-export carriers. Other possible mechanisms: passive diffusion and penetration by nanomicelles.

A new strategy to cause cell death in tumors might be based on increasing intracellular polyamine concentrations above physiological values to induce the production of oxidation metabolites at levels exceeding cell tolerance (Figure 1). However, the increase in intracellular polyamine concentration cannot be easily achieved by exogenous administration because the intracellular concentration of polyamines is tightly regulated by specific transporters that import or export polyamines depending on the cell necessity [13]. In addition, passive diffusion is hampered by the low concentration gradient outside-inside the cell and the massive protonation of polyamines in body fluids [14–16]. For this purpose, we prepared nanospermidine by spermidine encapsulation in nanomicelles and evaluated its ability to induce cell death in two neuroblastoma cell lines: NLF and BR6. Moreover, we evaluated the combination of nanospermidine with nanofenretinide to assess if the contribution of a drug able to increase intracellular ROS levels [17–20] could improve the cytotoxic effect of nanospermidine on tumor cells. Indeed, we had previously demonstrated that nanofenretinide was highly active in neuroblastoma and DIPG tumor models, and the effects were mediated by ROS increase [21–24].

To better understand the biological effect of this treatment, we employed a new powerful microscopic technique, the quantitative phase imaging (QPI), which employs various methods (e.g., holography, ptychography) to retrieve the phase shift of light waves passing through the cells. QPI techniques measure the extent of phase delay generated by the sample and record it as pixel values within the generated image. Pixel intensity is determined by the physical thickness and the refraction index of the cells, the latter depending on biomolecule composition and organization [25,26]. In this study, a Lifecyte microscope was employed to perform QPI based on ptychography. This microscope collects multiple diffraction patterns from spatially overlapping regions of the samples to form QPI images and estimate cell number, confluence, cell dry mass, cell morphology, and motility [27–30]. Furthermore, it is built within a cell CO_2 incubator that maintains the plate at 37 °C and 5% CO_2, thus allowing measures for long time periods without cell damage.

We used spermidine, among the other polyamines, as it has been proved to suppress tumorigenesis in healthy tissues and, in tumors, it contributed to the modulation of cancer-related functions, including immunoregulation, autophagy, and apoptosis [31]. In the tumor microenvironment, in particular, spermidine can induce the autophagy-dependent release of ATP, which, in turn, promotes immune surveillance [32,33]. Additionally, spermidine demonstrated the ability to alter macrophage immunometabolism and stimulate CD8+ T-cells [34] and memory B-cell responses [35–37].

2. Materials and Methods

2.1. Chemicals

N-4-hydroxyphenyl-retinamide (fenretinide, 4-HPR) was purchased from Olon Spa (Milan, Italy); spermidine, soy L-α-phosphatidylcholine, glyceryl tributyrate, 2-hydroxypropyl beta cyclodextrin (Mw 1460), and KOH from Sigma-Aldrich (Milan, Italy); ethanol absolute anhydrous from Carlo Erba Reagents (Milan, Italy). DMEM medium, dichlorofluorescein diacetate (DCHFDA), Hoechst 33342, glutamine, trypsine/EDTA solutions, and Human Serum were from Sigma-Aldrich.

2.2. Preparation of Spermidine Nanomicelles (NS) and Fenretinide Nanomicelles (NF)

Spermidine nanomicelles were prepared by mixing soy phosphatidylcholine (4 mmoles), glyceryl tributyrate (2 mmoles), spermidine (2 mmoles), 2-hydroxypropyl beta cyclodextrin (0.4 mmoles), and KOH 10 N (400 µL, 4 mmoles) to obtain a semisolid phase. Mixing to homogeneity was carried out by pressure and friction in a mortar grinder (RM 200 Retsch Verder, Italy) at 37 °C for 30 min at a 100 min^{-1} rate. The resultant semisolid phase was dispersed in PBS pH 7.4 to 50 mg/mL, filtered through 0.2 µm cellulose acetate filters (Fisher Scientific, Pittsburgh, PA, USA), and dialyzed for 72 h (dialysis membrane Mw cutoff 10 KD, Fisher Scientific) against PBS pH 7.4. The dialyzed phase was lyophilized. The dry residue was reconstituted with water and filtered again through 0.2 µm filters to obtain 50 mg/mL NS, representing the final formulative dispersion that was stored at −22 °C until use.

Fenretinide nanomicelles were prepared according to a method previously reported [21–24] by mixing soy phosphatidylcholine (4 mmoles), glyceryl tributyrate (2 mmoles), spermidine (2 mmoles), 2-hydroxypropyl beta cyclodextrin (0.4 mmoles), and KOH 10 N (400 µL, 4 mmoles) to obtain a semisolid phase. Mixing to homogeneity was carried out by pressure and friction in a mortar grinder at 37 °C for 30 min at a 100 min^{-1} rate. Fenretinide (1.2 mmoles) was dissolved in ethanol (300 µL) and KOH 10 N (120 µL, 1.2 mmole) and was subsequently added to the mixture. Homogenization was carried out in a mortar grinder for 30 min at a 100 min^{-1} rate. The resultant semisolid phase was dispersed in PBS pH 7.4 to 50 mg/mL, filtered through 0.2 µm filters, and dialyzed for 72 h (dialysis membrane Mw cutoff 10 KD) against PBS pH 7.4. The dialyzed phase was lyophilized. The dry residue was reconstituted with water and filtered again through 0.2 µm filters to obtain 50 mg/mL NF, representing the formulative dispersion that was stored at −22 °C until use.

2.3. Characterization of the Nanomicelles

Spermidine loading was evaluated in 50 mg/mL NS dispersions diluted (1:3, v/v) with an ethanol:water (1:1, v/v) mixture and analyzed by a fluorimetric method in comparison with the empty nanomicelles. The concentration obtained represented spermidine, both encapsulated in the nanomicelles and free in the aqueous phase. Therefore, to obtain the concentration of free spermidine, the nanomicelle dispersion was centrifuged in a 3.5 mL Ultra 5 KDa filter (Merck Millipore, Burlington, MA, USA) at 4000× g for 30 min and the ultrafiltrate was analyzed for spermidine content after dilution 1:3 with an ethanol:water (1:1, v/v) mixture. The difference between spermidine concentration in the nanomicelle dispersion and in the ultrafiltrate provided the concentration of the encapsulated spermidine. The nanomicelle loading was obtained as the ratio between the concentration (w/v) of the encapsulated spermidine and the concentration (w/v) of the nanomicelle dispersion.

The fluorimetric evaluation of spermidine in the nanomicelle dispersion and in the ultrafiltrate was carried out by a polyamine assay kit (Biovision, Milan, Italy) containing enzymes for the generation of hydrogen peroxide from spermidine and subsequent reaction of hydrogen peroxide with a fluorometric probe (Ex/Em = 535/587 nm) to yield a signal proportional to the amount of the polyamine present. The analysis was carried out according to the manufacturer's instructions.

Fenretinide loading was evaluated in 50 mg/mL NF dispersions diluted (1:3 v/v) with an ethanol:water (1:1, v/v) mixture and analyzed by UV spectroscopy (Shimadzu UV-1601) at 360 nm in comparison with the empty nanomicelles. The concentration obtained represented fenretinide, both encapsulated in the nanomicelles and free in the aqueous phase. To obtain the concentration of the free drug, the nanomicelle dispersion was centrifuged in a 3.5 mL Ultra 5 KDa filter at 4000× g for 30 min, and the ultrafiltrate was spectrophotometrically analyzed for drug content after dilution 1:3 with an ethanol:water (1:1, v/v) mixture. The difference between the drug concentration in the nanomicelle suspension and in the ultrafiltrate provided the concentration of the encapsulated drug. The nanomicelle loading was obtained as the ratio between the concentration (w/v) of the encapsulated drug and the concentration (w/v) of the nanomicelle dispersion.

Particle size, polydispersity, and zeta potential were measured at 37 °C (Malvern Nano-ZS Spectrometer, Malvern, UK) on the nanomicelle suspensions prepared in PBS and on the nanomicelle suspensions progressively diluted up to 1:100 (v/v) starting from 50 mg/mL concentration. Dilutions were made with PBS containing 10% (v/v) Human Serum (HS) to simulate the in-vivo dilution of the nanomicelles injected into the bloodstream. A minimum of 12 measurements per sample were made. Results were the combination of 3 10-min runs for a total accumulation correlation function time of 30 min. The results were volume-weighted.

Leakage of spermidine and fenretinide from NS and NF, respectively, was measured by dialysis at 37 °C, as previously described [21,22], with some minor changes. Briefly, the nanomicelles suspensions, prepared in NaCl 0.9% (w/v) at 50 mg/mL, were diluted to 10 mg/mL with pH 7.4 PBS containing 10% HS, and 1 mL of the diluted suspension was placed in a releasing chamber separated from a receiving compartment by a dialysis membrane (Mw cutoff 5KD, Fisher Scientific). The receiving compartment was filled with 10 mL pH 7.4 PBS containing 10% HS.

Leakage from the nanomicelles was determined by evaluating spermidine or fenretinide concentrations in the receiving phase at increasing time intervals. Spermidine concentration was obtained by the polyamine assay kit previously described, and fenretinide was evaluated spectrophotometrically by its maximum absorption wavelength (360 nm). Sink conditions were monitored throughout the experiment.

2.4. Cell Lines

NLF and BR6 neuroblastoma cells were kindly provided by Dr. Garrett Brodeur (Children's Hospital of Philadelphia, Philadelphia, PA, USA). These cell lines are routinely tested for integrity and authenticity, for endotoxins, mycoplasma, bacterial, and other viral contaminations, as well as genetic authenticity by multiplex PCR techniques. WS1 fibroblasts were used as a model for non-tumor cells. For the present study, the cells were grown in DMEM supplied with 10% HS, Penicillin, and Streptomycin at 37 °C in a 5% CO_2 humidified atmosphere. They were maintained in 25 cm^2 culture flasks (Corning, Tewksbury, MA, USA) and harvested using 0.25% Trypsin in 0.2 g/L EDTA solution.

2.5. MTT Assay

The tetrazolium salt 3-(4,5-dimethylthiazol-2-yl)-2,5-diphenyltetrazolium bromide (MTT) assay was used to detect cell proliferation and availability. Briefly, this assay is based on the reduction of MTT to the insoluble formazan salt by the cellular dehydrogenase. For this reason, the amount of formazan produced is considered a good indicator of the number of viable cells in the sample [38,39]. To perform the MTT assay, the cells were seeded at 10×10^3 cell/cm^2 in 96 multiwell plates, and, after 24 h, they were treated with free spermidine at concentrations varying from 10 μM to 2000 μM and with NS at nanoparticles concentrations ranging from 0.05 to 0.2 mg/mL corresponding to spermidine concentrations between 50 μM and 200 μM. Cells were also treated with fenretinide nanomicelles in combination with both free spermidine and nanospermidine. These studies were performed at constant fenretinide concentration (0.05 mg/mL nanoparticles corresponding to 10 μM fenretinide) and increasing concentrations of free spermidine (10 μM–2000 μM) or NS (nanoparticle concentrations 0.05–0.2 mg/mL corresponding to 50 μM–200 μM spermidine). After 24, 48, or 72 h, 10 μL of a 5 mg/mL MTT solution were added to each well to a final concentration of 0.5 mg/mL. After 4 h at 37 °C in the dark, 100 μL of a solution containing 10% (*w/v*) sodium dodecylsulfate (SDS) and HCl 0.01 mM were added to each well to dissolve the insoluble purple formazan crystals and left overnight on a shaker. The absorbance of each well was read on a TECAN plate reader (Männedorf, Switzerland) at 570 nm. To avoid the turbidity interference of biological samples, the read absorbance was normalized by a second reading at 690 nM.

2.6. Quantitative Phase Imaging (QPI) Microscopy

For QPI analysis, cells were seeded in a 96-well plate (Corning, Tewksbury, MA, USA) at 4×10^3 per well. After 24 h, the cells were treated at the same concentrations used for the MTT assays. QPI was performed by a Livecyte microscope (Phase Focus, Sheffield, UK). Images were acquired every 60 min for 3 days using a $10\times$ objective (0.25 NA) at 37 °C and 5% CO_2. QPI data were analyzed using Cell Analysis Toolbox software (Phase Focus, Sheffield, UK). We evaluated cell doubling time, displacement, dry mass, sphericity to assess cell vitality, and differentiation.

2.7. Measurement of Intracellular ROS Level

Reactive oxygen species were detected in intact cells, according to Bergamini et al. [40]. Briefly, NLF and BR6 were seeded at the density of 1.5×10^4 cells per well in a 96-well plate and incubated for 24 h to allow adhesion. Then, cells were treated for 4 h with NS at concentrations ranging from 0.05 to 0.2 mg/mL or NF at 0.05 mg/mL or with the combinations of NS and NF. Cells were then incubated with 10 μM DCFDA (Thermo Fisher Scientific, Waltham, MA, USA) for 1 h. To induce oxidative stress, cells were exposed to

150 µM tert-butyl hydroperoxide (TBH) in PBS for 30 min. Cells were then washed twice with PBS, and the fluorescence emission from each well was measured (λexc = 485 nm; λem = 535 nm) with a multi-plate reader (Enspire, Perkin Elmer, Monza, Italy). Data are reported as the mean $\pm$ SD of at least three independent experiments.

2.8. Confocal Laser-Scanning Fluorescence Microscopy

Confocal laser scanning microscopy (CLSM) is a type of high-resolution fluorescence microscopy that generates high-resolution images by superposition of photons emitted from the fluorescence sample and reaching the detector during one exposure period [41].

To image with CLSM, the cells were grown on glass coverslips. After 24 h, the samples were incubated at 37 °C for 30 min in the presence of 1 µg/mL Hoechst 33342 to stain cell nuclei, and, for the last 15 min, they were exposed to NS 0.2 mg/mL stained with Nile Red. The staining of NS was obtained by the addition of 1% (*w/w*) Nile Red to the semisolid mixture used for the preparation of NS. Reconstitution was done according to the preparation of NS nanomicelles followed by extensive dialysis to remove any residual unloaded dye. After cell exposure to Nile Red stained NS, the cells were washed with PBS 3 times, fixed with 3% formaldehyde for 10 min at room temperature, and washed repeatedly with 0.1 M glycine/PBS and PBS. As controls, the cells were exposed to Nile Red solution at the concentration corresponding to the NS treatment. Specimens were embedded in Mowiol and analyzed using a Nikon C1s confocal laser-scanning microscope, equipped with a Nikon PlanApo 40, 1.4-NA oil immersion lens. Excitation was performed at 405 nm with an argon laser and emission was recorded at 650 nm. The images were analyzed by Image J Software (version 1.53a, U. S. National Institutes of Health, Bethesda, MD, USA).

2.9. Statistical Analysis

All experiments were repeated at least three times on three independent samples. One-way analysis of variance (ANOVA) followed by Dunnett's multiple comparison test was used for repeated measurement values. Differences of $p < 0.05$ were considered significant. Statistical analysis was carried out using GraphPad Prism Software (version 6.0c, GraphPad Software, San Diego, CA, USA).

3. Results

3.1. Characterization of NS and NF

Spermidine and fenretinide nanomicelles were obtained by dispersion in PBS of the semisolid mixtures made by phospholipids, spermidine or fenretinide, glyceryl tributyrate, and 2-hydroxypropyl beta cyclodextrin. The spontaneous self-assembling, in the aqueous phase, of the mixture components triggered nanomicelles formation and inclusion of spermidine (Figure 2) or fenretinide [21,22] in the amphiphilic nanomicelle matrix.

Particle size, polydispersity, and zeta potential were measured on the nanomicelle suspensions starting from 50 mg/mL and progressively diluting with PBS containing 10% (*v/v*) HS to simulate nanoparticle dilution in vivo after injection.

The mean diameter of the nanomicelles resulted optimally sized for tumor accumulation by the Enhanced Permeability and Retention (EPR) effect [42–46], being 148.4 $\pm$ 3.4 nm for NS and 154.1 $\pm$ 10.3 nm for NF. The polydispersity was always lower than 0.3, indicating good dimensional homogeneity (Table 1).

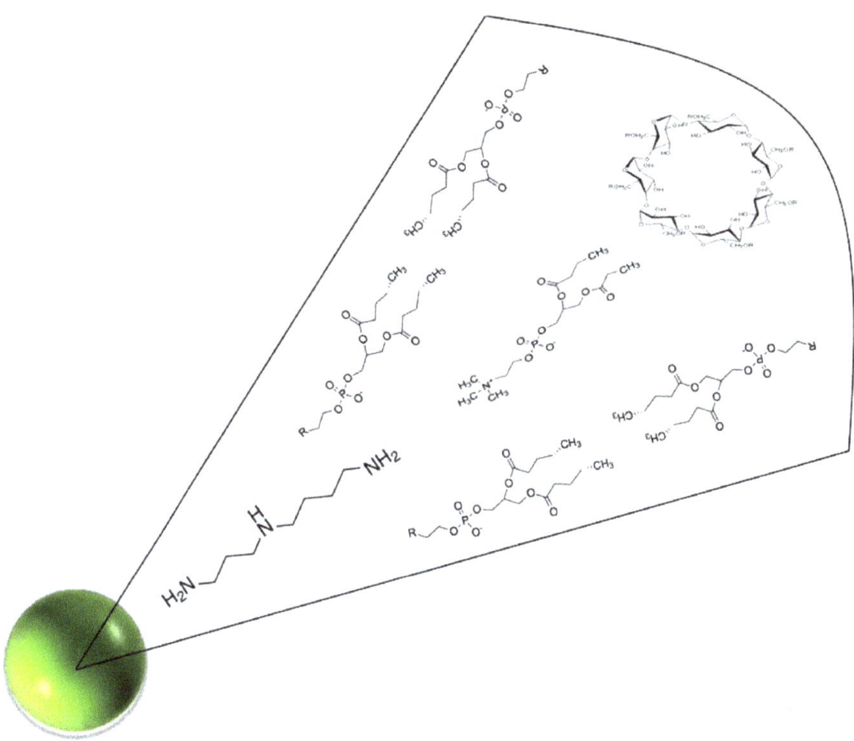

Figure 2. Schematic representation of the supramolecular organization of Nanospermidine main constituents: spermidine, phospholipids, 2-hydroxypropyl beta cyclodextrin.

Table 1. Physicochemical Characteristics of Nanospermidine (NS), Nanofenretinide (NF), and empty nanomicelles (No) in PBS at 50 mg/mL.

Nanomicelle Type	% Loading (*w:w*)	Mean Size (nm)	Polydispersity	Zeta Potential (mV)
NS	Spermidine 12.92 ± 2.37	148.4 ± 3.4	0.270 ± 0.013	−17.4 ± 0.57
NF	Fenretinide 7.82 ± 1.05	154.1 ± 10.3	0.258 ± 0.011	−27.2 ± 1.53
No	–	124.7 ± 3.5	0.183 ± 0.024	−21.7 ± 2.18

As is known, the EPR effect depends on the discontinuity of the tumor capillaries and the impaired lymphatic drainage in the tumor environment.

Capillary discontinuities, generally between 200 nm and 1.2 μm in diameter, have been shown to allow injected nanoparticles, smaller than 500 nm, to access tumor tissues by extravasation and accumulate due to reduced lymphatic drainage [47–50].

The surface of the nanomicelles was always negatively charged, as indicated by the zeta potential values (Table 1). Spermidine nanomicelles were characterized by the lowest absolute value of zeta potential in accordance with the alkaline character of spermidine which induces a decrease in the negative charge on the nanomicelle surface.

Dilution slightly increased the nanomicelle size (Figure 3A). The size increase was very low in the 50 to 1 mg/mL dilution range and higher between 0.5 and 0.05 mg/mL. The maximum size was obtained at 0.05 mg/mL, being 221.4 ± 23.7 nm for NF and

270.5 ± 16.1 nm for NS. This corresponded to diameter expansions of 43.67% for NF and 82.27% for NS. Size expansion at high dilutions could represent a favorable feature for nanomedicines because it could prevent retro-diffusion towards the venous circulation of the extravasated nanoparticles when their size increases due to the high dilutions provided by the tumor matrix.

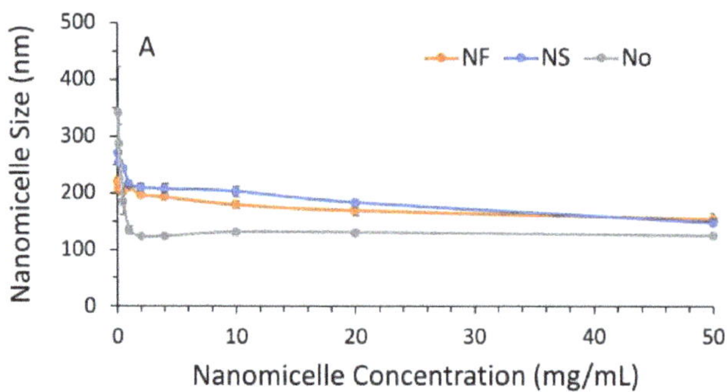

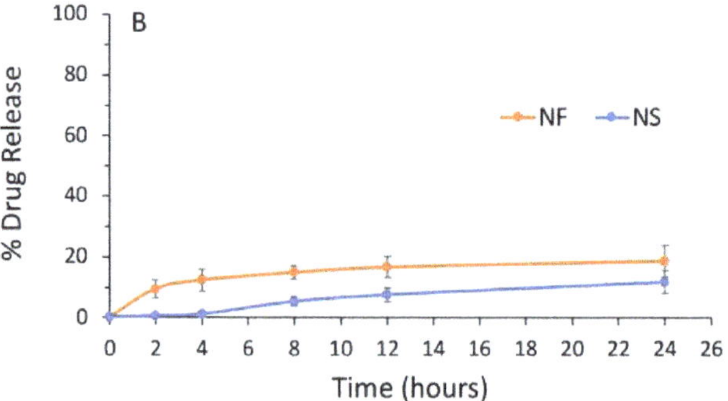

Figure 3. (**A**) Dimensional stability of the Nanomicelles to dilution in PBS containing 10% HS. (**B**) Stability of the Nanomicelles to leakage of their Spermidine (NS) or Fenretinide (NF) content in PBS containing 10% HS.

Spermidine and fenretinide release from NS and NF at 24 h was 18.83% ± 5.13 and 11.82% ± 2.33, respectively (Figure 3B), indicating stability towards drug leakage in circulation for time frames longer than those required for nanoparticle accumulation in solid tumors by extravasation in accordance with the EPR effect.

3.2. Effect of Free Spermidine and Nanospermidine on Cell Viability

Treatment with free spermidine in the concentration range of 10 μM–2000 μM did not elicit any detrimental effects up to 1200 μM (Figure 4). The higher concentration of 1500 μM induced a slowdown of cell proliferation in NLF and a complete inhibition in BR6. At 2000 μM, free spermidine significantly reduced the 72 h cell viability in both cell lines.

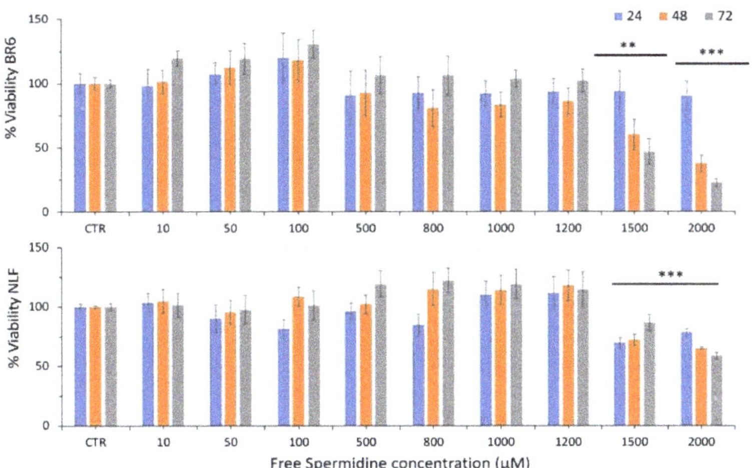

Figure 4. Relative viability of BR6 and NLF cells treated with increasing concentrations of Free Spermidine for 24, 48, and 72 h assessed by MTT assay. Data are presented as percentage versus control cells (100%) (mean ± SD, $n = 6$) (** $p < 0.01$, *** $p < 0.001$).

Nanospermidine had a tenfold higher effect than free spermidine in both tumor cell lines, as cell viability was significantly decreased at a nanospermidine concentration of 0.15 mg/mL corresponding to 150 μM spermidine (Figure 5).

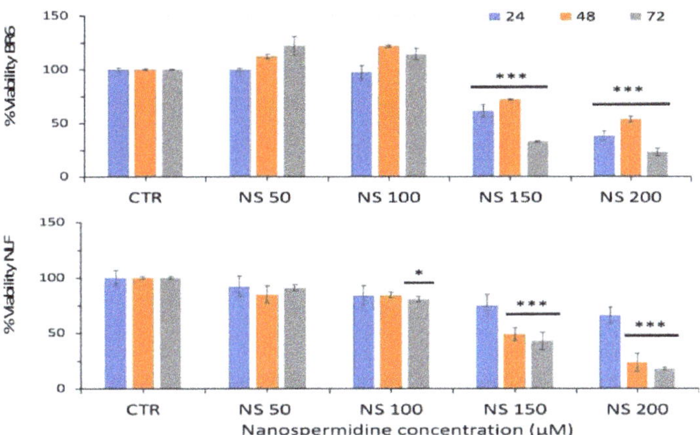

Figure 5. Relative viability of BR6 and NLF cells treated with increasing concentrations of Nanospermidine (NS) for 24, 48, and 72 h assessed by MTT assay. Data are presented as percentage versus control cells (100%) (mean ± SD, $n = 6$) (* $p < 0.05$, *** $p < 0.001$).

We had previously demonstrated the antitumor activity of nanofenretinide in these cell lines [21,22]; therefore, we evaluated the combination of nanospermidine with nanofenretinide to assess if the contribution of fenretinide could improve the cytotoxic effect of nanospermidine.

In combination with nanofenretinide, a slightly increased overall activity was obtained with nanospermidine (Figure 6) at the same concentrations that triggered cytotoxicity in single administrations: 0.15 mg/mL and 0.20 mg/mL nanospermidine, corresponding to 150 μM and 200 μM spermidine, respectively.

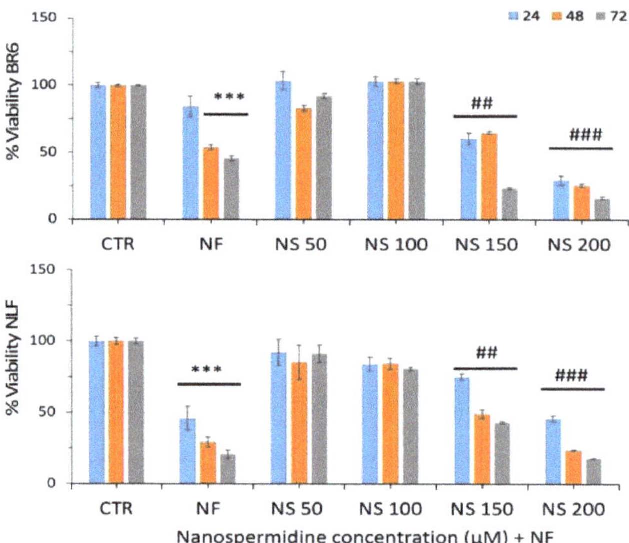

Figure 6. Relative viability of BR6 and NLF cells treated with increasing concentrations of Nanospermidine (NS) in the presence of Nanofenretinide (NF, 10 µM) for 24, 48, and 72 h assessed by MTT assay. Data are presented as percentage versus control cells (100%) (mean ± SD, $n = 6$) (*** $p < 0.001$) and versus NF (## $p < 0.01$, ### $p < 0.001$).

As a comparison, we evaluated the combination of free spermidine with nanofenretinide and, also, in this case, we observed increased activity at the concentrations that triggered cytotoxicity in single administrations: 1500 µM and 2000 µM spermidine (Figure 7).

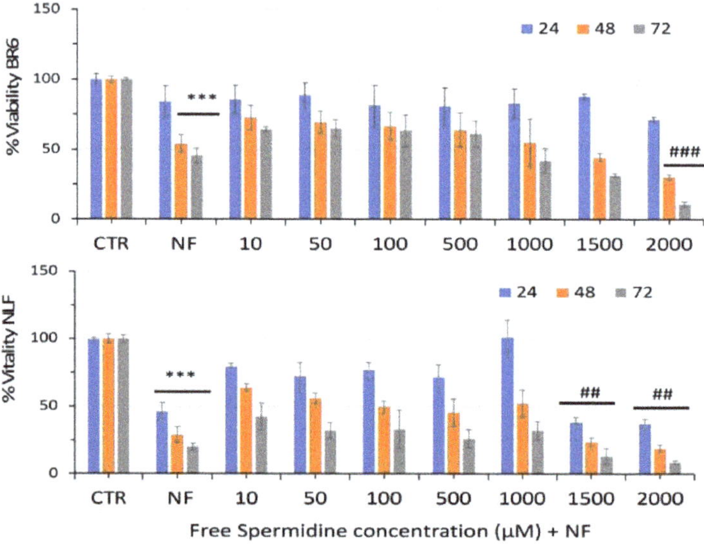

Figure 7. Relative viability of BR6 and NLF cells treated with increasing concentrations of Free Spermidine in the presence of Nanofenretinide (NF, 10 µM) for 24, 48, and 72 h assessed by MTT assay. Data are presented as percentage versus control cells (100%) (mean ± SD, $n = 6$) (*** $p < 0.001$) and versus NF (## $p < 0.01$, ### $p < 0.001$).

To exclude any contribution of the nanomicelles to the improved cytotoxicity of nanospermidine, we evaluated the effect of empty nanomicelles on cell viability. No decrease in cell vitality was observed up to 0.2 mg/mL nanomicelles, corresponding to the maximum concentration used in this study. However, higher concentrations induced a slight but significant decrease in cell viability in both cell lines (Figure 8).

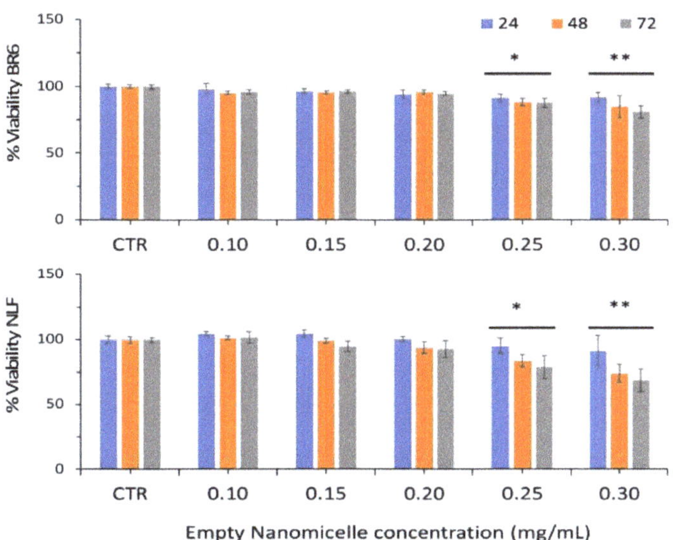

Figure 8. Relative viability of BR6 and NLF cells treated with Empty Nanomicelles at increasing concentrations for 24, 48, and 72 h assessed by MTT assay. Data are presented as percentage versus control cells (100%) (mean ± SD, $n = 6$) (* $p < 0.05$, ** $p < 0.01$).

Finally, the effect of nanospermidine was evaluated in normal WS1 fibroblasts at the same concentrations used in the tumor cells (Figure 9). No decrease in viability was obtained up to 150 µM spermidine, corresponding to the cytotoxic concentration in tumor cells. At 200 µM, a 28.36% viability decrease was observed after 72 h.

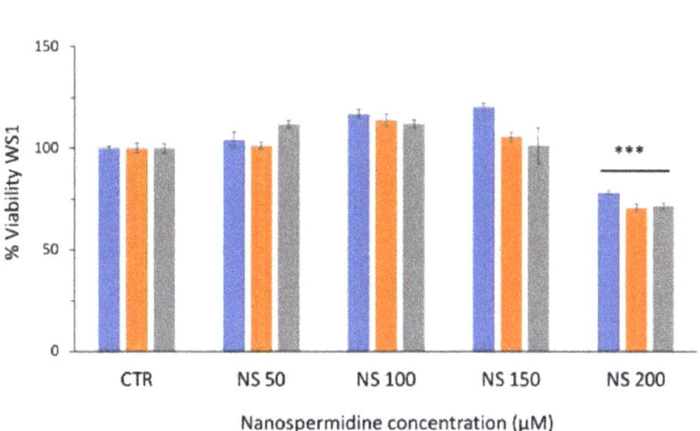

Figure 9. Relative viability of WS1 fibroblasts treated with increasing concentrations of Nanospermidine (NS) for 24, 48, and 72 h assessed by MTT assay. Data are presented as percentage versus control cells (100%) (mean ± SD, $n = 6$) (*** $p < 0.001$).

3.3. Quantitative Phase Imaging

QPI evaluated cell morphology over time, providing, at the same time, measures of several parameters such as shape, dry mass, and sphericity, as well as cellular proliferation and displacement in response to treatment [28,29].

The images (Figure 10A,B) were obtained at up to 72 h of treatment with nanospermidine, and the combination with nanofenretinide confirmed the MTT results, indicating cytotoxic effects at 200 µM with a higher activity of the combination than single nanospermidine.

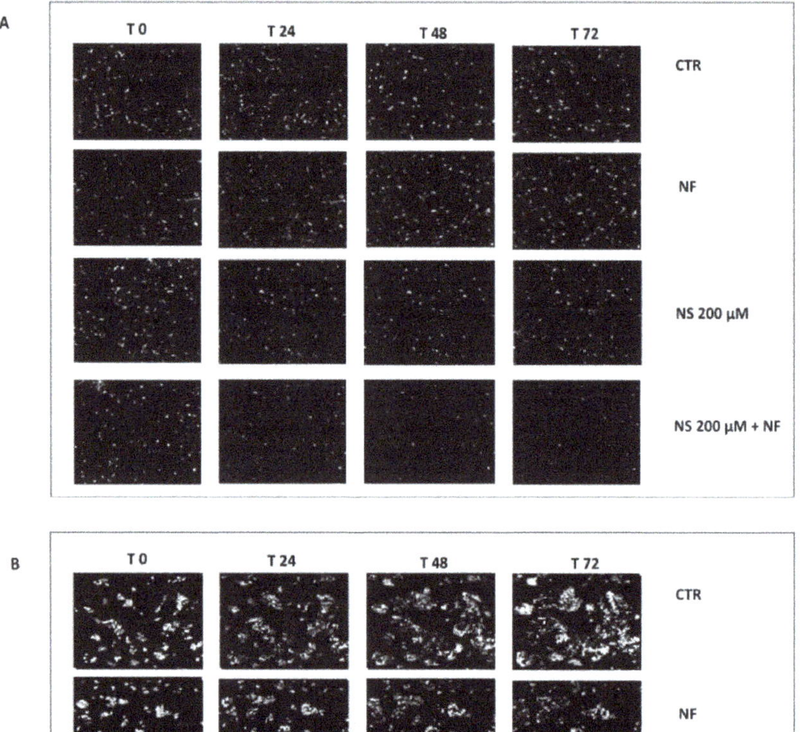

Figure 10. QPI. Representative image at 0, 24, 48, and 72 h of BR6 cells (**A**) and NLF cells (**B**) treated with Nanospermidine (NS, 200 µM spermidine) as a single agent and in combination with Nanofenretinide (NF, 10 µM fenretinide). As a comparison, the images of the untreated cells and the cells treated with NF are shown.

Cell doubling time, displacement, dry mass, and sphericity were analyzed over 72 h at the sub-cytotoxic concentrations 50 µM and 100 µM nanospermidine to exclude the interfering effect of cell death with detachment from the substrate.

The doubling time, which is a measure of cell growth, was increased by the combination in both cell lines, with a higher effect at 100 µM than 50 µM (Figure 11A), indicating a slowing down in cell proliferation.

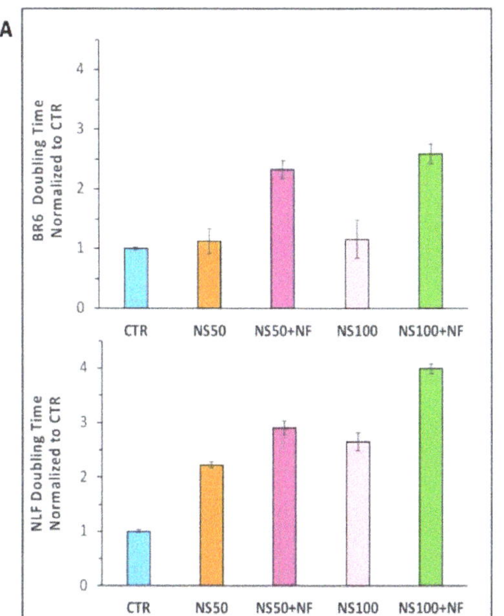
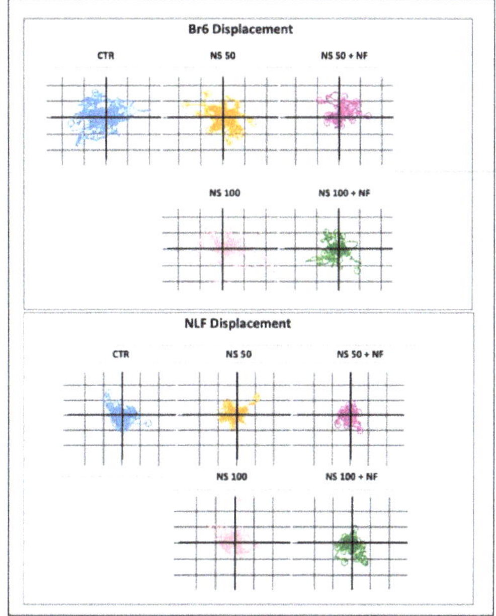

Figure 11. QPI data. (**A**) Histogram plot illustrating median cell doubling time of BR6 and NLF cells treated with nanospermidine (NS) and nanospermidine in combination with nanofenretinide (NS + NF) at the sub-cytotoxic concentrations of NS: 50 µM and 100 µM. (**B**) Displacement of BR6 and NLF cells treated with nanospermidine (NS) and nanospermidine in combination with nanofenretinide (NS + NF) at the sub-cytotoxic concentrations of NS: 50 µM and 100 µM.

Cell motility is obviously an important feature of tumor cells as it contributes to cancer invasion and metastasis [51,52]. Evaluation of cell displacement is now gaining increasing attention with improvements in time-lapse microscopy techniques [53].

In our study, we observed that cell displacement was decreased by nanospermidine, and the presence of fenretinide further limited cell displacement (Figure 11B).

In QPI, dry mass is measured from the phase delay considering that the refractive increment of biomolecules can be closely approximated by a constant [25].

A decrease in dry mass over time is indicative of a cytostatic or cytotoxic effect [54]. Sphericity quantifies the degree of cell roundness and is calculated as the ratio of the surface area of a sphere that has the same volume as the cell to the surface area of the cell. For adherent cells, such as those used in our study, sphericity provides a measure of the level of attachment of the cells to the substrate, since when the level of attachment increases the cells lose sphericity [55].

Treatment with the sub-cytotoxic concentrations of nanospermidine did not provide appreciable differences in dry mass (Figure 12A) and sphericity (Figure 12B) with respect to the controls. These results indicate that, despite the decrease in cell proliferation, the sub-cytotoxic concentrations of nanospermidine did not induce morphological variations.

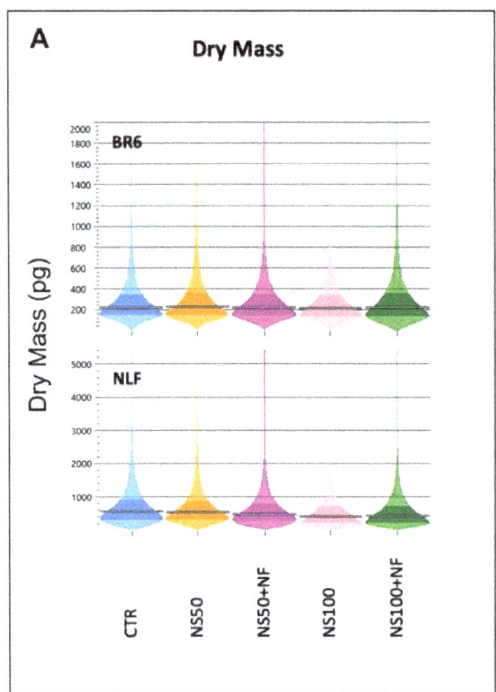

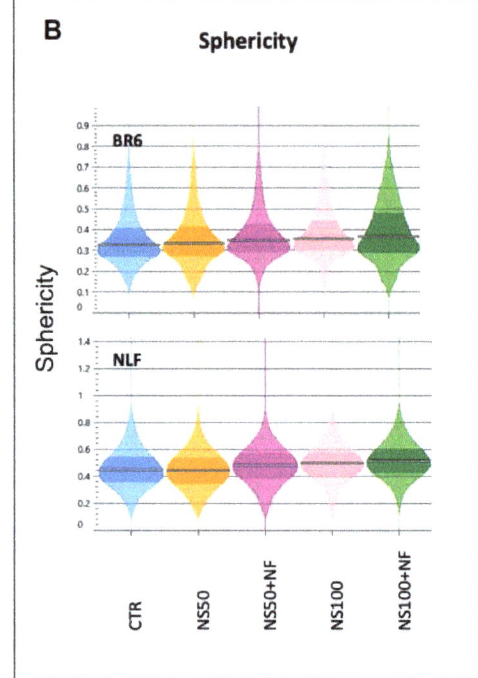

Figure 12. QPI data. (**A**) Dry mass at 72 h of BR6 and NLF cells treated with Nanospermidine (NS) and Nanospermidine in combination with Nanofenretinide (NS + NF) at the sub-cytotoxic concentrations of NS: 50 µM and 100 µM. (**B**) Sphericity at 72 h of BR6 and NLF cells treated with nanospermidine (NS) and nanospermidine in combination with nanofenretinide (NS + NF) at the sub-cytotoxic concentrations of NS: 50 µM and 100 µM. Analysis of cell doubling time, displacement, and dry mass was mediated on six wells for each treatment.

Combination treatments, on the contrary, slightly decreased dry mass and increased sphericity in both cell lines, indicating a cytotoxic effect induced by fenretinide.

3.4. ROS Increase in Treated Cells

Treatment with nanospermidine elicited ROS increases in both tumor cell lines. ROS increase was proportional to nanospermidine concentration (Figure 13). The highest ROS levels were obtained in NLF at 200 µM nanospermidine.

The presence of fenretinide further increased ROS at each nanospermidine concentration analyzed. At 200 µM nanospermidine, fenretinide provided comparable ROS levels in both cell lines.

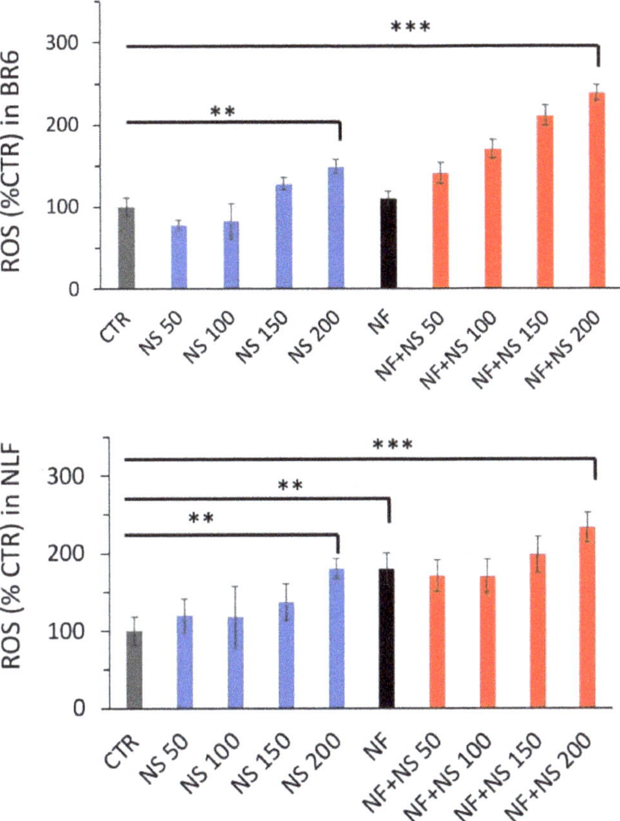

Figure 13. Effect of Nanospermidine (NS) and Nanospermidine in combination with Nanofenretinide (NS + NF) on ROS increase. ROS production was measured by H_2DClFDA fluorescence in BR6 and NLF cells treated 4 h with Nanospermidine at concentrations corresponding to 50, 100, 150, and 200 µM spermidine or with Nanospermidine in combination with Nanofenretinide at 0.05 mg/mL corresponding to 10 µM fenretinide. Data are presented as percentage versus control cells (100%) (mean $\pm$ SD, $n = 6$) (** $p < 0.01$, *** $p < 0.001$).

3.5. Confocal Laser-Scanning Fluorescence Microscopy

Cells treated with Red Nile stained nanospermidine were analyzed by confocal laser-scanning fluorescence microscopy. The images (Figure 14A) showed that nanospermidine can interact with the cells after short periods of time. The images reveal a uniform distribution of fluorescence in BR6 and a spotted-like dispersion in NLF. The quantitative analysis of fluorescence intensity (Figure 14B) showed that nanospermidine interaction was higher in BR6 than in NLF. WS1 did not provide fluorescent images after treatment with Red Nile stained nanospermidine for the same time period as the tumor cells.

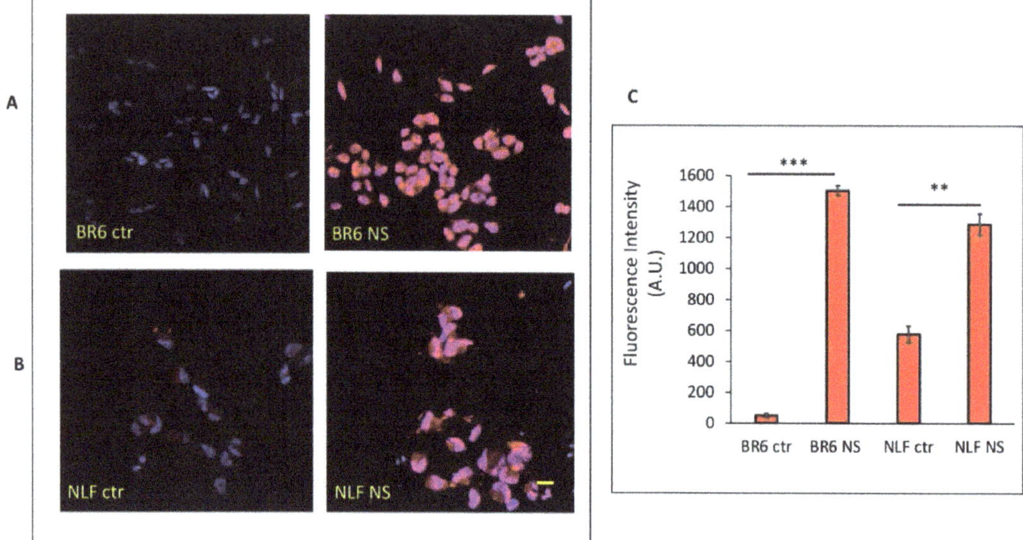

Figure 14. Confocal microscopy of BR6 (**A**) and NLF (**B**) after 15 min treatment with Nanospermidine (NS) stained with Nile Red and controls treated with Nile Red solution at the same concentration used with NS. Cell nuclei were evidenced by 1 μg/mL Hoechst 33348 staining. (**C**) Fluorescence intensities of the treated cells and controls. Photographs were taken at 40× magnification, bar = 20 μm. The images were analyzed by Image J Software and reported as fluorescence intensity per unit area. (mean ± SD, $n = 3$) (** $p < 0.01$, *** $p < 0.001$).

4. Discussion

The increased polyamine metabolism and its correlation with cancer have prompted many efforts at developing drugs that could inhibit polyamine biosynthesis or block the active, carrier-mediated transport of polyamine into the cells as a therapeutic approach in cancer treatment.

More recently, attention has been paid to molecules that, on the contrary, upregulate polyamine catabolism in cells as a means to generate toxic compounds that can induce cancer cell death [7]. In fact, it has been demonstrated that the production of H_2O_2 through polyamine catabolism is implicated in the cytotoxic responses of several types of tumors to different polyamine analogs [8].

Among the molecules currently used to upregulate polyamine catabolism, polyamine analogs, such as N1,N11-diethylnorspermine (DENSpm) and N1-ethyl-N11-[(cyclopropyl)m ethyl]-4,8- diazaundecane (CPENSpm), have demonstrated activity in several types of tumors [9,11,12]. However, these molecules showed some drawbacks, such as poor specificity for cancer cells, induction of epithelial-mesenchymal transition-dedifferentiation in non-cancerous cells, and limited activity in clinical trials [11,12].

Therefore, a new approach to stimulate polyamine catabolism might rely on increasing intracellular polyamine concentrations above their physiological values to trigger the production of oxidation metabolites at levels exceeding cell tolerance. However, exogenous administration of free polyamines cannot raise their intracellular concentrations over physiological levels because intracellular penetration is strictly regulated by import-export mechanisms based on specific membrane carriers, such as SLC22A16 and SLC3A2, that import or export polyamines depending on the cell necessity [13]. In addition, the carriers' activity is regulated by a feedback mechanism based on the immediate synthesis of antizyme, a regulatory protein that blocks the polyamine uptake in the presence of small increases in free cellular polyamine levels [14].

Other mechanisms of cell penetration, such as passive diffusion or endocytosis, are ineffective compared to the carrier-mediated active transport. Passive diffusion, in particular, is strongly hindered by the high intracellular polyamine concentrations in the mM range that sustain a negative outside-inside concentration gradient, being the physiological concentration of extracellular polyamines in the μM range. In addition, the extensive protonation of spermidine (pKa 10.9) at physiological pH reduces the concentration of the diffusible, unprotonated molecules to about 0.03% of the total extracellular concentration [14–16].

Furthermore, the extracellular concentration of polyamines is physiologically regulated to prevent accidental increases that could trigger an uncontrolled influx of polyamines into the cells driven by passive diffusion.

The control of extracellular polyamine concentration starts in the intestine where polyamines from the diet or produced by luminal bacteria are absorbed by a saturation regulated mechanism and are metabolized in the enterocytes before reaching the systemic circulation [56].

With the aim to overcome these limitations and obtain supra-physiological levels of spermidine in tumor cells, we prepared nanospermidine by encapsulation of spermidine in nanomicelles that we had previously demonstrated to be suitable carriers for fenretinide in tumor cells [21–24]. Next, we evaluated the effect of nanospermidine in two neuroblastoma cell lines: NLF (with MYCN amplification) and BR6 (without MYCN amplification, with TrKA expression).

A significant decrease in cell viability was obtained in both cell lines with nanospermidine concentrations of 0.15 mg/mL corresponding to 150 μM spermidine. With free spermidine, on the contrary, no cytotoxic effects were observed up to 1200 μM, and a slowing of cell proliferation in NLF and a complete inhibition in BR6 were obtained at 1500 μM.

The results were in agreement with the ability of the nanomicelles to transport the encapsulated spermidine within the cells, previously demonstrated for fenretinide encapsulated in the same nanomicelles [21,22] and supported by confocal laser-scanning fluorescence microscopy showing increased fluorescence in cells treated with Red Nile stained nanospermidine compared to controls treated with the free dye.

Cytotoxicity of empty nanomicelles was excluded by comparative experiments, indicating no cytotoxic effects at nanomicelle concentrations higher than those used in this study. Moreover, the cytotoxicity of nanospermidine was higher in tumor cells than in normal cells, such as WS1 fibroblasts, in accordance with the increased sensitivity of tumor cells to ROS increase compared to their normal counterparts. In fact, it has been demonstrated that cancer cells have ROS levels higher than normal cells, and their redox system is easily overwhelmed by additional ROS production that induces oxidative stress leading to cell death [57,58].

Based on this feature, many strategies are being developed to raise ROS levels and realize targeted therapies by induction of cell death in tumors without relevant toxicity to normal cells [59].

The tenfold higher cytotoxicity of nanospermidine than free spermidine suggested the suitability of the nanomicelles as polyamine transporters and revealed the efficiency of this mechanism to provide the cell with high amounts of spermidine owing to the favorable spermidine loading (12.92% *w/w*) in the nanomicelles. On the contrary, the much higher concentration of free spermidine required to trigger cytotoxicity correlates with the difficulty for the free molecule to enter the cells in amounts exceeding physiological values, evading cell transport regulation. These results denote the unsuitability of free spermidine as an antitumor agent because the high concentrations required at the tumor site would be hardly achieved even by the administration of large doses.

Moreover, the results obtained indicate that nanospermidine can be regarded as a new tool to increase intracellular polyamines levels and induce tumor cell death by using spermidine at biocompatible doses. In addition, the in-vivo administration of nanospermidine could exploit the mechanism of nanomedicines accumulation at the tumor site by extravasation through the discontinuities of the capillaries generated by angiogenesis, providing on-target activity. The ability of nanospermidine to increase its size at high dilutions could represent an additional favorable feature for targeted localization. Indeed, in the presence of high dilutions, such as those provided by the tumor matrix, the size expansion of the extravasated nanoparticles could prevent their retro-diffusion to the venous circulation that, at present, is recognized as a drawback in passive drug targeting of nanomedicines [60].

We found a correlation between nanospermidine cytotoxicity and increased intracellular ROS levels. This prompted us to evaluate the combination of nanospermidine with nanofenretinide, as we had previously demonstrated that nanofenretinide was endowed with high antitumor activity towards these same cell lines [21,22] due to ROS increase.

Nanofenretinide improved the effect of nanospermidine at both cytotoxic and sub-cytotoxic nanospermidine concentrations. It increased cell death at the cytotoxic concentrations while, at the sub-cytotoxic concentrations, it decreased cell proliferation, as indicated by prolonged duplication times and decreased cell motility.

Furthermore, cotreatment with nanofenretinide at the sub-cytotoxic nanospermidine concentrations slightly increased sphericity and decreased biomass. These slight but important modifications were disclosed by time-lapse QPI techniques that measured several parameters at the same time and provided a multifaced image of the cells. They indicated that, while treatment with nanospermidine at sub-cytotoxic concentrations did not influence cell morphology, cotreatment with nanofenretinide may induce apoptosis, even at such low nanospermidine concentrations, as suggested by the reduced biomass and the increased sphericity indicative of decreased cell attachment. This information could not have been obtained by other techniques such as MTT assay.

5. Conclusions

Nanospermidine, obtained by spermidine encapsulation in nanomicelles, can represent a new tool in antitumor therapy being endowed with both the pharmacokinetics characteristics of nanomedicines and the activity of polyamines catabolism modulators.

As a nanocarrier, nanospermidine can undergo accumulation at the tumor site by extravasation through the discontinuities of the tumor capillaries, thus allowing spermidine localization at the pathological site.

At the cellular level, nanospermidine can provide spermidine penetration into the cells in amounts exceeding physiological values, thus increasing ROS production over limiting values for cell tolerance and triggering cell death.

Administration of nanospermidine in combination with nanofenretinide further increases the antitumor activity of nanospermidine.

Hence, nanospermidine may represent a new approach in tumor treatment because it provides a high intracellular increase of the polyamine and tumor cell death by administration of micromolar, physiological doses of spermidine that make the treatment tolerable and biocompatible.

Finally, the proven ability of spermidine to activate macrophages and memory B cell responses makes nanospermidine worth further evaluation to assess if the antitumor effect demonstrated in this study can be further improved by inhibition of tumor immunosuppression in in vivo settings.

Author Contributions: P.L. and M.R. contributed equally to this work. Conceptualization, P.L., M.R., I.O., P.B. and G.F.; Methodology, P.L., M.R., P.B. and G.F.; Investigation, P.L. and M.R.; Formal Analysis, P.L., M.R., P.B., G.F. and I.O.; Writing, P.B., G.F. and I.O.; Funding acquisition, I.O., P.B. and G.F. All authors have read and agreed to the published version of the manuscript.

Funding: The project work was partially funded by the "Ministero dell'Università e della Ricerca" (Targeting Hedgehog pathway: Virtual screening identification and sustainable synthesis of novel Smo and Gli inhibitors and their pharmacological drug delivery strategies for improved therapeutic effects in tumors. PRIN 2017, project n. 20175XBSX4).

Data Availability Statement: The data presented in this study are available upon request from the corresponding author, I.O.

Acknowledgments: The authors wish to acknowledge the support of the Center for Applied Biomedical Research (CRBA) of the University of Bologna.

Conflicts of Interest: The authors declare no conflict of interest.

References

1. Gobert, A.P.; Latour, Y.L.; Asim, M.; Barry, D.P.; Allaman, M.M.; Finley, J.L.; Smith, T.M.; McNamara, K.M.; Singh, K.; Sierra, J.C.; et al. Protective Role of Spermidine in Colitis and Colon Carcinogenesis. *Gastroenterology* **2022**, *162*, 813–827.e8. [CrossRef] [PubMed]
2. McNamara, K.M.; Gobert, A.P.; Wilson, K.T. The role of polyamines in gastric cancer. *Oncogene* **2021**, *40*, 4399–4412. [CrossRef] [PubMed]
3. Li, H.; Wu, B.K.; Kanchwala, M.; Cai, J.; Wang, L.; Xing, C.; Zheng, Y.; Pan, D. YAP/TAZ drives cell proliferation and tumour growth via a polyamine-eIF5A hypusination-LSD1 axis. *Nat. Cell Biol.* **2022**, *24*, 373–383. [CrossRef] [PubMed]
4. Arruabarrena-Aristorena, A.; Zabala-Letona, A.; Carracedo, A. Oil for the cancer engine: The cross-talk between oncogenic signaling and polyamine metabolism. *Sci. Adv.* **2018**, *4*, eaar2606. [CrossRef] [PubMed]
5. Flynn, A.T.; Hogarty, M.D. Myc, Oncogenic Protein Translation, and the Role of Polyamines. *Med. Sci.* **2018**, *6*, 41. [CrossRef]
6. Li, J.; Meng, Y.; Wu, X.; Sun, Y. Polyamines and related signaling pathways in cancer. *Cancer Cell Int.* **2020**, *20*, 539. [CrossRef]
7. Pledgie, A.; Huang, Y.; Hacker, A.; Zhang, Z.; Woster, P.M.; Davidson, N.E.; Casero, R.A., Jr. Spermine oxidase SMO(PAOh1), Not N1-acetylpolyamine oxidase PAO, is the primary source of cytotoxic H_2O_2 in polyamine analogue-treated human breast cancer cell lines. *J. Biol. Chem.* **2005**, *280*, 39843–39851. [CrossRef]
8. Murray Stewart, T.; Dunston, T.T.; Woster, P.M.; Casero, R.A., Jr. Polyamine catabolism and oxidative damage. *J. Biol. Chem.* **2018**, *293*, 18736–18745. [CrossRef]
9. McCloskey, D.E.; Yang, J.; Woster, P.M.; Davidson, N.E.; Casero, R.A., Jr. Polyamine analogue induction of programmed cell death in human lung tumor cells. *Clin. Cancer Res.* **1996**, *2*, 441–446.
10. Casero, R.A., Jr.; Wang, Y.; Stewart, T.M.; Devereux, W.; Hacker, A.; Wang, Y.; Smith, R.; Woster, P.M. The role of polyamine catabolism in anti-tumour drug response. *Biochem. Soc. Trans.* **2003**, *31*, 361–365. [CrossRef]
11. Ivanova, O.N.; Snezhkina, A.V.; Krasnov, G.S.; Valuev-Elliston, V.T.; Khomich, O.A.; Khomutov, A.R.; Keinanen, T.A.; Alhonen, L.; Bartosch, B.; Kudryavtseva, A.V.; et al. Activation of Polyamine Catabolism by N^1,N^{11}-Diethylnorspermine in Hepatic HepaRG Cells Induces Dedifferentiation and Mesenchymal-Like Phenotype. *Cells* **2018**, *7*, 275. [CrossRef]
12. Thomas, T.J.; Thomas, T. Cellular and Animal Model Studies on the Growth Inhibitory Effects of Polyamine Analogues on Breast Cancer. *Med. Sci.* **2018**, *6*, 24. [CrossRef]
13. Bae, D.H.; Lane, D.J.R.; Jansson, P.J.; Richardson, D.R. The old and new biochemistry of polyamines. *Biochim. Biophys. Acta Gen. Subj.* **2018**, *1862*, 2053–2068. [CrossRef]
14. Mitchell, J.L.; Thane, T.K.; Sequeira, J.M.; Thokala, R. Unusual aspects of the polyamine transport system affect the design of strategies for use of polyamine analogues in chemotherapy. *Biochem. Soc. Trans.* **2007**, *35*, 318–321. [CrossRef]
15. Cohen, S.S. *A Guide to the Polyamines*; Oxford University Press: New York, NY, USA, 1998.
16. Pegg, A.E.; Casero, R.A., Jr. Current status of the polyamine research field. *Methods Mol. Biol.* **2011**, *720*, 3–35. [CrossRef]
17. Asumendi, A.; Morales, M.C.; Alvarez, A.; Arechaga, J.; Perez-Yarza, G. Implication of mitochondria-derived ROS and cardiolipin peroxidation in N-(4-hydroxyphenyl)retinamide-induced apoptosis. *Br. J. Cancer* **2002**, *86*, 1951–1956. [CrossRef]
18. O'Donnell, P.H.; Guo, W.X.; Reynolds, C.P.; Maurer, B.J. N-(4-hydroxyphenyl)retinamide increases ceramide and is cytotoxic to acute lymphoblastic leukemia cell lines, but not to non-malignant lymphocytes. *Leukemia* **2002**, *16*, 902–910. [CrossRef]
19. Makena, M.R.; Koneru, B.; Nguyen, T.H.; Kang, M.H.; Reynolds, C.P. Reactive Oxygen Species-Mediated Synergism of Fenretinide and Romidepsin in Preclinical Models of T-cell Lymphoid Malignancies. *Mol. Cancer Ther.* **2017**, *16*, 649–661. [CrossRef]
20. Kim, Y.K.; Hammerling, U. The mitochondrial PKCdelta/retinol signal complex exerts real-time control on energy homeostasis. *Biochim. Biophys. Acta Mol. Cell Biol. Lipids* **2020**, *1865*, 158614. [CrossRef]
21. Orienti, I.; Farruggia, G.; Nguyen, F.; Guan, P.; Calonghi, N.; Kolla, V.; Chorny, M.; Brodeur, G.M. Nanomicellar Lenalidomide-Fenretinide Combination Suppresses Tumor Growth in an MYCN Amplified Neuroblastoma Tumor. *Int. J. Nanomed.* **2020**, *15*, 6873–6886. [CrossRef]
22. Orienti, I.; Nguyen, F.; Guan, P.; Kolla, V.; Calonghi, N.; Farruggia, G.; Chorny, M.; Brodeur, G.M. A Novel Nanomicellar Combination of Fenretinide and Lenalidomide Shows Marked Antitumor Activity in a Neuroblastoma Xenograft Model. *Drug Des. Dev. Ther.* **2019**, *13*, 4305–4319. [CrossRef] [PubMed]
23. Upton, D.; Valvi, S.; Liu, J.; Yeung, N.; George, S.; Ung, C.; Khan, A.; Franshaw, L.; Ehteda, A.; Shen, H.; et al. DIPG-07. High throughput drug screening identifies potential new therapies for diffuse intrinsic pontine gliomas (DIPGs). *Neuro Oncol.* **2020**, *22* (Suppl. S3), iii288. [CrossRef]
24. Upton, D.; Valvi, S.; Liu, J.; Yeung, N.; George, S.; Ung, C.; Khan, A.; Franshaw, L.; Ehteda, A.; Shen, H.; et al. RARE-08. Potential new therapies for diffuse intrinsic pontine gliomas identified through high throughput drug screening. *Neuro Oncol.* **2021**, *23* (Suppl. S1), i42. [CrossRef]

25. Barer, R. Interference microscopy and mass determination. *Nature* **1952**, *169*, 366–367. [CrossRef] [PubMed]
26. Zangle, T.A.; Teitell, M.A. Live-cell mass profiling: An emerging approach in quantitative biophysics. *Nat. Methods* **2014**, *11*, 1221–1228. [CrossRef] [PubMed]
27. Maiden, A.M.; Rodenburg, J.M.; Humphry, M.J. Optical ptychography: A practical implementation with useful resolution. *Opt. Lett.* **2010**, *35*, 2585–2587. [CrossRef] [PubMed]
28. Kasprowicz, R.; Suman, R.; O'Toole, P. Characterising live cell behaviour: Traditional label-free and quantitative phase imaging approaches. *Int. J. Biochem. Cell Biol.* **2017**, *84*, 89–95. [CrossRef]
29. Marrison, J.; Raty, L.; Marriott, P.; O'Toole, P. Ptychography—A label free, high-contrast imaging technique for live cells using quantitative phase information. *Sci. Rep.* **2013**, *3*, 2369. [CrossRef]
30. Suman, R.; Smith, G.; Hazel, K.E.; Kasprowicz, R.; Coles, M.; O'Toole, P.; Chawla, S. Label-free imaging to study phenotypic behavioural traits of cells in complex co-cultures. *Sci. Rep.* **2016**, *6*, 22032. [CrossRef]
31. Chen, Y.; Zhuang, H.; Chen, X.; Shi, Z.; Wang, X. Spermidine-induced growth inhibition and apoptosis via autophagic activation in cervical cancer. *Oncol. Rep.* **2018**, *39*, 2845–2854. [CrossRef]
32. Fan, J.; Feng, Z.; Chen, N. Spermidine as a target for cancer therapy. *Pharmacol. Res.* **2020**, *159*, 104943. [CrossRef]
33. McCubbrey, A.L.; McManus, S.A.; McClendon, J.D.; Thomas, S.M.; Chatwin, H.B.; Reisz, J.A.; D'Alessandro, A.; Mould, K.J.; Bratton, D.L.; Henson, P.M.; et al. Polyamine import and accumulation causes immunomodulation in macrophages engulfing apoptotic cells. *Cell Rep.* **2022**, *38*, 110222. [CrossRef]
34. Puleston, D.J.; Zhang, H.; Powell, T.J.; Lipina, E.; Sims, S.; Panse, I.; Watson, A.S.; Cerundolo, V.; Townsend, A.R.; Klenerman, P.; et al. Autophagy is a critical regulator of memory CD8$^+$ T cell formation. *eLife* **2014**, *3*, e03706. [CrossRef]
35. Puleston, D.J.; Buck, M.D.; Klein Geltink, R.I.; Kyle, R.L.; Caputa, G.; O'Sullivan, D.; Cameron, A.M.; Castoldi, A.; Musa, Y.; Kabat, A.M.; et al. Polyamines and eIF5A Hypusination Modulate Mitochondrial Respiration and Macrophage Activation. *Cell Metab.* **2019**, *30*, 352–363.e8. [CrossRef]
36. Latour, Y.L.; Gobert, A.P.; Wilson, K.T. The role of polyamines in the regulation of macrophage polarization and function. *Amino Acids* **2020**, *52*, 151–160. [CrossRef]
37. Zhang, H.; Alsaleh, G.; Feltham, J.; Sun, Y.; Napolitano, G.; Riffelmacher, T.; Charles, P.; Frau, L.; Hublitz, P.; Yu, Z.; et al. Polyamines Control eIF5A Hypusination, TFEB Translation, and Autophagy to Reverse B Cell Senescence. *Mol. Cell* **2019**, *76*, 110–125.e9. [CrossRef]
38. Mosmann, T. Rapid colorimetric assay for cellular growth and survival: Application to proliferation and cytotoxicity assays. *J. Immunol. Methods* **1983**, *65*, 55–63. [CrossRef]
39. Green, L.M.; Reade, J.L.; Ware, C.F. Rapid colorimetric assay for cell viability: Application to the quantitation of cytotoxic and growth inhibitory lymphokines. *J. Immunol. Methods* **1984**, *70*, 257–268. [CrossRef]
40. Bergamini, C.; Moruzzi, N.; Sblendido, A.; Lenaz, G.; Fato, R. A water soluble CoQ10 formulation improves intracellular distribution and promotes mitochondrial respiration in cultured cells. *PLoS ONE* **2012**, *7*, e33712. [CrossRef]
41. He, T.; Sun, Y.; Qi, J.; Hu, J.; Huang, H. Image deconvolution for confocal laser scanning microscopy using constrained total variation with a gradient field. *Appl. Opt.* **2019**, *58*, 3754–3766. [CrossRef]
42. Fang, J.; Nakamura, H.; Maeda, H. The EPR effect: Unique features of tumor blood vessels for drug delivery, factors involved, and limitations and augmentation of the effect. *Adv. Drug Deliv. Rev.* **2011**, *63*, 136–151. [CrossRef] [PubMed]
43. Gao, X.; Cui, Y.; Levenson, R.M.; Chung, L.W.; Nie, S. In vivo cancer targeting and imaging with semiconductor quantum dots. *Nat. Biotechnol.* **2004**, *22*, 969–976. [CrossRef] [PubMed]
44. Jain, R.K.; Stylianopoulos, T. Delivering nanomedicine to solid tumors. *Nat. Rev. Clin. Oncol.* **2010**, *7*, 653–664. [CrossRef] [PubMed]
45. Matsumura, Y.; Maeda, H. A new concept for macromolecular therapeutics in cancer chemotherapy: Mechanism of tumoritropic accumulation of proteins and the antitumor agent smancs. *Cancer Res.* **1986**, *46*, 6387–6392.
46. Maeda, H.; Takeshita, J.; Kanamaru, R. A lipophilic derivative of neocarzinostatin. A polymer conjugation of an antitumor protein antibiotic. *Int. J. Pept. Protein Res.* **1979**, *14*, 81–87. [CrossRef]
47. Nagy, J.A.; Chang, S.H.; Dvorak, A.M.; Dvorak, H.F. Why are tumour blood vessels abnormal and why is it important to know? *Br. J. Cancer* **2009**, *100*, 865–869. [CrossRef]
48. Maeda, H.; Nakamura, H.; Fang, J. The EPR effect for macromolecular drug delivery to solid tumors: Improvement of tumor uptake, lowering of systemic toxicity, and distinct tumor imaging in vivo. *Adv. Drug Deliv. Rev.* **2013**, *65*, 71–79. [CrossRef]
49. Prabhakar, U.; Maeda, H.; Jain, R.K.; Sevick-Muraca, E.M.; Zamboni, W.; Farokhzad, O.C.; Barry, S.T.; Gabizon, A.; Grodzinski, P.; Blakey, D.C. Challenges and key considerations of the enhanced permeability and retention effect for nanomedicine drug delivery in oncology. *Cancer Res.* **2013**, *73*, 2412–2417. [CrossRef]
50. Greish, K. Enhanced permeability and retention effect for selective targeting of anticancer nanomedicine: Are we there yet? *Drug Discov. Today Technol.* **2012**, *9*, e71–e174. [CrossRef]
51. De la Fuente, I.M.; Lopez, J.I. Cell Motility and Cancer. *Cancers* **2020**, *12*, 2177. [CrossRef]
52. Stuelten, C.H.; Parent, C.A.; Montell, D.J. Cell motility in cancer invasion and metastasis: Insights from simple model organisms. *Nat. Rev. Cancer* **2018**, *18*, 296–312. [CrossRef]
53. Paul, C.D.; Mistriotis, P.; Konstantopoulos, K. Cancer cell motility: Lessons from migration in confined spaces. *Nat. Rev. Cancer* **2017**, *17*, 131–140. [CrossRef]

54. Rappaz, B.; Breton, B.; Shaffer, E.; Turcatti, G. Digital holographic microscopy: A quantitative label-free microscopy technique for phenotypic screening. *Comb. Chem. High Throughput Screen* **2014**, *17*, 80–88. [CrossRef]
55. Girshovitz, P.; Shaked, N.T. Generalized cell morphological parameters based on interferometric phase microscopy and their application to cell life cycle characterization. *Biomed. Opt. Express* **2012**, *3*, 1757–1773. [CrossRef]
56. Larque, E.; Sabater-Molina, M.; Zamora, S. Biological significance of dietary polyamines. *Nutrition* **2007**, *23*, 87–95. [CrossRef]
57. Trachootham, D.; Alexandre, J.; Huang, P. Targeting cancer cells by ROS-mediated mechanisms: A radical therapeutic approach? *Nat. Rev. Drug. Discov.* **2009**, *8*, 579–591. [CrossRef]
58. Halliwell, B. The antioxidant paradox. *Lancet* **2000**, *355*, 1179–1180. [CrossRef]
59. Perillo, B.; Di Donato, M.; Pezone, A.; Di Zazzo, E.; Giovannelli, P.; Galasso, G.; Castoria, G.; Migliaccio, A. ROS in cancer therapy: The bright side of the moon. *Exp. Mol. Med.* **2020**, *52*, 192–203. [CrossRef]
60. Durymanov, M.O.; Rosenkranz, A.A.; Sobolev, A.S. Current Approaches for Improving Intratumoral Accumulation and Distribution of Nanomedicines. *Theranostics* **2015**, *5*, 1007–1020. [CrossRef]

 *pharmaceutics*

Article

Surface Functionalization of Silica Nanoparticles: Strategies to Optimize the Immune-Activating Profile of Carrier Platforms

Benjamin Punz [†], Litty Johnson [†], Mark Geppert, Hieu-Hoa Dang, Jutta Horejs-Hoeck, Albert Duschl and Martin Himly *

Department of Biosciences and Medical Biology, University of Salzburg, 5020 Salzburg, Austria; benjamin.punz@plus.ac.at (B.P.); litty.johnson@plus.ac.at (L.J.); mark.geppert@plus.ac.at (M.G.); hieu-hoa.dang@plus.ac.at (H.-H.D.); jutta.horejs-hoeck@plus.ac.at (J.H.-H.); albert.duschl@plus.ac.at (A.D.)
* Correspondence: martin.himly@plus.ac.at; Tel.: +43-662-8044-5713
† These authors contributed equally to this work.

Abstract: Silica nanoparticles (SiNPs) are generally regarded as safe and may represent an attractive carrier platform for nanomedical applications when loaded with biopharmaceuticals. Surface functionalization by different chemistries may help to optimize protein loading and may further impact uptake into the targeted tissues or cells, however, it may also alter the immunologic profile of the carrier system. In order to circumvent side effects, novel carrier candidates need to be tested thoroughly, early in their development stage within the pharmaceutical innovation pipeline, for their potential to activate or modify the immune response. Previous studies have identified surface functionalization by different chemistries as providing a plethora of modifications for optimizing efficacy of biopharmaceutical (nano)carrier platforms while maintaining an acceptable safety profile. In this study, we synthesized SiNPs and chemically functionalized them to obtain different surface characteristics to allow their application as a carrier system for allergen-specific immunotherapy. In the present study, crude natural allergen extracts are used in combination with alum instead of well-defined active pharmaceutical ingredients (APIs), such as recombinant allergen, loaded onto (nano)carrier systems with immunologically inert and stable properties in suspension. This study was motivated by the hypothesis that comparing different charge states could allow tailoring of the binding capacity of the particulate carrier system, and hence the optimization of biopharmaceutical uptake while maintaining an acceptable safety profile, which was investigated by determining the maturation of human antigen-presenting cells (APCs). The functionalized nanoparticles were characterized for primary and hydrodynamic size, polydispersity index, zeta potential, endotoxin contamination. As potential candidates for allergen-specific immunotherapy, the differently functionalized SiNPs were non-covalently coupled with a highly purified, endotoxin-free recombinant preparation of the major birch pollen allergen Bet v 1 that functioned for further immunological testing. Binding efficiencies of allergen to SiNPs was controlled to determine uptake of API. For efficacy and safety assessment, we employed human monocyte-derived dendritic cells as model for APCs to detect possible differences in the particles' APC maturation potential. Functionalization of SiNP did not affect the viability of APCs, however, the amount of API physisorbed onto the nanocarrier system, which induced enhanced uptake, mainly by macropinocytosis. We found slight differences in the maturation state of APCs for the differently functionalized SiNP–API conjugates qualifying surface functionalization as an effective instrument for optimizing the immune response towards SiNPs. This study further suggests that surface-functionalized SiNPs could be a suitable, immunologically inert vehicle for the efficient delivery of biopharmaceutical products, as evidenced here for allergen-specific immunotherapy.

Keywords: NH_2; COOH; APCs; moDCs; uptake; maturation

Citation: Punz, B.; Johnson, L.; Geppert, M.; Dang, H.-H.; Horejs-Hoeck, J.; Duschl, A.; Himly, M. Surface Functionalization of Silica Nanoparticles: Strategies to Optimize the Immune-Activating Profile of Carrier Platforms. *Pharmaceutics* **2022**, *14*, 1103. https://doi.org/10.3390/pharmaceutics14051103

Academic Editors: João Paulo Figueiró Longo and Lúis Alexandre Muehlmann

Received: 20 April 2022
Accepted: 18 May 2022
Published: 21 May 2022

Publisher's Note: MDPI stays neutral with regard to jurisdictional claims in published maps and institutional affiliations.

1. Introduction

The past years have witnessed a boost in the importance and growth of novel biopharmaceuticals which include peptides, proteins and antibodies [1]. When we consider the USFDA approvals from the year 2017 to 2019, more than 30% of the approved drugs were biologics (2017: 12 of 34; 2018: 17 of 42; 2019: 10 of 48) [2]. Compared to the traditional small drug molecules, biologics provide high target specificity and pharmacokinetics resulting in minimal off-target effects, making them more favorable [3,4]. Even though they exhibit increased effectiveness for a wide range of diseases including cancer and metabolic disorders, they face major challenges. One of the biggest challenges is the delivery of biologics to achieve their maximum therapeutic potential. This is attributed to their structural complexity, decreased stability, distribution, and permeability across biological barriers [5]. Nanoparticulate systems constitute an effective strategy to improve the delivery of biologics. The biologics can be either encapsulated inside the nanoparticles (NPs) or loaded onto their surface by covalent conjugation or by physical adsorption [6].

Silica NPs (SiNP) are an attractive platform especially for protein delivery due to their excellent biocompatibility, potential for surface modification, safety, tunability, and stability. They have been studied previously for intra- and extracellular protein delivery and enzyme mobilization [7–9]. The functionalization of SiNPs offers diverse prospects, especially in the field of protein delivery. Hollow-type mesoporous SiNPs with surface amino functionalization were used for the delivery of bovine viral diarrhea virus protein, and show promising potential for the development of nanoparticle-based recombinant subunit vaccines [10]. Wolley et al. successfully demonstrated the conjugation of platelet activation-specific antibodies to polyamido-amine dendrimer-functionalized SiNPs, facilitating the development of a diagnostic tool in cardiovascular disease [11]. Furthermore, carboxylic acid functionalization of SiNPs has been reported to improve protein-loading efficacy and sustained release, and to improve the particles' thermal stability and adaptability [12,13]. However, all these benefits remain of limited value if the functionalized nanomaterial exerts an enhanced or altered immunological effect, specifically when directed against the cargo (adaptive immunity). Immunological inertness is thus, for many applications, desired when considering functionalized nanoparticles for drug delivery. Ideally, they should remain undetectable by the innate immune system, display a low propensity to activate the adaptive immune system, and serve the purpose of delivering their cargo effectively to the target site. In this regard, surface functionalization of particulate systems should be scrutinized, as the functional groups or molecules associated with the surface of NPs can cause immunomodulation or induce an unexpected immune recognition resulting in an unwanted response. For example, polyethylene glycol (PEG) can induce an anti-PEG immune response and immunological memory, thereby resulting in reduced clinical efficacy and increased adverse effects [14]. Furthermore, even if the free molecule in solution exhibits weak immunogenicity, it can become immunogenic when associated with a carrier [15,16]. For instance, while PEG by itself is considered as immunologically inert and safe, PEGylation of nanoparticles results in the induction of anti-PEG antibodies, leading to the enhanced clearance of NPs from the blood and hypersensitivity reactions in patients [16]. Moreover, the physicochemical properties of nanomaterials, including their coatings, have been reported to determine the outcome of an immune response [17]. Thus, it is essential to elucidate the immune effects of functionalized nanoparticles at an early stage during development of nanobiopharmaceutics.

Dendritic cells are the most potent antigen-presenting cells (APCs) that can prompt the initiation of a primary immune response. Hence, they are highly relevant cells to investigate the immune effects of functionalized nanomaterials. Several reports have revealed the influence of the physicochemical properties of nanoparticles on the molecular mechanisms leading to an immune response. This includes the recognition by immune cells, efficiency of uptake, APC maturation, antigen processing, presentation, and T cell differentiation [18–20]. For example, nanoparticles with an optimum size of 50 nm were found to be taken up more effectively compared smaller or larger particles [21].

In this study, we compared the immune effects of aminopropyltriethoxysilane- (SiNP_A-conferring amino (NH_2) group) and Meldrum's acid- (SiNP_M-conferring a carboxy (COOH) group) functionalized silica particles with uncoated, i.e., SiNPs, silica NPs in APCs. SiNP_M was used as an alternative to previously investigated undecanoic acid-functionalized particles due to increased stability in suspension and, thus better performance in immunological assays involving APCs. Dendritic cells derived from human peripheral blood monocytes, termed monocyte-derived dendritic cells (moDCs) were used as APCs to study immune effects. We have previously observed alterations in the antigen processing patterns when Bet v 1 was conjugated to SiNPs due to the structural alteration of the protein at the nanomaterial interface [22]. Thus, the immune effects of the particles with and without conjugation to Bet v 1 (the model protein) were also explored here. Bet v 1, the major birch pollen allergen is a well-characterized, well-studied protein; its properties as an allergen facilitate specific immunological assessments in follow-up studies to determine its suitability as a candidate for allergen-specific immunotherapy. Here, we investigated the two major molecular mechanisms contributing to the initiation or modulation of an immune response, including the uptake of antigen and the maturation of the APCs.

2. Materials and Methods

2.1. Synthesis and Functionalization of SiNPs

2.1.1. Synthesis of SiNPs

SiNPs used in this study were synthesized by the Stöber method as previously described by Liberman et al. [23]. Briefly, 200 mL ethanol (96%) and 36 mL deionized water were heated to 75 °C in a two-neck round-bottom flask attached to a reflux condenser. To this mixture, 10 mL of 25% aqueous ammonia was added and the system was equilibrated for 15 min. This was followed by the addition of 15 mL tetraethylorthosilicate (TEOS), and the reaction was vigorously stirred for 2 h at 500 rpm. The obtained silica dispersion was centrifuged for 1 h at $6000\times g$ at 4 °C. The pellet was then washed three times with deionized water by centrifugation at $6000\times g$ for 30 min. Then, the washed particles were suspended in ethanol (96%) to circumvent extensive particle agglomeration, and dried at 80 °C for 12 h.

2.1.2. Functionalization of SiNP with Amino (NH_2) Groups

In order to attach functional amino groups to the surface of the SiNP, a slightly modified version of the protocol from Avella et al. [24] was followed. In the beginning of this procedure, 200 mg 3-aminopropyltrethoxysilane (APTES) was dissolved in 20 mL of water and kept at room temperature for 2 h under continuous magnetic stirring to start the hydrolysis reaction. SiNPs (1 g) dissolved in 25 mL ethanol and sonicated for around 30 min was added to the reaction mix and kept at 70 °C for 23 h under continuous magnetic stirring. The functionalized SiNPs were then dried at 80 °C for 24 h and stored at room temperature. To confirm if the APTES functionalization worked, the samples were tested with the salicylaldehyde test and subjected to FTIR and zeta potential measurements.

2.1.3. Functionalization of SiNPs with Carboxy (COOH) Groups

The modification of SiNPs with carboxy functional groups was performed by following a slightly modified protocol by Barczak et al. [25]. First, 0.1 g of SiNPs, 0.024 g of Meldrum's acid (MA), and 33.3 mL of toluene were added to a round-bottom flask attached to a reflux condenser and heated to 110 °C in an oil bath with continuous stirring for 3 h. Then, the yellowish reaction mixture was centrifuged at $6000\times g$ and washed once with 1% chloroform (diluted in Millipore water) and twice with absolute ethanol to assure that the unreacted species derived from the Meldrum's acid decomposition were removed. The washed mixture was then placed into a clean round-bottom flask with 33.3 mL of water and boiled for another 3 h. Finally, the contents of the flask were poured into a crystallization cup and dried overnight at 70 °C. The SiNP_M-functionalized particles obtained were then stored in an airtight container.

2.1.4. Efficiency of Functionalization Reaction

The efficient functionalization of SiNPs with both amino and carboxy groups was confirmed by FTIR, Schiff base reaction and zeta potential measurement. The aqueous NP dispersions were dried overnight at 80 °C, and the dried samples were used for FTIR measurements. The FTIR spectra were recorded using an FTIR spectrometer (Tensor 27, Bruker Optics, Ettlingen, Germany) equipped with an ATR (MIRacle ATR, Pike technologies, Fitchburg, WI, USA) accessory and Opus Software (Bruker Optics Version 6.5, Billerica, MA, USA). The Schiff base reaction was performed with salicylaldehyde to prove the presence of NH_2 groups with modified concentrations from Cuoq et al. (2012) [26]. Hereby, 10 µg of functionalized particles was added to 1 mL of absolute ethanol and 16 µL salicylaldehyde.

2.2. Characterization of SiNP

Size, Surface Charge, and Morphology of NPs

The hydrodynamic particle size of the SiNP was determined using nanoparticle tracking analysis (NTA) and dynamic light scattering (DLS). For NTA measurements, the different SiNPs were diluted to a concentration of 10 µg/mL with deionized water and analyzed using the NanoSight LM10 instrument (Malvern, Kassel, Germany) and the NTA 3.2 Dev Build 3.2.16 software (Malvern Panalytical, Malvern, UK). For the measurements, the standard measurement protocol was followed with 5 captures per measurement and 15 s capture duration. The hydrodynamic sizes and zeta potentials of the SiNPs were further confirmed by DLS. A concentration of 100 µg/mL of NPs was prepared in the buffer used for conjugation and the measurements were performed using the ZetaSizer Nano ZS (Malvern Panalytical, Malvern, UK) and the ZetaSizer Software (Malvern, 7.03, Malvern Panalytical, Malvern, UK), modulating the settings for refractive index of the NP composition and dispersant (1.45 for silica).

The primary particle size of the SiNPs was determined by transmission electron microscopy (TEM). For measurement, 2 µL of a 10 µg/mL NP dispersion was dried overnight on a lacy carbon-coated copper TEM grid and imaged using the JEM F200 (JEOL, Freising, Germany) electron microscope in TEM mode operated at 200 kV. Primary particle size was determined by calculating the mean ± SD of minimum 10 particles via image processing with the ImageJ software (NIH, Bethesda, MD, USA) and manual measuring.

2.3. Quantification of Endotoxin Contamination

The lipopolysaccharide (LPS) contamination in the synthesized and functionalized particle systems was tested by two different methods in agreement with recommendation for proper endotoxin testing of nanomaterials [27].

2.3.1. Monocyte Activation Test (MAT)

All studies involving human cells were conducted in accordance with the guidelines of the World Medical Association's Declaration of Helsinki. According to the national regulations, no additional approval by the local ethics committee was required in the case of anonymous blood cells discarded after plasmapheresis (buffy coats). Peripheral blood mononuclear cells (PBMCs) were isolated from blood samples of anonymous donors by density gradient centrifugation. From the PBMCs, monocytes were then purified using magnetically-activated cell sorting by CD14[+] MicroBeads UltraPure human kit (Miltenyi Biotec, Bergisch-Gladbach, Germany) according to the manufacturer's protocol. A total concentration of 1.5×10^5 cells/mL was seeded in the wells at a volume of 270 µL per well in cell culture medium containing RPMI, 10% heat-inactivated FCS, 1% 2 mM L-Glutamine, 1% Pen–Strep, and 50 mM 2-mercaptoethanol. The cells were stimulated with 30 µL (final sample concentration 100 µg/mL) of sample and incubated at 37 °C for 24 h. Endotoxin-free water was used as the diluent. After incubation, the supernatants were collected and tested for cytokine (IL-6, TNF-α) release by ELISA (Peprotech, London, UK), since these are the prominent cytokines induced by LPS. An LPS standard curve ranging from 1500 to 0.7 pg/mL on the differently coated cytokine ELISA plates was used to determine the quantitative amount of LPS in the samples.

2.3.2. HEK Blue™ LPS Detection Assay

The HEK Blue™ LPS detection assay, obtained from InvivoGen (San Diego, CA, USA), is a very simple, sensitive, and reliable assay to detect biologically active LPS. Briefly, 20 µL samples were added and a LPS standard curve ranging from 12.5 to 0 ng/mL was made. Endotoxin-free water was used as the diluent. hTLR4 cells were prepared at a concentration of 140,000 cells per mL and 180 µL were then added to the samples, incubated at 37 °C, 5% CO_2 for 16 h. Afterwards, NF-κB activation was measured in detection medium containing Quanti Blue™ by reading the absorbance at 650 nm. The remaining protocol was followed in accordance with the manufacturer's instructions.

2.4. Binding Efficency

To determine the binding efficiency, 500 µg/mL of NPs were incubated with 100 µg/mL of protein (Bet v 1) in the presence of 0.9% NaCl and 10 mM HEPES buffer pH 7.4 (binding buffer) at a volume of 500 µL at 4 °C for 17 h on a test tube rotator. The recombinant birch pollen allergen Bet v 1 used in this study was prepared as previously described [28]. To determine the total amount of protein bound to the surface of the NPs, the samples were centrifuged twice for 1 h with $16,000 \times g$ at 4 °C and pellet and supernatant was thoroughly separated. The two-step collection of the supernatant reduces the risk of contaminating the supernatant with the pellet. The bound protein was quantified directly from the pellet using 15% SDS-PAGE by relating it to the concentration of standards using the ImageLab software (version 6.0.1, 2017, Bio-Rad Laboratories, Hercules, CA, USA). In parallel, the unbound protein in the supernatant was quantified using BCA (bicinchoninic acid) assay. From this the percentage of bound protein was estimated indirectly. Hence, testing the amount of proteins left in the supernatant with the BCA assay also allows determine whether the amount of protein in the pellet and supernatant add up to the total amount of Bet v 1 present in the reaction mix during coupling.

2.5. Generation of Human Monocyte-Derived Dendritic Cells (moDCs)

PBMCs were isolated from buffy coats by density gradient centrifugation using histopaque-1077 (Sigma, St. Louis, MO, USA). The buffy coats from donors were kindly provided by the SALK (Salzburger Landesklinikum). The monocytes were then separated from the PBMCs using the adherence method. The adherent monocytes were then cultured in dendritic cell medium containing RPMI (Sigma), 10% heat-inactivated FCS (Biowest, Nuaillé, France), 1% 2 mM L-glutamine (Sigma), 1% Pen–Strep (Sigma), 50 mM 2-mercaptoethanol (Sigma) supplemented with 50 ng/mL granulocyte macrophage colony-stimulating factor (GM-CSF) (Life Technologies, Carlsbad, CA, USA) and interleukin 4 (IL-4) (Life Technologies, Carlsbad, CA, USA) for 6 days. The cells were refed with 100 ng/mL of GM-CSF and IL-4 on day 3. The protocol for generation of moDCs is described in detail by Posselt et al. [29].

2.6. Viability of Human Monocyte-Derived Dendritic Cells (moDCs)

The effect of functionalized SiNPs on the cytotoxicity of moDCs was assessed by the commonly used lactate dehydrogenase (LDH) assay, which determines the release of the cytoplasmic enzyme LDH. For analysis, 90 µL of LDH buffer (200 mM NaCl, 80 mM Tris/HCl, pH 7.2) was mixed with 90 µL of the media samples of moDCs stimulated for 24 h. Then, 180 µL of the LDH reaction mixture (0.4 mM NADH and 4 mM sodium pyruvate in LDH buffer) were added and the absorption at 340 nm was immediately recorded and followed over a timeframe of 15 min. Cell viability was calculated by comparing extracellular LDH activities of samples with extracellular LDH activities of moDCs treated with 0.1% Triton X-100 (total LDH release).

Additionally, the viability was determined using the Fixable Viability Dye eFluor506 (eBioscience, Waltham, MA, USA) by flow cytometric analysis using a 510/50 band pass filter. The percentage of viable cells was calculated by gating CD1a on the x-axis, indicating the well-differentiated moDCs and viability stain on the y-axis.

2.7. Kinetics and Mechanism of Uptake

To assess the impact of functionalization on the uptake by moDCs, the protein Bet v 1 was fluorescently labelled with pHrodo™ Red, succinimidyl ester (ThermoFisher Scientific, Waltham, MA, USA) according to the manufacturer's instruction. The detailed protocol for labelling can be found in the supplementary materials file. The labelled protein was bound to SiNPs as described in 2.4. moDCs were seeded in a 24-well plate at a density of 1×10^5 cells per mL, stimulated with 1 µg/mL of Bet v 1 bound to 100 µg/mL of different SiNPs and incubated for different time points (1, 2, 4, 6, 8, 24 h) to assess the kinetics of uptake. The mechanism of uptake was investigated using four different inhibitors, i.e., 2 µM Cytochalasin D (macropinocytosis and phagocytosis) (Sigma); 20 µM chlorpromazine-hydrochloride (clathrin-mediated endocytosis) (Sigma); 1 µM filipin (caveolin-dependent endocytosis) (Sigma); and 10 µM rottlerin (macropinocytosis) (Merck, Darmstadt, Germany) [30–33].

The concentration of inhibitors for the study was determined by testing its effect on the viability of moDCs. The inhibitors were preincubated with the moDCs for the desired time (cytochalasin D for 90 min and chlorpromazine hydrochloride, filipin, and rottlerin for 30 min). The samples were incubated with the cells for a period of 24 h. After the desired incubation time, the amount of pHrodo inside the cells was quantitatively determined by flow cytometry. The gating strategy presented is mean fluorescence intensity (MFI) and this has been obtained by excluding the dead cells, nanoparticles, and doublets. For the inhibition experiments, the Bet v 1 control was considered as 100% uptake control and all other samples were placed in relation to that. The gating strategy is shown in Figure S2 in Supplementary Materials.

2.8. Flow Cytometry

The expression of co-stimulatory molecules was assessed by flow cytometry. moDCs were seeded at a density of 1×10^5 cells per mL in a 24-well plate and incubated with either 1 µg/mL of Bet v 1 bound to 100 µg/mL of SiNP or 100 µg/mL of SiNP without Bet v 1 for 24 h at 37 °C and 5% CO_2. The cells were then collected and stained for α-HLA-DR APC (Invitrogen, Waltham, MA, USA), Fixable Viability Dye eFluor506 (eBioscience, Waltham, MA, USA), α-CD1a BV421 (Biolegend, San Diego, CA, USA), α-CD86 PE (eBioscience), α-CD40 FITC (Biolegend), and α-CD83 PE-Cy™7 (BD Bioscience, Heidelberg, Germany). The cells were then fixed by adding 4% PFA (Sigma) solution in PBS, before sample acquisition using the FACS Canto II flow cytometer (BD Biosciences). The data thus obtained were analyzed using the FlowJo X 10.0.7r2 software. The data are presented in relation to the controls, which has been obtained by excluding the dead cells, nanoparticles, and doublets. The gating strategy is shown in Figure S2, in addition to the fluorescence minus one (FMO) control in Figure S3.

2.8.1. Cytokine Multiplexing

The 45-Plex Human Procarta-Plex™ (ThermoFisher, Waltham, MA, USA) was used to measure the release of cytokines and chemokines from moDCs activated with SiNPs following manufacturer's instructions. Shortly, beads were rinsed (PBS, 0.05% Tween-20) and resuspended in the assay buffer (PBS, 0.05% Tween-20, 1% heat-inactivated FCS) (Biowest) before adding 8.34 µL per well into a 96 V-bottom well plate. Afterwards, 15 µL of samples were added, and the plate was incubated on a 500 rpm shaking orbital shaker at 4 °C overnight. The following day, samples were washed three times and resuspended in 15 µL of detection antibody solution before being incubated at room temperature for 30 min. After three further washes, 20 µL of Streptavidin-PE solution (1:1 in assay buffer) was added to each well and incubated at room temperature for 30 min. Lastly, the samples were rinsed three times before being resuspended in drive fluid for testing. The data were processed using Procarta Plex Analyst Software (ThermoFisher, Waltham, MA, USA) and measured on a Luminex Magpix device.

2.8.2. ELISA

ELISA was performed in accordance with the manufacturer's protocol (Peprotech, London, UK).

2.9. Statistical Analysis

Statistical analyses were accomplished with GraphPad Prism 9. For multiple comparisons, one-way ANOVA (Analysis of Variance) followed by a Tukey's post hoc test was performed; p-values ≤ 0.05 were considered as statistically significant (* $p \leq 0.05$; ** $p \leq 0.01$; *** $p \leq 0.001$; **** $p \leq 0.0001$).

3. Results and Discussion

3.1. Characterization of SiNPs

SiNPs synthesized wibyth the Stöber method were chemically functionalized with negatively charged COOH (SiNP_M) and positively charged NH_2 (SiNP_A) functional groups. As the initial step, the SiNPs and functionalized SiNPs were characterized for their primary size and morphology. The TEM images of the particles displayed a uniform spherical shape with a primary size of 51.02 ± 3.80 nm (SiNP), 47.31 ± 4.72 nm (SiNP_A) and 50.42 ± 4.57 nm (SiNP_M) (Figure 1) [22].

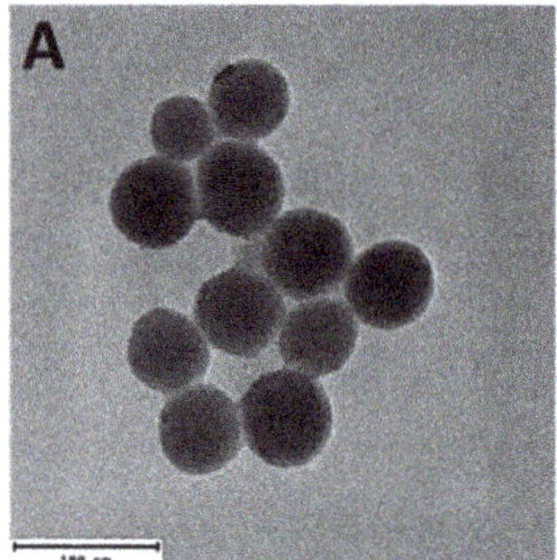
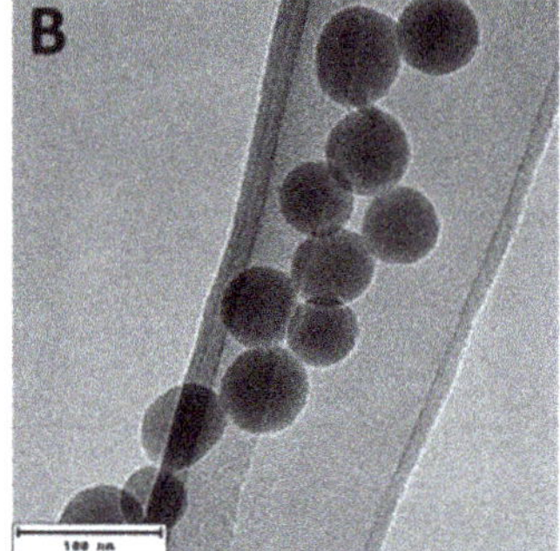
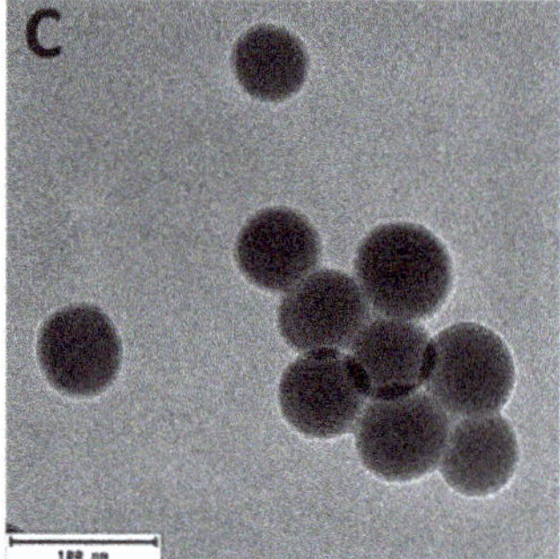

Figure 1. TEM images of the differently functionalized SiNPs for determination of primary particle size (**A**) SiNP, (**B**) SiNP_A, (**C**) SiNP_M, Size bar: 100 nm.

Thereafter, the efficiency of functionalization was tested by FTIR, zeta potential measurements, and the salicylaldehyde test. The FTIR spectra of all the samples exhibited the characteristic peak of silica at 1070 cm^{-1} [34] (Figure S4A–C). SiNP_A showed an additional peak at 1547 cm^{-1} due to NH_2 bending, which indicated that the functionalization was effective [35] (Figure S4D). For SiNP_M we observed the "C=O" stretch vibration at a wavelength of 1730–1700 cm^{-1}; however, the "C–O" stretch and "O–H" bend vibrations at 1320–1310 and 960–900 cm^{-1} were not evident [36] (Figure S4C,D). This might be due to some partial degree of esterification of the carboxyl group when washed with ethanol [37].

To further confirm effective particle functionalization, we determined the pH values of their aqueous suspensions and the surface charge of the particles with the ZetaSizer Nano ZS. SiNP_A exhibited a pH of 8.5 and positive zeta potential (33.4 $\pm$ 1.1 mV, whereas SiNP_M exhibited a pH of 4.5 and a negative zeta potential (-26.2 ± 1.8 mV) when compared to the SiNPs that displayed a pH of 7.0 and a negative zeta potential (-25.7 ± 0.5 mV) (Table 1). Moreover, the amino functionalization was tested by a Schiff base reaction. The addition of salicylaldehyde to SiNP_A led to the formation of a bright yellow-colored Schiff base. Centrifugation of the sample resulted in the formation of a yellow-colored pellet and a clear supernatant (Figure S1). This further proved effective amino functionalization.

Table 1. Physicochemical properties of SiNP, SiNP_A and SiNP_M. Size (NTA): Average ± SD of the hydrodynamic diameter mode values of five measurements. Size (DLS): Average ± SD of mean values from intensity- and number-weighted distribution analyses of three measurements. Zeta potential: average ± SD in mV determined when resuspended in buffer (HEPES). PDI: Polydispersity index ranging from 0 (perfectly uniform sample with respect to particle size) to 1 (highly polydisperse sample with multiple particle size populations).

Type	Size [nm] (NTA)	Size [nm] (DLS) Intensity	Size [nm] (DLS) Number	Zeta Potential [mV]	PDI
SiNP	112.2 ± 38.7	99.0 ± 42.4	53.1 ± 0.6	−25.7 ± 0.5	0.34
SiNP_A	206.2 ± 53.5	108.0 ± 46.4	56.3 ± 1.6	33.4 ± 1.1	0.41
SiNP_M	136.8 ± 62.1	114.5 ± 58.9	50.2 ± 7.4	−26.2 ± 1.8	0.38

Finally, the hydrodynamic sizes and distributions of the particles were determined by NTA and DLS (weighted for number and intensity, Figure S5). The SiNPs displayed an intensity-weighted mean hydrodynamic size of 99.0 ± 42.4 nm (SiNPs), 114.5 ± 58.9 nm (SiNP_M), and 108.0 ± 46.4 nm (SiNP_A) by DLS, which were in good agreement with NTA (Table 1). The intensity-weighted values and distributions (Figure S5A–F) show that for most of the particles, small agglomerates were abundant in suspension, whereas the number-weighted distributions (Figure S5G–I) indicate the presence of still non-agglomerated particles, since their sizes match the values obtained from TEM (Figure 1). From all the characterization data, it can be concluded that the functionalization worked efficiently and both the functionalized and SiNPs exhibited a hydrodynamic size of about 100 nm in dispersion.

3.2. Lipopolysaccharide/Endotoxin Content in SiNP

Lipopolysaccharide (LPS) induces the expression of pro-inflammatory cytokines and surface activation markers at a concentration as low as 20 pg/mL in primary human monocytes [38]. A potential contamination of LPS can cause a health threat in pharmaceutical products or compromise experimental results. Thus, it is necessary to determine the LPS contents in the particle systems, proteins and reagents involved in the in vitro experiments. These were tested by the most sensitive monocyte activation test [39,40] and the HEK Blue™ LPS Detection Kit 2. The results from both the tests indicate that the LPS content in the samples is below 20 pg/mL (Figure 2). Bet v 1 was previously tested after the original preparation and showed LPS values of less than 0.044 pg/μg [41].

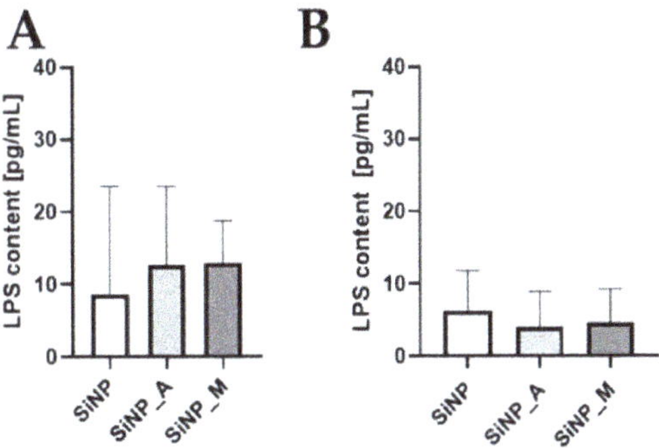

Figure 2. Quantification of LPS content by (**A**) HEK blue assay and (**B**) monocyte activation test (MAT) to analyze LPS contamination in the SiNP samples and Bet v 1. The LPS content was quantified based on LPS standard curves. The data are presented as mean ± SD (*n* = 4).

3.3. Impact of Particle Functionalization on the Binding Efficiency

Nanoparticle surfaces spontaneously adsorb proteins and form a nanoparticle protein corona [42,43]. Allergens are known to form a stable corona on the particle interface [43]. We have previously demonstrated that silica nanoparticles both mesoporous and nonporous adsorb Bet v 1, effectively forming a biocorona [22,44]. The efficiency of binding was determined by SDS-PAGE and BCA assay. From the SDS-PAGE analysis, 21.75 ± 6.50% of Bet v 1 was bound to SiNPs at pH 7.4, whereas the SiNP_M and SiNP_A showed binding efficiencies of 11.41 ± 4.55% and 0.1 ± 0.1%, respectively (Figures 3A and S6). The BCA assay displayed similar yet statistically significant results (Figure 3B). Both SiNP_M and SiNP_A functionalization decreased the binding efficiency when compared to the SiNPs. A similar decrease in serum protein adsorption to COOH-modified mesoporous SiNPs was previously reported by Lin et al. [45] and Beck et al. [46]. While a decreased protein binding was reported to be due to the increasing negative charge on the particles surface, the influence of the particles' charge on protein binding was not evident in this study. Furthermore, SiNP_A showed almost negligible binding capacity. The pKa of the used SiNP_A is 7.6, based on the components' ratios during functionalization [47]. Consequently, Bet v 1 displays an isoelectric point (IEP) in the similar range [48]. This would result in hindered binding due to the similarities in their isoelectric points. Here, it is clear that functionalization of SiNPs is altering the binding efficiency.

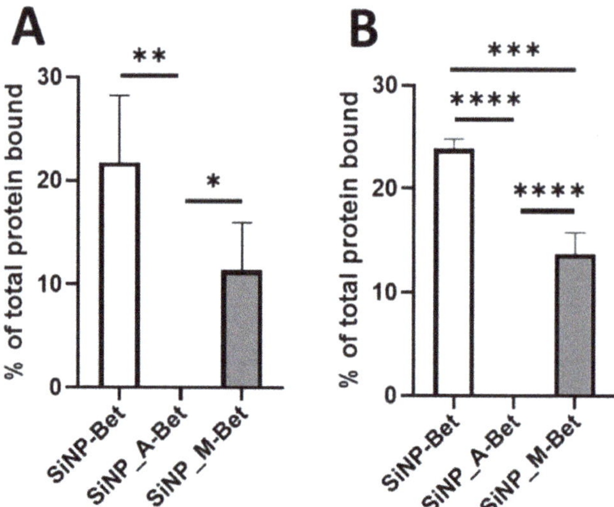

Figure 3. The binding efficiencies of differently functionalized SiNP samples were determined (**A**) directly from the pellet by SDS-PAGE and (**B**) indirectly from the supernatant by BCA assay. The data are presented as mean ± SD ($n = 4$) * $p \leq 0.05$; ** $p \leq 0.01$; *** $p \leq 0.001$; **** $p \leq 0.0001$.

3.4. SiNPs Do Not Affect the Viability of Antigen-Presenting Cells

Monocytes were isolated from peripheral blood mononuclear cells (PBMCs) by adherence method and artificially differentiated into immature monocyte-derived dendritic cells (moDCs) by the addition of IL-4 and GM-CSF. The immature moDCs were then stimulated with the samples (100 µg/mL of SiNP, 1 µg/mL of Bet v 1 bound to SiNP, and 1 µg/mL of Bet v 1) for 24 h. The impact of surface functionalization on the viability of moDCs was assessed by LDH assay and flow cytometry after staining the cells with Fixable Viability Dye eFluor506 (live/dead staining). The data from the LDH assay show that the functionalization did not alter the viability of the cells significantly, although we observed a small decrease in viability of moDCs when incubated with SiNP-Bet (Figure 4A). Still, the moDCs proved to be more than 80% viable. The negligible impact of functionalization on the viability was further proved by live/dead staining using flow cytometry analysis (Figure 4B). The gating strategy used for the analysis of the flow cytometry data is represented in Figure S2.

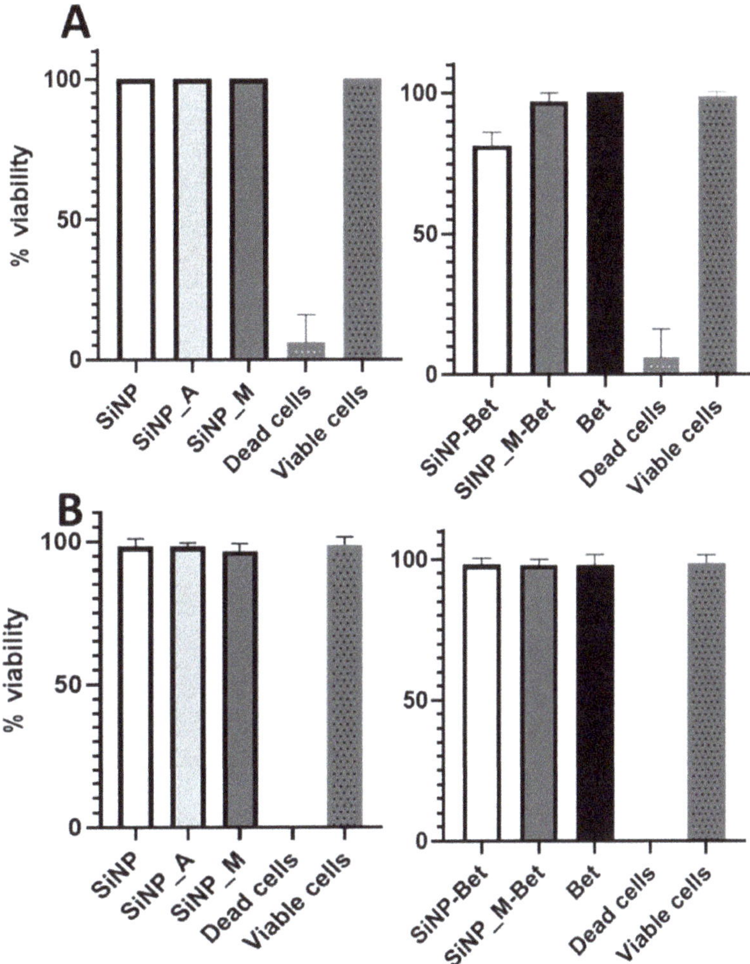

Figure 4. Viability of the moDCs tested with (**A**) LDH assay based on the release of LDH and (**B**) Live/dead staining via flow cytometry after stimulation of moDCs with the samples after 24 h. The negative control in the experiment was dead cells (incubated at 95 °C for 10 min) and the positive control (viable cells) was unstimulated moDCs.

3.5. SiNP Adsorption Induces Enhanced Allergen Uptake into Antigen-Presenting Cells Preferentially by Macropinocytosis

The first and foremost mechanism contributing to an immune response involves the recognition and internalization of antigens. The internalization of nanoparticles can change based on their physicochemical properties such as size, shape and surface chemistry [6]. We studied the kinetics of uptake using pHrodo-labeled Bet v 1. The labeled Bet v 1 was bound to differently functionalized SiNPs and the samples (Bet v 1, SiNP_M-Bet, and SiNP-Bet) were incubated with moDCs for 1, 2, 4, 6, 8, and 24 h with a concentration of 1 µg/mL of Bet v 1. All particle systems were stable in the suspension for 24 h as controlled by sedimentation analysis employing the silicomolybdic assay (Figure S7) and the protein labeling did not interfere with SiNP adsorption of Bet v 1, as confirmed by SDS-PAGE (Figure S8). Furthermore, the viability of moDCs was not affected when stimulated with the samples (Figure S9). We observed a significant increase in the uptake of allergen when bound to SiNPs at almost all time points (Figure 5A). This increased tendency for uptake was previously reported by Lu et al. (2009) and Heller et al. (2009) where it was shown that uptake is mostly dependent on size, and the maximum uptake by cells occurred when the size of NPs are about 50 nm [21,49]. Similar to Holzapfel et al. (2006), we found an enhanced uptake of protein bound to COOH-functionalized nanoparticles at early time points (up to 8 h) [50]. To clarify further the potential differences in the uptake behavior, we investigated the mechanism of uptake by using specific inhibitors of the major endocytosis mechanism of antigen uptake. The moDCs were pre-incubated with the inhibitors for their required times (Table S1), stimulated with the samples for 24 h and uptake was monitored by flow cytometry. The viability of moDCs remained 90 to 95% when incubated with the inhibitors (Figure S10). Chloropormazine hydrochoride (CPZ) inhibits clathrin-mediated endocytosis, and cytochalasin D (CytoD) inhibits both phagocytosis and macropinocytosis. Filipin inhibits caveolin-dependent endocytosis and rottlerin inhibits macropinocytosis [30–33]. The results revealed that there was a significant decrease in uptake of SiNP-Bet when incubated with rottlerin (around 75% inhibition), cytochalasin D (around 50% inhibition), and chlorpromazine HCl (40% inhibition). Thus, Bet v 1 molecules adsorbed to SiNPs are taken up mostly by macropinocytosis and Clathrin mediated endocytosis. It has previously been reported by Sahay et al. (2010) [51] that silica-based nanoparticles are internalized with clathrin-mediated endocytosis mechanisms. However, in SiNP_M-Bet we observed inhibition with rottlerin (around 70%) and cytochalasin D (around 50%). Macropinocytosis was thus revealed as the dominant mechanism of uptake for SiNP_M-Bet. Similarly, in Bet, about 40% uptake was inhibited with rottlerin and cytochalasin D, suggesting macropinocytosis as the preferred mechanism of uptake (Figure 5B) [32,52]. Overall, we observed an efficient uptake of Bet v 1 when adsorbed to SiNPs and that macropinocytosis is the preferred endocytosis mechanism for the internalization of Bet v 1. However, we did not observe any significant differences in either the kinetics or mechanisms of uptake between the functionalized SiNP samples, putatively due to negligible differences in size.

3.6. SiNP Adsorption and Functionalization Do Not Alter the Maturation of Antigen-Presenting Cells

In the final part of the study, we investigated the impact of functionalization on the maturation of APCs, which is characterized by the expression of co-stimulatory molecules like HLA-DR, CD86, CD83, and more, along with the release of soluble mediators such as cytokines. These co-stimulatory molecules and cytokines mediate the presentation of antigen to the T cells to modulate an immune response. To study the maturation of APCs, immature moDCs were stimulated with the samples for 24 h and the expression of CD40, CD80, CD83, CD86, and HLA-DR was measured by flow cytometry. The gating strategy is presented in Figure S2. The results obtained clearly showed negligible deviations in the expression of co-stimulatory molecules for Bet v 1 when compared to the untreated moDCs (Figure 6). This is in line with the study of Aglas et al. [53]. Furthermore, as reported before, the positive control (100 ng/mL LPS) showed a significant upregulation of all measured costimulatory molecules [54]. SiNPs showed the tendency to upregulate CD40, CD86, and HLA-DR, particularly when protein was conjugated onto them. In contrast, SiNP_M and SiNP_A caused no upregulation of costimulatory molecules and can thus be considered as a more inert particle system than silica without further chemical functionalization (Figure 6). The release of soluble mediators upon stimulation of moDCs was investigated using the human Procarta Plex Multiplex assay, where 45 different cytokines, chemokines, and growth factors were analyzed; raw data for the multiplex assay are available at https://doi.org/10.5281/zenodo.6473305 (accessed on 20 April 2022). For the control, LPS was used; it induced the release of the cyto-/chemokines and growth factors (IFN-gamma, IL12p70, IL-13, IL-2,

IL-4, IL-6, TNF-alpha, IL-18, IL-10, IL-1alpha, IL-1RA, Eetaxin, GRO-alpha, IL-8, IP-10, MCP-1, MIP-1alpha, MIP-1beta, SDF-1alpha, RANTES, HGF, and VEGF-A) in agreement with previous results [55]. When testing the SiNP samples and Bet v 1, we observed the release of the cyto-/chemokines IL-4, IL-6, IL-8, TNF-α, IL-1RA, MCP-1, MIP-1α, and MIP-1β (Figure 7). Notably, we found statistical significance between the Bet v 1-conjugated SiNP samples in IL-8, IL1-RA, and MIP-1α release, suggesting that functionalized particles are immunologically more inert compared to non-functionalized SiNPs when they are used as biopharmaceutical carrier system. To sum up, SiNPs tended to cause the upregulation of costimulatory molecules and increased release of pro-inflammatory cytokines, while these tendencies were not present in the functionalized samples. This indicates that surface functionalization of SiNPs can further decrease the maturation potential of moDCs and can thus be instrumental in preventing the activation and proliferation of T cells, rendering them safer in the context of inducing of an unwanted immune response. It further implies that functionalized SiNPs may remain undetectable as they showed no propensity to activate the innate immune system.

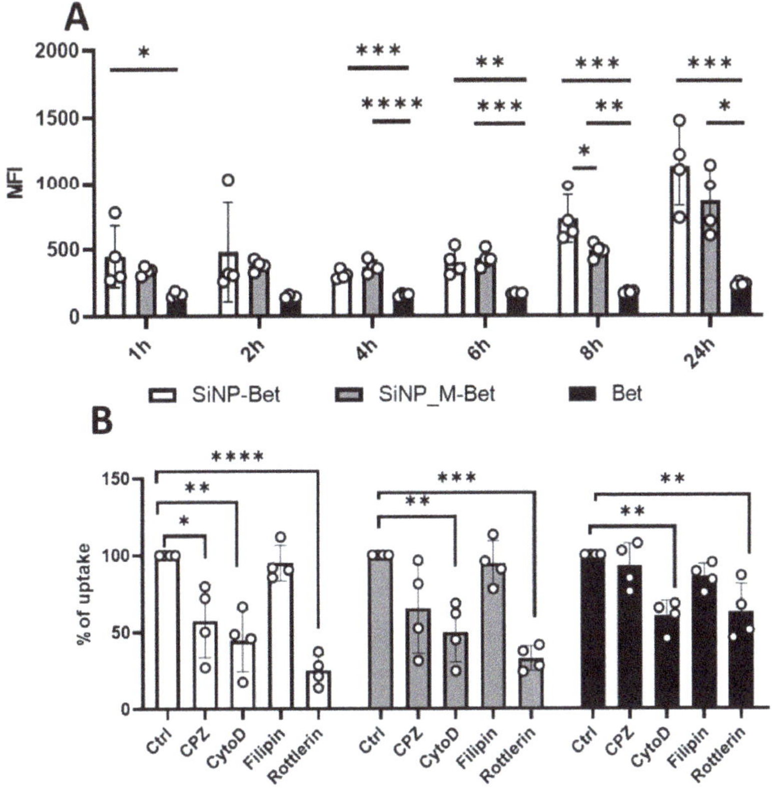

Figure 5. Impact on the uptake of allergen by moDCs. (**A**) Kinetics of uptake represented as mean fluorescence intensity (MFI) over the course of 24 h with SiNP-Bet, SiNP_M-Bet, and Bet. (**B**) Inhibition of uptake with the endocytosis inhibitors chlorpromazine hydrochloride (CPZ), cytochalasin D (CytoD), filipin, and rottlerin after stimulation with moDCs for 24 h ($n = 4$ individual donors). The data are presented as mean $\pm$ SD (* $p \leq 0.05$; ** $p \leq 0.01$; *** $p \leq 0.001$; **** $p \leq 0.0001$).

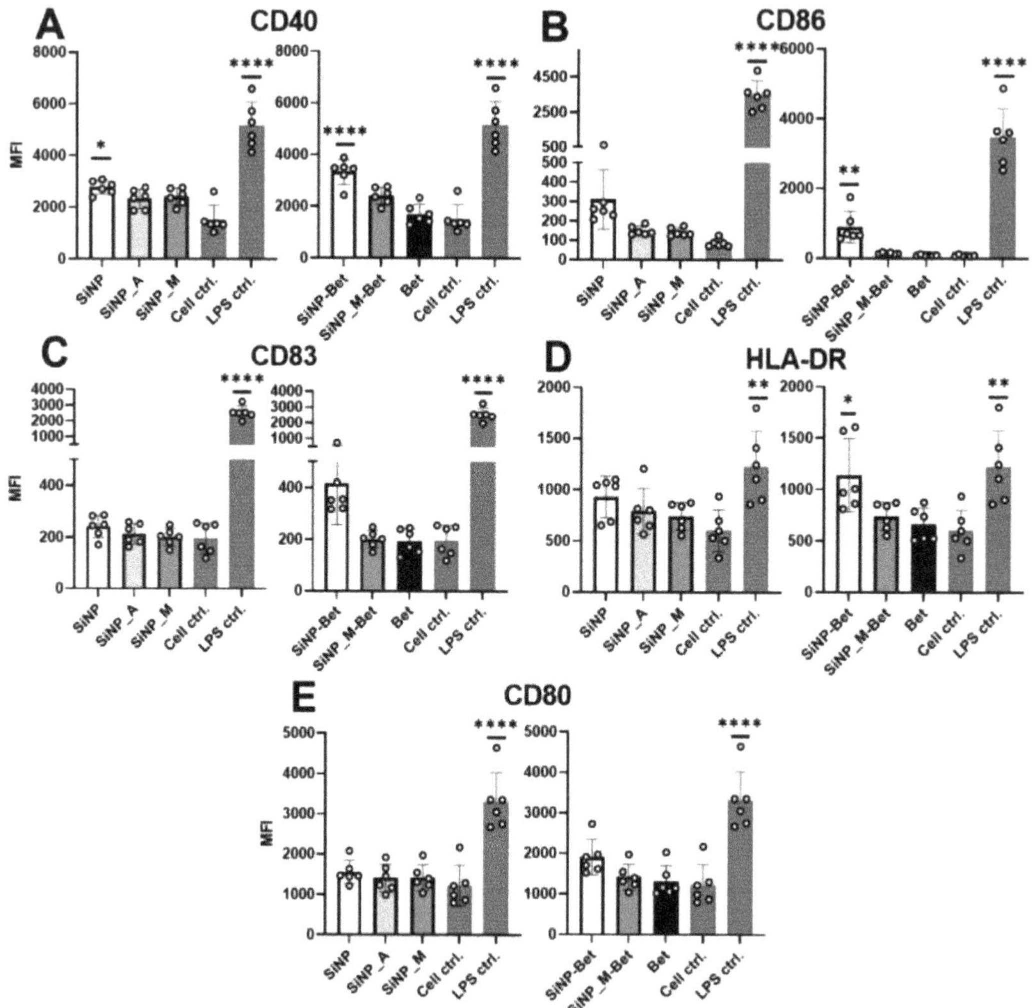

Figure 6. Co-stimulatory molecule expression by stimulated moDCs with functionalized SiNPs and Bet. (**A**) CD40, (**B**) CD86, (**C**) CD83, (**D**) HLA-DR, and (**E**) CD80 were analyzed. LPS was used as the positive control and a statistical significance was observed in comparison to the cell control indicating a successful stimulation. All particles and conjugates were compared to Bet v 1 for statistical analysis (n = 6 individual donors) (* $p \leq 0.05$; ** $p \leq 0.01$; **** $p \leq 0.0001$).

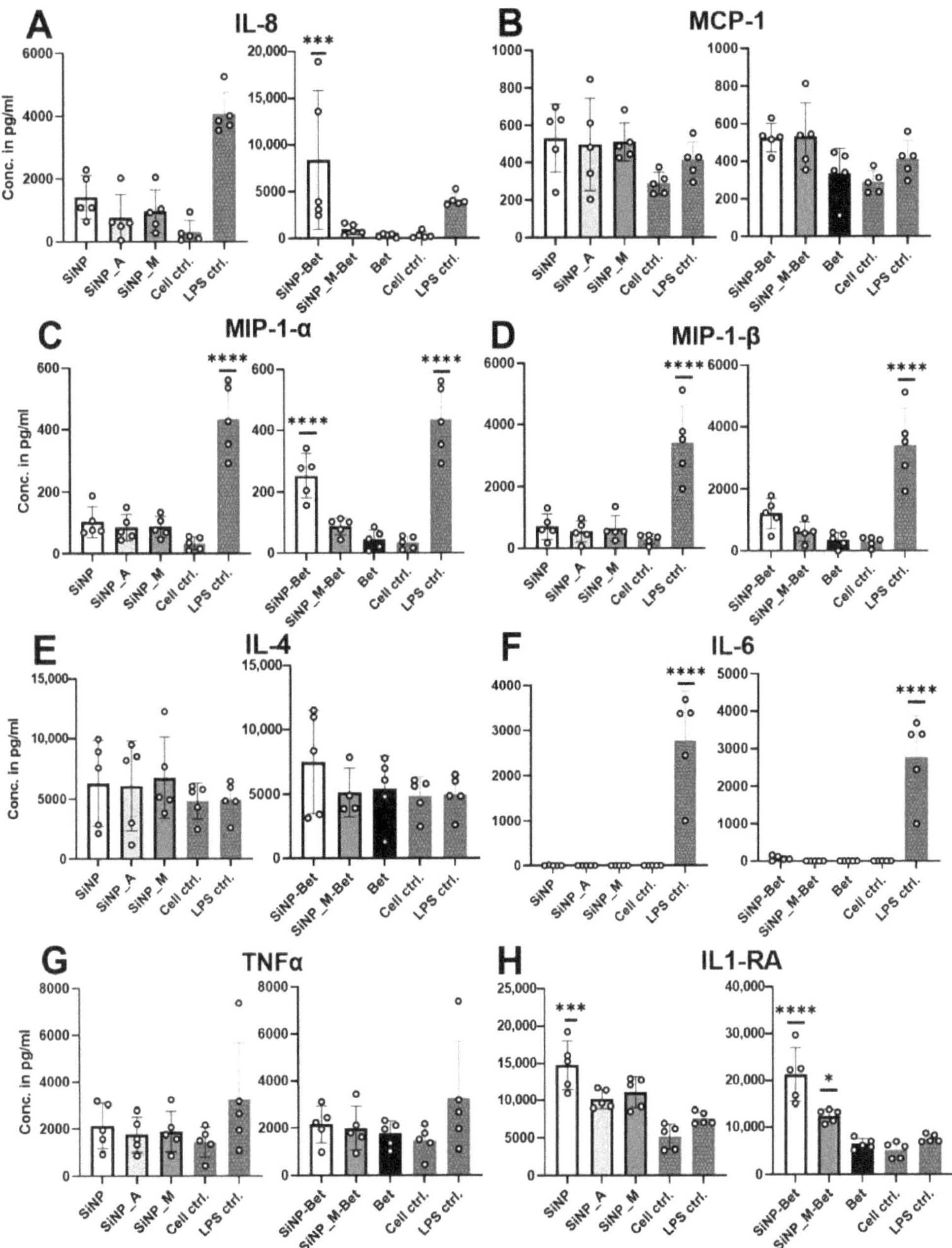

Figure 7. Cyto-/chemokine release of APCs. (**A**) IL8, (**B**) MCP-1, (**C**) MIP-1α, (**D**) MIP-1 β, (**E**) IL-4, (**F**) IL-6, (**G**) TNFα, (**H**) IL1-RA. LPS was used as the positive control and a statistical significance was observed in comparison to the cell control, indicating successful stimulation. The particles were compared to Bet v 1 for statistical analysis (*n* = 5 individual donors) (* $p \leq 0.05$, *** $p \leq 0.001$; **** $p \leq 0.0001$).

4. Conclusions

In this study, we have investigated whether different types of surface functionalization of SiNP impact immune effects by differential activation, i.e., the maturation, of APCs. In addition to first-line safety assessment, early insight on efficacy (second-line) is crucial for decision making in pharmaceutical development; in the context of immunotherapy, this represents the investigation of the immune activating and modulating capacities of administered substances, including nanomaterial platforms. Both the immune activating as well as modulating potentials involve APCs as a crucial player. As a suitable surrogate for human APCs, monocyte-derived dendritic cells were used here. We observed negligible changes induced by the differently functionalized SiNPs in the viability and maturation state of APCs, which is an indication of their immunological inertness. Nevertheless, a significant impact on the binding of proteins to the differently functionalized SiNPs was evident, and could not be simply explained by considering the isoelectric point of protein the net charge of particles upon functionalization. Enhanced uptake was noted for the particle system used here, irrespective of functionalization, while surface functionalization even optimized their immunological inertness, suggesting that SiNPs may be a suitable vehicle for the delivery of biopharmaceutical products, in particular for allergen-specific immunotherapy. Inert (nano)carriers afford the opportunity to intentionally load additional immune modifiers for fine-tuning of desired immunological outcome (e.g., immunomodulation towards a regulatory T helper cell response in allergy or immunosuppression in autoimmunity). As protein binding can be modulated by various types of chemical functionalization, functionalized SiNPs could be useful to deliver a range of biopharmaceutical drugs and lessen the potential immune response. Nevertheless, in vivo studies about the safety and efficacy profiles are required.

Supplementary Materials: The following supporting information can be downloaded at: https://www.mdpi.com/article/10.3390/pharmaceutics14051103/s1, Reference [56] is cited in the supplementary materials. Figure S1: Schiff base reaction for testing amino functionalization of SiNP; Figure S2. Gating strategy for moDC experiments; Figure S3. Fluorescence minus one (FMO) controls of moDCs; Figure S4. FTIR spectra of differently functionalized SiNP; Figure S5. NTA and DLS size distribution graphs; Figure S6. SDS-PAGE for estimation of protein binding; Figure S7. Suspension stability test; Table S1. Uptake inhibitors; Figure S8. SDS-PAGE analysis with pHrodo-labelled Bet v 1; Figure S9. Viability of moDCs on exposure to labelled allergen; Figure S10. Viability of moDCs on exposure to selected inhibitors.

Author Contributions: Conceptualization: L.J. and M.H. Methodology: B.P., L.J., J.H.-H., M.G. and M.H. Investigation: B.P., L.J., M.G. and H.-H.D. Visualization: B.P. and L.J. Writing—original draft: B.P., L.J. and M.H. Writing—review & editing: B.P., L.J., M.G., J.H.-H., A.D. and M.H. Supervision: L.J. and M.H. Funding acquisition: A.D. and M.H. All authors have read and agreed to the published version of the manuscript.

Funding: This work was funded by the international PhD program "Immunity in Cancer and Allergy" of the Austrian Science Fund (Open Access Funding by the Austrian Science Fund (FWF), grant number W01213), the EU H2020 NanoCommons project (grant number 731032), the SmartCERIALS project of the Austrian Research Promotion Agency (FFG, grant number 890610) and the Allergy Cancer Bio-Nano Research Centre (ACBN) of the University of Salzburg.

Institutional Review Board Statement: Not applicable.

Informed Consent Statement: Not applicable.

Data Availability Statement: All data generated or analyzed during this study are included in this manuscript and its supplementary information files. Raw data of the cyto-/chemokine secretion multiplex assay are available at https://doi.org/10.5281/zenodo.6473305 (accessed on 20 April 2022).

Acknowledgments: The authors thank Robert Mills-Goodlet; Ingrid Hasenkopf; Alexandra Fux; Angela Schmiedlechner; Sabine Hofer; Norbert Hofstätter; and Nanett Mokran for provision of materials for experiments, conduct of supporting experiments and scientific inputs.

Conflicts of Interest: There are no conflict to declare.

References

1. Begley, A. (Bio)pharma Continuous Manufacturing Market to Reach \$2.3 Billion by 2027. EPR. 2021. Available online: https://www.europeanpharmaceuticalreview.com/news/163519/pharma-and-biopharma-continuous-manufacturing-market-to-reach-2-3-billion-by-2027/ (accessed on 17 November 2021).
2. De la Torre, B.G.; Albericio, F. The pharmaceutical industry in 2019. An analysis of fda drug approvals from the perspective of molecules. *Molecules* **2020**, *25*, 745. [CrossRef] [PubMed]
3. Anselmo, A.C.; Gokarn, Y.; Mitragotri, S. Non-invasive delivery strategies for biologics. *Nat. Rev. Drug Discov.* **2018**, *18*, 19–40. [CrossRef] [PubMed]
4. Geynisman, D.M.; De Velasco, G.; Sewell, K.L.; Jacobs, I. Biosimilar biologic drugs: A new frontier in medical care. *Postgrad. Med.* **2017**, *129*, 460–470. [CrossRef] [PubMed]
5. Mitragotri, S.; Burke, P.A.; Langer, R. Overcoming the challenges in administering biopharmaceuticals: Formulation and delivery strategies. *Nat. Rev. Drug Discov.* **2014**, *13*, 655–672. [CrossRef]
6. Johnson, L.; Duschl, A.; Himly, M. Nanotechnology-based vaccines for allergen-specific immunotherapy: Potentials and challenges of conventional and novel adjuvants under research. *Vaccines* **2020**, *8*, 237. [CrossRef]
7. Niu, Y.; Yu, M.; Meka, A.; Liu, Y.; Zhang, J.; Yang, Y.; Yu, C. Understanding the contribution of surface roughness and hydrophobic modification of silica nanoparticles to enhanced therapeutic protein delivery. *J. Mater. Chem. B* **2016**, *4*, 212–219. [CrossRef]
8. Kalantari, M.; Yu, M.; Jambhrunkar, M.; Liu, Y.; Yang, Y.; Huang, X.; Yu, C. Designed synthesis of organosilica nanoparticles for enzymatic biodiesel production. *Mater. Chem. Front.* **2018**, *2*, 1334–1342. [CrossRef]
9. Xu, C.; Lei, C.; Huang, L.; Zhang, J.; Zhang, H.; Song, H.; Yu, M.; Wu, Y.; Chen, C.; Yu, C. Glucose-responsive nanosystem mimicking the physiological insulin secretion via an enzyme–polymer layer-by-layer coating strategy. *Chem. Mater.* **2017**, *29*, 7725–7732. [CrossRef]
10. Mahony, D.; Cavallaro, A.; Mody, K.; Xiong, L.; Mahony, T.; Qiao, S.; Mitter, N. In vivo delivery of bovine viral diahorrea virus, e2 protein using hollow mesoporous silica nanoparticles. *Nanoscale* **2014**, *6*, 6617–6626. [CrossRef]
11. Woolley, R.; Roy, S.; Prendergast, Ú.; Panzera, A.; Basabe-Desmonts, L.; Kenny, D.; McDonagh, C. From particle to platelet: Optimization of a stable, high brightness fluorescent nanoparticle based cell detection platform. *Nanomed. Nanotechnol. Biol. Med.* **2013**, *9*, 540–549. [CrossRef]
12. Tu, J.; Boyle, A.L.; Friedrich, H.; Bomans, P.H.; Bussmann, J.; Sommerdijk, N.A.; Jiskoot, W.; Kros, A. Mesoporous silica nanoparticles with large pores for the encapsulation and release of proteins. *ACS Appl. Mater. Interfaces* **2016**, *8*, 32211–32219. [CrossRef] [PubMed]
13. Saikia, D.; Deka, J.R.; Wu, C.-E.; Yang, Y.-C.; Kao, H.-M. Ph responsive selective protein adsorption by carboxylic acid functionalized large pore mesoporous silica nanoparticles sba-1. *Mater. Sci. Eng. C* **2019**, *94*, 344–356. [CrossRef] [PubMed]
14. Kozma, G.; Shimizu, T.; Ishida, T.; Szebeni, J. Anti-peg antibodies: Properties, formation and role in adverse immune reactions to pegylated nano-biopharmaceuticals. *Adv. Drug Deliv. Rev.* **2020**, *154–155*, 163–175. [CrossRef] [PubMed]
15. Richter, A.W.; Åkerblom, E. Polyethylene glycol reactive antibodies in man: Titer distribution in allergic patients treated with monomethoxy polyethylene glycol modified allergens or placebo, and in healthy blood donors. *Int. Arch. Allergy Immunol.* **1984**, *74*, 36–39. [CrossRef] [PubMed]
16. Shiraishi, K.; Yokoyama, M. Toxicity and immunogenicity concerns related to pegylated-micelle carrier systems: A review. *Sci. Technol. Adv. Mater.* **2019**, *20*, 324–336. [CrossRef]
17. Jia, J.; Zhang, Y.; Xin, Y.; Jiang, C.; Yan, B.; Zhai, S. Interactions between nanoparticles and dendritic cells: From the perspective of cancer immunotherapy. *Front. Oncol.* **2018**, *8*, 404. [CrossRef]
18. Blank, F.; Gerber, P.; Rothen-Rutishauser, B.; Sakulkhu, U.; Salaklang, J.; De Peyer, K.; Gehr, P.; Nicod, L.P.; Hofmann, H.; Geiser, T. Biomedical nanoparticles modulate specific cd4+ t cell stimulation by inhibition of antigen processing in dendritic cells. *Nanotoxicology* **2011**, *5*, 606–621. [CrossRef]
19. Est-Witte, S.E.; Livingston, N.K.; Omotoso, M.O.; Green, J.J.; Schneck, J.P. Nanoparticles for generating antigen-specific t cells for immunotherapy. *Semin. Immunol.* **2021**, *56*, 101541. [CrossRef]
20. Jarrett, R. The activation of mature dendritic cells using nanoparticle drug delivery system. In Proceedings of the Georgia Undergraduate Research Conference, Statesboro, GA, USA, 14 November 2014.
21. Lu, F.; Wu, S.H.; Hung, Y.; Mou, C.Y. Size effect on cell uptake in well-suspended, uniform mesoporous silica nanoparticles. *Small* **2009**, *5*, 1408–1413. [CrossRef]
22. Johnson, L.; Aglas, L.; Soh, W.T.; Geppert, M.; Hofer, S.; Hofstätter, N.; Briza, P.; Ferreira, F.; Weiss, R.; Brandstetter, H. Structural alterations of antigens at the material interface: An early decision toolbox facilitating safe-by-design nanovaccine development. *Int. J. Mol. Sci.* **2021**, *22*, 10895. [CrossRef]
23. Liberman, A.; Mendez, N.; Trogler, W.C.; Kummel, A.C. Synthesis and surface functionalization of silica nanoparticles for nanomedicine. *Surf. Sci. Rep.* **2014**, *69*, 132–158. [CrossRef] [PubMed]
24. Avella, M.; Bondioli, F.; Cannillo, V.; Di Pace, E.; Errico, M.E.; Ferrari, A.M.; Focher, B.; Malinconico, M. Poly (ε-caprolactone)-based nanocomposites: Influence of compatibilization on properties of poly (ε-caprolactone)–silica nanocomposites. *Compos. Sci. Technol.* **2006**, *66*, 886–894. [CrossRef]

25. Barczak, M. Functionalization of mesoporous silica surface with carboxylic groups by meldrum's acid and its application for sorption of proteins. *J. Porous Mater.* **2019**, *26*, 291–300. [CrossRef]
26. Cuoq, F.; Masion, A.; Labille, J.; Rose, J.; Ziarelli, F.; Prelot, B.; Bottero, J.-Y. Preparation of amino-functionalized silica in aqueous conditions. *Appl. Surf. Sci.* **2013**, *266*, 155–160. [CrossRef]
27. Himly, M.; Geppert, M.; Hofer, S.; Hofstätter, N.; Horejs-Höck, J.; Duschl, A. When would immunologists consider a nanomaterial to be safe? Recommendations for planning studies on nanosafety. *Small* **2020**, *16*, 1907483. [CrossRef]
28. Hoffmann-Sommergruber, K.; Susani, M.; Ferreira, F.; Jertschin, P.; Ahorn, H.; Steiner, R.; Kraft, D.; Scheiner, O.; Breiteneder, H. High-level expression and purification of the major birch pollen allergen, bet v 1. *Protein Expr. Purif.* **1997**, *9*, 33–39. [CrossRef]
29. Posselt, G.; Schwarz, H.; Duschl, A.; Horejs-Hoeck, J. Suppressor of cytokine signaling 2 is a feedback inhibitor of tlr-induced activation in human monocyte-derived dendritic cells. *J. Immunol.* **2011**, *187*, 2875–2884. [CrossRef]
30. Kuhn, D.A.; Vanhecke, D.; Michen, B.; Blank, F.; Gehr, P.; Petri-Fink, A.; Rothen-Rutishauser, B. Different endocytotic uptake mechanisms for nanoparticles in epithelial cells and macrophages. *Beilstein J. Nanotechnol.* **2014**, *5*, 1625–1636. [CrossRef]
31. Hsiao, I.-L.; Gramatke, A.M.; Joksimovic, R.; Sokolowski, M.; Gradzielski, M.; Haase, A. Size and cell type dependent uptake of silica nanoparticles. *J. Nanomed. Nanotechnol.* **2014**, *5*, 1000248.
32. Smole, U.; Radauer, C.; Lengger, N.; Svoboda, M.; Rigby, N.; Bublin, M.; Gaier, S.; Hoffmann-Sommergruber, K.; Jensen-Jarolim, E.; Mechtcheriakova, D. The major birch pollen allergen bet v 1 induces different responses in dendritic cells of birch pollen allergic and healthy individuals. *PLoS ONE* **2015**, *10*, e0117904. [CrossRef]
33. Sarkar, K.; Kruhlak, M.J.; Erlandsen, S.L.; Shaw, S. Selective inhibition by rottlerin of macropinocytosis in monocyte-derived dendritic cells. *Immunology* **2005**, *116*, 513–524. [CrossRef] [PubMed]
34. Osswald, J.; Fehr, K. Ftir spectroscopic study on liquid silica solutions and nanoscale particle size determination. *J. Mater. Sci.* **2006**, *41*, 1335–1339. [CrossRef]
35. Martínez-Carmona, M.; Ho, Q.P.; Morand, J.; García, A.; Ortega, E.; Erthal, L.C.; Ruiz-Hernandez, E.; Santana, M.D.; Ruiz, J.; Vallet-Regí, M. Amino-functionalized mesoporous silica nanoparticle-encapsulated octahedral organoruthenium complex as an efficient platform for combatting cancer. *Inorg. Chem.* **2020**, *59*, 10275–10284. [CrossRef] [PubMed]
36. Smith, B.C. The C=O bond, part iii: Carboxylic acids. *Spectroscopy* **2018**, *33*, 14–20.
37. Seré, S.; Vounckx, U.; Seo, J.W.; Lenaerts, I.; Van Gool, S.; Locquet, J.-P. Proof of concept study: Mesoporous silica nanoparticles, from synthesis to active specific immunotherapy. *Front. Nanotechnol.* **2020**, *13*, 584233. [CrossRef]
38. Schwarz, H.; Gornicec, J.; Neuper, T.; Parigiani, M.A.; Wallner, M.; Duschl, A.; Horejs-Hoeck, J. Biological activity of masked endotoxin. *Sci. Rep.* **2017**, *7*, 44750. [CrossRef]
39. Hacine-Gherbi, H.; Denys, A.; Carpentier, M.; Heysen, A.; Duflot, P.; Lanos, P.; Allain, F. Use of toll-like receptor assays for the detection of bacterial contaminations in icodextrin batches released for peritoneal dialysis. *Toxicol. Rep.* **2017**, *4*, 566–573. [CrossRef]
40. Hartung, T. The human whole blood pyrogen test: Lessons learned in twenty years. *ALTEX: Altern. Anim. Exp.* **2015**, *32*, 79–100. [CrossRef]
41. Machado, Y.; Freier, R.; Scheiblhofer, S.; Thalhamer, T.; Mayr, M.; Briza, P.; Grutsch, S.; Ahammer, L.; Fuchs, J.E.; Wallnoefer, H.G. Fold stability during endolysosomal acidification is a key factor for allergenicity and immunogenicity of the major birch pollen allergen. *J. Allergy Clin. Immunol.* **2016**, *137*, 1525–1534. [CrossRef]
42. Saptarshi, S.R.; Duschl, A.; Lopata, A.L. Interaction of nanoparticles with proteins: Relation to bio-reactivity of the nanoparticle. *J. Nanobiotechnol.* **2013**, *11*, 26. [CrossRef]
43. Radauer-Preiml, I.; Andosch, A.; Hawranek, T.; Luetz-Meindl, U.; Wiederstein, M.; Horejs-Hoeck, J.; Himly, M.; Boyles, M.; Duschl, A. Nanoparticle-allergen interactions mediate human allergic responses: Protein corona characterization and cellular responses. *Part. Fibre Toxicol.* **2015**, *13*, 3. [CrossRef] [PubMed]
44. Mills-Goodlet, R.; Schenck, M.; Chary, A.; Geppert, M.; Serchi, T.; Hofer, S.; Hofstätter, N.; Feinle, A.; Huesing, N.; Gutleb, A.C. Biological effects of allergen-nanoparticle conjugates: Uptake and immune effects determined on haelvi cells under submerged vs. Air-liquid interface conditions. *Environ. Sci. Nano* **2020**, *7*, 2073–2086. [CrossRef]
45. Lin, C.-Y.; Yang, C.-M.; Lindén, M. Influence of serum concentration and surface functionalization on the protein adsorption to mesoporous silica nanoparticles. *RSC Adv.* **2019**, *9*, 33912–33921. [CrossRef] [PubMed]
46. Beck, M.; Mandal, T.; Buske, C.; Linden, M. Serum protein adsorption enhances active leukemia stem cell targeting of mesoporous silica nanoparticles. *ACS Appl. Mater. Interfaces* **2017**, *9*, 18566–18574. [CrossRef] [PubMed]
47. Zhang, H.; He, H.-X.; Wang, J.; Mu, T.; Liu, Z.-F. Force titration of amino group-terminated self-assembled monolayers using chemical force microscopy. *Appl. Phys. A* **1998**, *66*, S269–S271. [CrossRef]
48. Breitenbach, M.; Ferreira, F.; Jilek, A.; Swoboda, I.; Ebner, C.; Hoffmann-Sommergruber, K.; Briza, P.; Scheiner, O.; Kraft, D. Biological and immunological importance of bet v 1 isoforms. In *New Horizons in Allergy Immunotherapy*; Springer: Berlin/Heidelberg, Germany, 1996; pp. 117–126.
49. Jin, H.; Heller, D.A.; Sharma, R.; Strano, M.S. Size-dependent cellular uptake and expulsion of single-walled carbon nanotubes: Single particle tracking and a generic uptake model for nanoparticles. *ACS Nano* **2009**, *3*, 149–158. [CrossRef]
50. Holzapfel, V.; Lorenz, M.; Weiss, C.K.; Schrezenmeier, H.; Landfester, K.; Mailänder, V. Synthesis and biomedical applications of functionalized fluorescent and magnetic dual reporter nanoparticles as obtained in the miniemulsion process. *J. Phys. Condens. Matter* **2006**, *18*, S2581. [CrossRef]

51. Sahay, G.; Alakhova, D.Y.; Kabanov, A.V. Endocytosis of nanomedicines. *J. Control. Release* **2010**, *145*, 182–195. [CrossRef]
52. Kankaanpää, P.; Tiitta, S.; Bergman, L.; Puranen, A.-B.; von Haartman, E.; Linden, M.; Heino, J. Cellular recognition and macropinocytosis-like internalization of nanoparticles targeted to integrin $\alpha2\beta1$. *Nanoscale* **2015**, *7*, 17889–17901. [CrossRef]
53. Aglas, L.; Gilles, S.; Bauer, R.; Huber, S.; Araujo, G.R.; Mueller, G.; Scheiblhofer, S.; Amisi, M.; Dang, H.-H.; Briza, P. Context matters: Th2 polarization resulting from pollen composition and not from protein-intrinsic allergenicity. *J. Allergy Clin. Immunol.* **2018**, *142*, 984–987.e986. [CrossRef]
54. Ardeshna, K.M.; Pizzey, A.R.; Devereux, S.; Khwaja, A. The pi3 kinase, p38 sap kinase, and nf-κb signal transduction pathways are involved in the survival and maturation of lipopolysaccharide-stimulated human monocyte–derived dendritic cells. *Blood J. Am. Soc. Hematol.* **2000**, *96*, 1039–1046.
55. Li, W.; Yang, S.; Kim, S.O.; Reid, G.; Challis, J.R.; Bocking, A.D. Lipopolysaccharide-induced profiles of cytokine, chemokine, and growth factors produced by human decidual cells are altered by lactobacillus rhamnosus gr-1 supernatant. *Reprod. Sci.* **2014**, *21*, 939–947. [CrossRef] [PubMed]
56. Coradin, T.; Eglin, D.; Livage, J. The silicomolybdic acid spectrophotometric method and its application to silicate/biopolymer interaction studies. *Spectroscopy* **2004**, *18*, 567–576. [CrossRef]

Article

Induction of Immunogenic Cell Death by Photodynamic Therapy Mediated by Aluminum-Phthalocyanine in Nanoemulsion

Mosar Corrêa Rodrigues [1,2], Wellington Tavares de Sousa Júnior [2], Thayná Mundim [2], Camilla Lepesqueur Costa Vale [2], Jaqueline Vaz de Oliveira [1], Rayane Ganassin [1,2], Thyago José Arruda Pacheco [1], José Athayde Vasconcelos Morais [1,2], João Paulo Figueiró Longo [2], Ricardo Bentes Azevedo [2] and Luis Alexandre Muehlmann [1,2,*]

[1] Laboratory of Nanoscience and Immunology, Faculty of Ceilandia, University of Brasilia, Brasilia 72220-900, Brazil; mosarcr@gmail.com (M.C.R.); jaquelinevazdeoliveira@gmail.com (J.V.d.O.); rayaneganassin@hotmail.com (R.G.); thyagojap@gmail.com (T.J.A.P.); joseathayde_9@hotmail.com (J.A.V.M.)
[2] Department of Genetics and Morphology, Institute of Biological Sciences, University of Brasilia, Brasilia 70910-900, Brazil; wellingtonjunior123@hotmail.com (W.T.d.S.J.); thaynamundim@hotmail.com (T.M.); camilla_costavale@hotmail.com (C.L.C.V.); jplongo82@gmail.com (J.P.F.L.); razevedo@unb.br (R.B.A.)
* Correspondence: luisalex@unb.br

Abstract: Photodynamic therapy (PDT) has been clinically employed to treat mainly superficial cancer, such as basal cell carcinoma. This approach can eliminate tumors by direct cytotoxicity, tumor ischemia, or by triggering an immune response against tumor cells. Among the immune-related mechanisms of PDT, the induction of immunogenic cell death (ICD) in target cells is to be cited. ICD is an apoptosis modality distinguished by the emission of damage-associated molecular patterns (DAMP). Therefore, this study aimed to analyze the immunogenicity of CT26 and 4T1 treated with PDT mediated by aluminum-phthalocyanine in nanoemulsion (PDT-AlPc-NE). Different PDT-AlPc-NE protocols with varying doses of energy and AlPc concentrations were tested. The death mechanism and the emission of DAMPs–CRT, HSP70, HSP90, HMGB1, and IL-1β–were analyzed in cells treated in vitro with PDT. Then, the immunogenicity of these cells was assessed in an in vivo vaccination-challenge model with BALB/c mice. CT26 and 4T1 cells treated in vitro with PDT mediated by AlPc IC_{50} and a light dose of 25 J/cm^2 exhibited the hallmarks of ICD, i.e., these cells died by apoptosis and exposed DAMPs. Mice injected with these IC_{50} PDT-treated cells showed, in comparison to the control, increased resistance to the development of tumors in a subsequent challenge with viable cells. Mice injected with 4T1 and CT26 cells treated with higher or lower concentrations of photosensitizer and light doses exhibited a significantly lower resistance to tumor development than those injected with IC_{50} PDT-treated cells. The results presented in this study suggest that both the photosensitizer concentration and light dose affect the immunogenicity of the PDT-treated cells. This event can affect the therapy outcomes in vivo.

Keywords: nanobiotechnology; apoptosis; damage associated molecules patterns; immunotherapy

Citation: Rodrigues, M.C.; de Sousa Júnior, W.T.; Mundim, T.; Vale, C.L.C.; de Oliveira, J.V.; Ganassin, R.; Pacheco, T.J.A.; Vasconcelos Morais, J.A.; Longo, J.P.F.; Azevedo, R.B.; et al. Induction of Immunogenic Cell Death by Photodynamic Therapy Mediated by Aluminum-Phthalocyanine in Nanoemulsion. *Pharmaceutics* **2022**, *14*, 196. http://doi.org/10.3390/pharmaceutics14010196

Academic Editor: Maria Nowakowska

Received: 3 December 2021
Accepted: 12 January 2022
Published: 14 January 2022

Publisher's Note: MDPI stays neutral with regard to jurisdictional claims in published maps and institutional affiliations.

1. Introduction

Photodynamic therapy (PDT) is generally based on three harmless components: molecular oxygen, photosensitizer, and light [1]. Once combined, they yield a robust production of reactive oxygen species (ROS), which can be lethal to the target cell [2–4]. As a cancer treatment modality, PDT can directly kill tumor cells due to its photocytotoxicity, cause infarction of the cancerous tissue by its effects on the tumor microvasculature and activate the immune system against tumor antigens [5]. The activation of the immune system by PDT has been the subject of intense research in recent years. In a murine experimental model of ectopic 4T1 mammary adenocarcinoma, PDT reduced the incidence of metastatic foci in the lungs, even when applied to the primary tumor [6]. Importantly, different parameters of PDT can affect its ability to induce antitumor responses, such as the type [5] and concentration of photosensitizer, as well as the irradiation regimen [7], a fact that has to be taken into account in the design of PDT-based immunotherapy.

The cell death mechanism triggered by PDT seems to play a key role on its immune effects. For instance, PDT mediated by the photosensitizer aluminum-phthalocyanine (AlPc) can induce both necrosis and apoptosis in murine melanoma B16F10 cells [7]. In this PDT setting, necrosis predominates at higher concentrations of AlPc, while apoptosis is the main cell death mechanism elicited at lower concentrations of this photosensitizer [7]. In the context of antitumor immune responses induced by PDT, the induction of immunogenic cell death (ICD) in target cells is to be cited [8]. According to Garg et al. [5], ICD is an apoptosis modality distinguished by the emission of DAMPs, which are potent immune activators. As agonists of various receptors involved in the immune response, DAMPs can attract and activate different immune cells [9]. They are also capable of promoting proinflammatory events, such as the maturation and activation of antigen-presenting cells, such as dendritic cells and T-cell activating macrophages [8,10].

In this study, the immunogenicity of two murine cancer cell lines–colorectal carcinoma (CT26) and mammary adenocarcinoma (4T1) cells–submitted to different PDT protocols mediated by a nanoemulsion containing aluminum-phthalocyanine (PDT-AlPc-NE) was evaluated. AlPc was chosen as the model photosensitizer because of its high singlet oxygen photogeneration yield and for its efficacy against both primary tumors and metastasis of murine cancer cells, such as 4T1 cells [1,6], in in vivo models. However, the possible immune-related, cellular mechanisms behind the efficacy of AlPc have not been studied in those models. The results show that both CT26 and 4T1 cells emitted different DAMP (calreticulin-CRT, heat shock proteins (HSP)-70 and -90, interleukin 1 beta-IL-1B, and high mobility group-box 1 (HMGB1) after specific PDT-AlPc-NE protocols in vitro. In an in vivo vaccination-challenge model, these PDT-treated cells rendered mice more resistant against the development of experimental tumors.

2. Materials and Methods

2.1. Materials

Roswell Park Memorial Institute medium (RPMI 1640), fetal bovine serum (FBS), penicillin, streptomycin, and trypsin were all purchased from Gibco, Carlsbad, CA, USA. Ethanol (99.3° GL) and glucose were all purchased from J.T. Trypan blue and dimethylsulfoxide (DMSO 99.5%); Trypan blue, $4'$,6-diamidino-e-fenilindol (DAPI), Annexin V (AnV), and propidium iodide (PI) were purchased from Sigma-Aldrich (St. Louis, MO, USA). Aluminum-phthalocyanine (AlPc) was purchased from Aldrich Chemical Company (St. Louis, MO, USA). PBS was purchased from Pinhais, Paraná, Brazil. Enzyme-linked immunosorbent assay (ELISA) kit for high mobility group box-1 (HMGB1) was purchased in IBL International GmbH (Hamburg, Germany). ELISA kit for heat shock proteins (HSP)-70 and −90, primary anti-mouse/human antibody against CRT, and Alexa Fluor® 488 Goat Anti-Mouse (IgG) secondary antibodies (IgG488) were purchased from ABCAM (Cambridge, UK). ELISA kit for interleukin-1beta (IL-1β) was purchased from Life Technologies (Carlsbad, CA, USA).

2.2. Light Source

A LED array system was used for irradiation of cells in vitro. This system was composed of a 20-LED-lamp-array model XL001WP01NRC660 (Shenzhen Sealand Optoelectronics, Ltd., Shenzhen, Guangdong, China) attached to a metal cooling unit and controlled by a constant current LED driver (Recom Power, Inc., Dietzenbach, Germany) model RCD-24-0.35/W. The energy fluence (J/cm^2) was adjusted according to the following formula: [power (W) $\times$ irradiation time (s)]/area of the light sensor (cm^2). LED spectral emission was recorded with a portable spectrometer (Ocean Optics, Inc., USA) with a spectral resolution of 0.2 nm in a range of 600–700 nm. A light power meter (Fieldmax II, Coherent Inc., Santa Clara, CA, USA) with a 1.9 cm diameter circular light-sensing area was used to measure maximum power and power as a function of the distance between the illumination system and the target. This system has a maximum power of 800 mW and a maximum fluence rate of 55 mW/cm^2 inside the 2.5-cm-radius illuminated area. The actual power used was equivalent to 122 mW. The spectral output is limited to the 660 nm red band.

2.3. Nanoemulsion Containing Aluminum-Phthalocyanine (AlPc-NE)

The nanoemulsion containing aluminum-phthalocyanine (AlPc-NE) was prepared by the spontaneous nanoemulsification method described by Muehlmann et al. [11]. Briefly, 9 g Kolliphor® ELP and 3 g castor oil were mixed with 40 mL AlPc 100 μM in ethanol. This solution was stirred at 50 °C for 15 min. Ethanol was then removed at 80 °C under magnetic stirring. Next, 70 mL of distilled water was added, and the dispersion was left under stirring at room temperature until a transparent nanoemulsion was obtained. The volume was then completed to 100 mL with PBS. The final concentration of AlPc in this formulation was 40 μM.

2.4. Cell Culture

Both the murine colorectal carcinoma (CT26) and the murine mammary adenocarcinoma cell line (4T1) were purchased from American Type Culture Collection (ATCC, Manassas, VA, USA). CT26 cells and 4T1 cells were cultured in RPMI supplemented with 10% (v/v) FBS, 100 units penicillin/mL, and 100 mg streptomycin/mL and maintained in an incubator under a humidified atmosphere with 5% CO$_2$ at 37 °C. The different in vitro tests in this study were performed using either 12- or 96-well microplates, with 4×10^4 CT26 or 1×10^4 4T1 cells per well.

2.5. Animals

Immunocompetent 12-week-old female BALB/c mice were kept in an animal facility, with free access to Purina and water and were handled according to procedures previously approved by the Ethics Committee on Animal Use of the University from Brasilia, Brazil (UnB/Doc n. 5529/2015, approved on 3 March 2015).

2.6. PDT-AlPc-NE In Vitro

CT26 or 4T1 cells were incubated in the dark at 37 °C with different concentrations of AlPc-NE for 15 min. Then, the cells were washed with PBS and maintained in an incubator for further 15 min with complete RPMI 1640 medium. Next, the cells were irradiated with LED 660 nm for 10 min, with a final energy dose of 25 J/cm^2. The AlPc-NE concentrations that reduced cell viability by 50% (IC$_{50}$) and 90% (IC$_{90}$) were then calculated. In subsequent experiments, the cells were treated with PDT protocols, with their respective AlPc-NE IC$_{50}$ and IC$_{90}$, maintained in an incubator and analyzed for cell death pathways and DAMP exposure.

2.7. Cell Death Pathways

Two methods were used to verify the cell death pathway triggered in the CT26 cells and 4T1 cells by PDT-AlPc-NE protocols: (i) fluorescence microscopy (microscope EVOS-FL, Thermo Fisher Scientific InC., Waltham, MA, USA) of cells stained with acridine orange and propidium iodide (AO/PI) at 4 h after the PDT-AlPc-NE protocols described by Kasibhatla et al. [12] and (ii) flow cytometry of cells stained with Alexa Fluor 488-annexin V and propidium iodide (AnV/PI) at 24 h after the PDT-AlPc-NE. Mitoxantrone (1.5 μM) was used as a positive control for ICD and apoptosis. The necrosis-positive control group consisted of cells frozen and thawed three times, as described by Garg et al. [13].

2.8. Immunofluorescence

The detection of CRT, HSP70, and HSP90 was performed by immunofluorescence. After the treatments, the cells were fixed with ethanol 70% (v/v, in water) at room temperature for 30 min. Next, the cells were incubated with an anti-CRT (1:75), anti-HSP70 (1:75) or anti-HSP90 (1:150) primary antibody in cold blocking buffer (2% BSA in PBS) for 1 h in an incubator (37 °C), followed by washing with PBS and incubation with IgG488 secondary antibody (1:250) for 30 min at room temperature. The nuclei of the cells were labeled with DAPI. The stained cells were visualized with a fluorescence microscope (EVOS-FL, Thermo Fisher Scientific Inc., Waltham, MA, USA). The fluorescence of the stained cells was quantified with the software Photoshop CC 2015 (Adobe Systems, San Jose, CA, USA) and ImageJ 1.8.0_172 (NIH, Madison, WI, USA).

2.9. ELISA

ELISA kits were used to quantify IL-1β (Life Technologies, Carlsbad, CA, USA) and HMGB1 (IBL International GmbH, Hamburg, Germany) in the culture supernatants according to the manufacturer's protocol 24 h after the difference treatment with PDT-AlPc-NE, MTX, and F-T.

2.10. Vaccination-Challenge Assay

The immunogenicity of the treated cells was assessed with an in vivo vaccination-challenge model. Briefly, CT26 cells and 4T1 cells were treated as follows, respectively: (i) PDT1 12.2 nM or 9.01 nM with 25 J/cm^2; (ii) PDT2 31.5 nM or 19.4 nM with 25 J/cm^2; (iii) PDT3 12.2 nM or 9.01 nM with 67 J/cm^2; (iv) PDT4 31.5 nM or 19.4 nM with 67 J/cm^2; and (v) mitoxantrone (1.5 μM); and (vi) frozen-thawed. Then, 100 μL of the suspension of cells (4 × 10^5 CT26 or 1 × 10^5 4T1 cells/mL) were subcutaneously injected into the right flank of mice. This process was repeated a second time, with a ten-day interval between vaccinations. Seven days after the 2nd vaccine, the mice were challenged with a subcutaneous graft of viable CT26 or 4T1 cells on the left flank. Following the challenge with viable CT26 or 4T1 cells, the animals were monitored for tumor onset, tumor volume evolution, and survival.

2.11. Computed Tomography

The mice were anesthetized with ketamine and xylazine (80 and 10 mg/kg, respectively) and subjected to computed tomography (PET-SPECT and CT-Bruker, Ettlingen, Germany). It used low-resolution (CT single good, low dose (200 μA), low voltage (35 μA)), and standard CT quality. Two hundred and fifty image projections were collected per animal with three minutes of exposure. Then, images were reconstructed with standard software reconstruction modes to evaluate the pulmonary radiopacity profile and bone structure. The pulmonary radiopacity profile was assessed in all planes of the image by delimiting the pulmonary region using the inner side of the rib cage as a reference. The software generates an automatic Hounsfield (HU) scale, and the voxel (3D pixel) intensity values create a Gaussian distribution in this HU scale. The less-dense voxels on the scale are positioned on the left, while the denser ones are placed on the right. Increased pulmonary density, which is correlated with lung metastasis [6], was evaluated by the Gaussian curve displacement to the right. The area under the curve of each experimental animal was calculated as the integral of the curve in a bar graph.

2.12. Statistical Analyses

All statistical analyses were performed with GraphPad Prism 7.0 software (San Diego, CA, USA). Correlation between variables was analyzed with the Pearson and Spearman test. Significant differences between groups were assessed by one-way analysis of variance (ANOVA) followed by Tukey or Bonferroni's post-tests ($\alpha = 0.05$). Results are expressed as mean ± standard error of the mean.

3. Results and Discussion

Several studies have shown that PDT is capable of triggering immune responses against tumor antigens [5,14,15]. One of the candidate mechanisms underlying this immune effect is the occurrence of ICD in cancer cells exposed to PDT [5,7]. Thus, the main goal of the present work was to verify whether two critical variables of PDT protocols, namely photosensitizer concentration, and light dose, affect the ability of this approach to induce ICD in two different murine cancer cell lines, 4T1 and CT26, in vitro. CT26 and 4T1 cells were subjected to different PDT-AlPc-NE protocols to obtain IC50 and IC90. The concentrations obtained for CT26 can be found in the Supplementary Materials (Figures S1 and S2). The data for 4T1 cells were published in Rodrigues et al. [1].

ICD is characterized by cell death by apoptosis with a well-defined pattern of DAMP exposure. It is well described that PDT can preferentially trigger apoptosis or necrosis depending on protocol parameters such as the concentration of photosensitizer and the energy dose applied [1,7]. The results in Figure 1 corroborate findings in the literature, as the cell mechanism triggered by PDT depended on the protocol parameters. Both CT26 and 4T1 cells succumbed to apoptosis when submitted to PDT with their respective AlPc-NE IC_{50}, 12.2 nM, and 9.01 nM, respectively, and the same energy dose—25 J/cm^2. As expected, with the same AlPc-NE concentrations, but under a higher energy dose—67 J/cm^2, a significant increase in the percentage of necrotic cells was observed. Moreover, necrosis was predominant in both CT26 and 4T1 cells exposed to their respective AlPc-NE IC_{90}–31.5 and 19.4 nM–at both 25 J/cm^2 and 67 J/cm^2 energy doses (Figure 1).

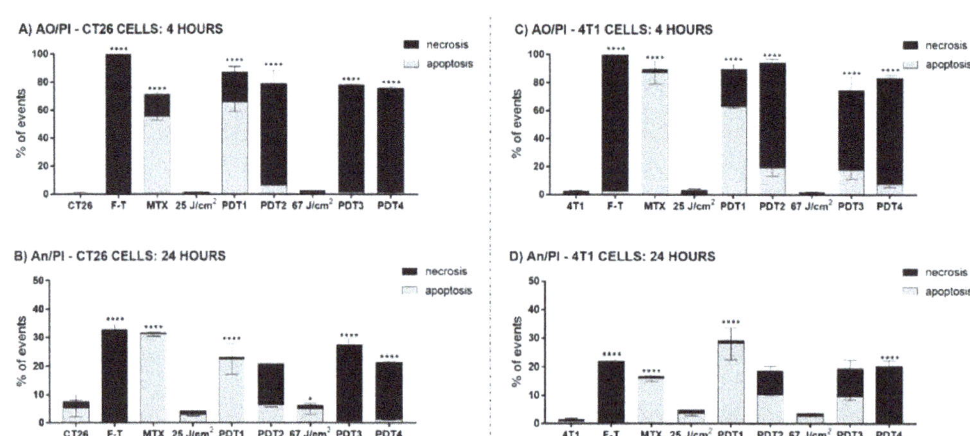

Figure 1. The induction of necrosis and apoptosis by PDT-AlPc-NE is affected by both the concentration of photosensitizer and the energy dose. (**A**) CT26 cells analyzed by the AO/PI method after 4 h of treatment; (**B**) CT26 cells analyzed by AnV/PI method after 24 h of treatments; (**C**) 4T1 cells analyzed by the AO/PI method after 4 h of treatment, and (**D**) 4T1 cells analyzed by the AnV/PI method after 24 h of treatments. Light gray bars represent the results of apoptotic cells, and dark gray bars represent necrotic cells. Untreated cells are represented with CT26 and 4T1. PDT protocols for CT26 cells: PDT1 = 12.2 nM and 25 J/cm^2; PDT2 = 31.5 nM and 25 J/cm^2; PDT3 = 12.2 nM and 67 J/cm^2; and PDT4 = 31.5 nM and 67 J/cm^2. PDT protocols for 4T1 cells: PDT1 = 9.01 nM and 25 J/cm^2; PDT2 = 19.4 nM and 25 J/cm^2; PDT3 = 9.01 nM and 67 J/cm^2; and PDT4 = 19.4 nM and 67 J/cm^2. MTX: mitoxantrone; F-T: three cycles of freeze-thawing; AO/PI: acridine orange and propidium iodide; AnV/PI: Annexin V and propidium iodide. Superscripts * and **** represent $p < 0.05$ and $p < 0.0001$ relatives to apoptotic and necrotic cells of the same group. Data are presented as mean $\pm$ SEM for triplicates.

It is noteworthy that the results profiles observed for both cell lines treated with IC_{50} and 25 J/cm^2 were similar to those obtained with the ICD-positive control MTX. As expected, F-T induced necrosis in both cells studied. The treatments with MTX and F-T are often used as positive controls for apoptosis and necrosis, respectively [7,13]. Thus, variations in the PDT-AlPc-NE protocol parameters affect the type of cell death induced in the studied cells, which can be correlated to the intensity of the oxidative stress in the target cell following PDT.

The profile of DAMPs released by cells succumbing to PDT was also investigated. The exposure of CRT on the plasma membrane is an essential feature of ICD, as this DAMP facilitates the recognition and phagocytosis of the target cell by antigen-presenting cells [13,16,17]. These, in turn, will present the processed tumor cell antigens, potentially inducing an antitumor immune response mediated by $CD8^+$ T cells [18,19]. Moreover, the exposure of HSP70 and HSP90 can increase the immunogenicity of the cells [20,21]. As shown in Figure 2, the exposure of HSP70, HSP90, and CRT were affected by variations in PDT-NE-AlFtCl parameters in both CT26 and 4T1 cells. The images used for fluorescence quantification can be found in the Supplementary Materials (Figures S3–S5). When these cells were submitted to PDT with AlPc-NE IC_{50} and an energy dose of 25 J/cm^2, a more intense CRT exposure, HSP70, and HSP90 was observed. A significantly lower exposure of these DAMPs was observed with the PDT protocols based on higher AlPc concentrations and higher energy doses. Other studies have also shown this same dose-dependency concerning DAMP exposure [22,23]. As expected, MTX induced HSP70, HSP90, and CRT exposure, while the F-T process did not cause the exposure of these DAMPs on the plasma membrane. The apparent increased exposure of HSP70 and HSP90 in CT26 cells treated with PDT4 is most probably due to the disruption of the plasma membrane that occurs in necrotic cells, which enables for the staining of these intracellular proteins by immunofluorescence.

Another important hallmark of ICD is the release of HMGB1 to the extracellular medium [24–26]. HMGB1 is a nuclear protein associated with nucleosomes [26]. HMGB1 protein is released during the late phase of ICD, since long after the onset of apoptosis, and the chromatin becomes deconcentrated and, consequently, HMGB1 release occurs [24,26,27]. This DAMP attracts DC cells and macrophages; upon recognition, these cells become mature and responsible for activating T cells [13,28]. The results presented in Figure 2 show that all the tested PDT protocols induced the release of HMGB1 by both CT26 and 4T1 cells, with a more intense release being observed with the protocol with AlPc IC_{50} and energy dose of 25 J/cm^2.

The release of the proinflammatory cytokine IL-1β was also assessed. For both the 4T1 and CT26 lines, only PDT with AlPc IC_{50} and 25 J/cm^2, and MTX, promoted a significant release of IL-1β compared to control (Figure 2G). IL-1β can modify the activity of many immune cell types, such as monocytes, macrophages, neutrophils, and lymphocytes, and can induce the release of other essential cytokines and chemokines involved in the activation of adaptive immune responses [29,30].

Thus, the in vitro experiments suggest that PDT-AlPc-NE can induce apoptosis and DAMPs exposure, a death pattern characteristic of ICD, in CT26 cells and 4T1 cells. Moreover, the results evidence that both the percentage of apoptotic cells and the DAMPs release profile are affected by the PDT parameters, specifically the AlPc-NE concentration and the energy dose.

As suggested by the literature [31], the immunogenicity of cells undergoing ICD can be assessed in in vivo vaccination-challenge models. Thus, CT26 cells and 4T1 cells subjected to different in vitro PDT protocols were used as prophylactic vaccines injected subcutaneously into the flank of the animals, as shown in Figure 3. The results show that 50% and 40% of mice vaccinated with CT26 cells treated with PDT1 [12.2 nM and 25 J/cm^2] and MTX, respectively, did not develop tumors up to 250 days after the challenge (Figure 3B). This result further shows that PDT can elicit ICD, an event already described in the literature. According to Garg et al. [21], 70% of mice vaccinated with CT26 cells treated with hypericin-mediated PDT were tumor-free.

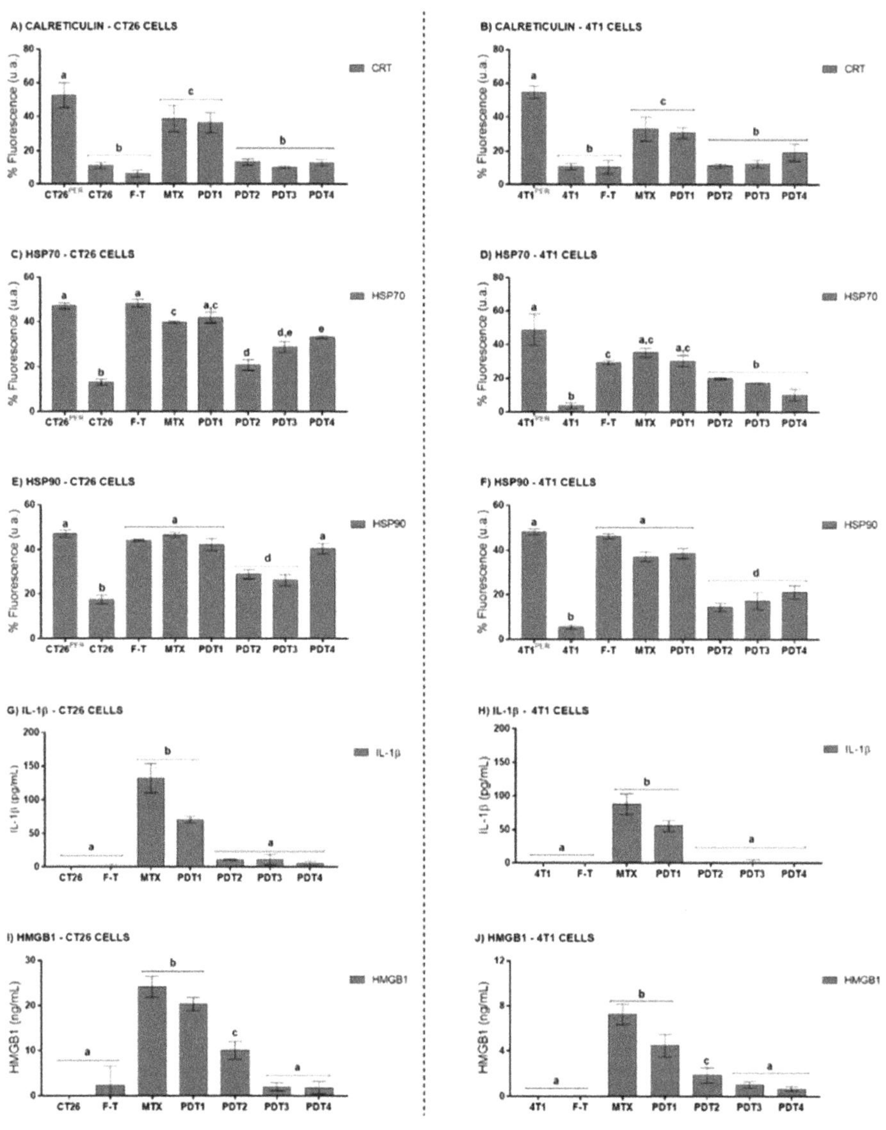

Figure 2. PDT-AlPc-NE induces the release of DAMPs by murine colorectal carcinoma (CT26) and murine mammary adenocarcinoma (4T1) cells. Surface CRT (**A,B**), surface HSP70 (**C,D**), and surface HSP90 (**E,F**). Supernatant IL-1β (**G,H**) and HMGB1 (**I,J**) letters refer to CT26 and 4T1 cells, respectively. Untreated cells are representing with CT26 and 4T1. PDT protocols for CT26 cells: PDT1 = 12.2 nM and 25 J/cm^2; PDT2 = 31.5 nM and 25 J/cm^2; PDT3 = 12.2 nM and 67 J/cm^2; and PDT4 = 31.5 nM and 67 J/cm^2. PDT protocols for 4T1 cells: PDT1 = 9.01 nM and 25 J/cm^2; PDT2 = 19.4 nM and 25 J/cm^2; PDT3 = 9.01 nM and 67 J/cm^2; and PDT4 = 19.4 nM and 67 J/cm^2. MTX: mitoxantrone; F-T: three cycles of freeze-thawing. Superscripts ^{PER} mean the cells submitted to permeation with Triton X-100 0.1%. Equal letters represent results without significant differences between groups. Data are presented as mean ± SEM for triplicates.

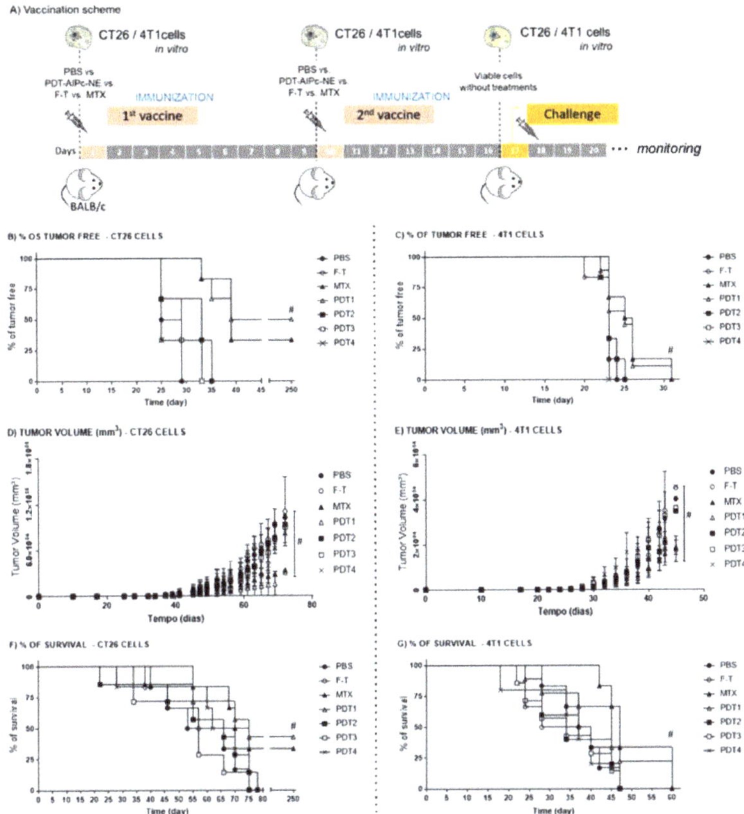

Figure 3. Assessment of the in vivo immunogenicity of cells treated with different protocols of photodynamic therapy. (**A**) Representation of the vaccination-challenge schedule using CT26 cells and 4T1 cells treated with different PDT-AlPc-NE protocols vs F-T vs MTX. CNTR represent the animal that received just PBS without cells. After the challenge the animals were monitored for: the onset of tumors (**B,C**); tumor volume (**D,E**); and survival (**F,G**) referring CT26 and 4T1, respectively. PDT protocols for CT26 cells—: PDT1 = 12.2 nM and 25 J/cm^2; PDT2 = 31.5 nM and 25 J/cm^2; PDT3 = 12.2 nM and 67 J/cm^2; and PDT4 = 31.5 nM and 67 J/cm^2. PDT protocols for 4T1 cells: PDT1 = 9.01 nM and 25 J/cm^2; PDT2 = 19.4 nM and 25 J/cm^2; PDT3 = 9.01 nM and 67 J/cm^2; and PDT4 = 19.4 nM and 67 J/cm^2. 1st vaccine day 0; 2nd vaccine day 10 and the challenge day 17. MTX (mitoxantrone); F-T (three cycles of freeze-thawing); PBS: phosphate buffered saline. Superscript # means $p < 0.05$ in comparison to control group (PBS) at the endpoint. For all data, $n = 6$ mice, mean $\pm$ SEM.

Interestingly, although 4T1 cells exhibited an ICD-related pattern of DAMPs exposure in the in vitro tests described above, they were less immunogenic in vivo than the CT26 cells (Figure 3C). The 4T1 cells treated with PDT or MTX did not elicit a fully protective immunization of the mice; all the animals presented tumors during the experiment. However, there was a significant delay in tumor development in animals vaccinated with MTX- or PDT1-treated cells compared to the other groups (Figure 3C,D). Moreover, these mice showed a prolonged survival time (Figure 3F).

The CT26 and 4T1 cells have distinct characteristics, such as their origin, immunogenic profile, and dissemination pattern. The 4T1 murine cells come from a spontaneously originated stage IV breast adenocarcinoma and exhibit a high metastization potential to the lungs, bones, liver, spleen, lymph nodes, and brain [6,32,33]. The CT26 murine cell line is derived from a chemically induced tumor, experimentally developed by the administration of N-nitrous-N-methylurethane [34]. Although CT26 cells can generate lung metastasis, they are less aggressive than the 4T1 cells, which can be a consequence of differences in the efficacy of their respective immunoevasion strategies [6,35,36]. The literature suggests that CT26 cells are highly immunogenic [37].

Phenotypically, the 4T1 cells can create an immune-suppressive environment that avoids T cell surveillance, thus protecting tumor cells [6]. The mechanisms are related to the abnormal hematopoiesis process, which is driven to the myeloid lineage that produces more myeloid cells in this unusual event. Moreover, due to an excess of growth factors produced by the 4T1 cells, the produced cells are primarily immature cells with immunosuppressive activity. The consequence of this process is an imbalance between the collective memory lymphocytic response, triggered by the vaccination, and the immunosuppressive actions coordinated by the immature immunosuppressive myeloid cells generated after the 4T1 cells stimuli. All these immunological conditions created by the 4T1 cells can explain why the vaccination is less effective in this model compared to the CT26 tumor model.

This study also evaluated the appearance of metastasis foci in the lungs of animals submitted to vaccines with CT26 or 4T1 cells (Figure 4 and Supplementary Materials Figures S6 and S7, respectively). It was found that animals vaccinated with CT26 cells treated with PDT1 had a lung density similar to that of healthy animals, suggesting a low incidence of metastasis (Figure 4). It has already been shown a reduction in CT26 cells metastatic foci in the lungs of mice treated with bacteriochlorin-mediated PDT, an event associated with an immune response induced by PDT against these cells [38]. Regarding the 4T1 cells, a high incidence of metastasis was found in tumor-bearing control mice. However, the radiopacity of the lungs of animals vaccinated with cells treated with PDT1 or MTX was not statistically different from that presented by control, healthy animals (Figure 4B). This result further suggests that, even though these mice (PDT1 and MTX) presented tumors, as discussed earlier, the development of both the grafted tumor and the metastatic foci was somehow reduced. This could be due to an immune response induced in mice by the 4T1 cells succumbing to ICD. Previously, Longo et al. [6] showed that PDT significantly prolonged the survival of mice bearing grafted 4T1 cells tumor, an event associated with a drastic reduction in the number of metastatic foci in the lungs. Moreover, the authors showed that PDT reduced the count of myeloid-derived suppressor cells (MDSC) in the spleen, which can be linked to a reduced ability of 4T1 cells to escape the immunosurveillance and to establish metastasis in PDT-treated mice [6]. That finding could result from the PDT-induced ICD in 4T1 cells, affecting the systemic immune response against tumor cells.

A) % OF PULMONARY DENSITY IN HU - CT26 CELLS

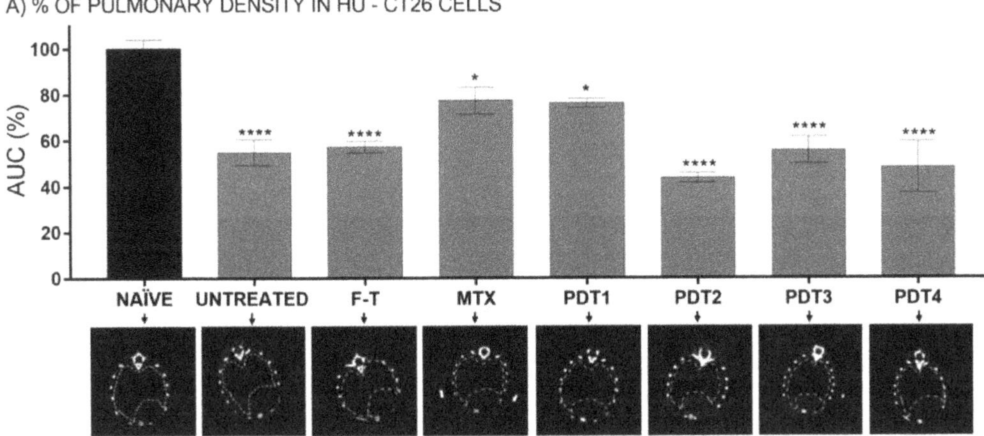

B) % OF PULMONARY DENSITY IN HU - 4T1 CELLS

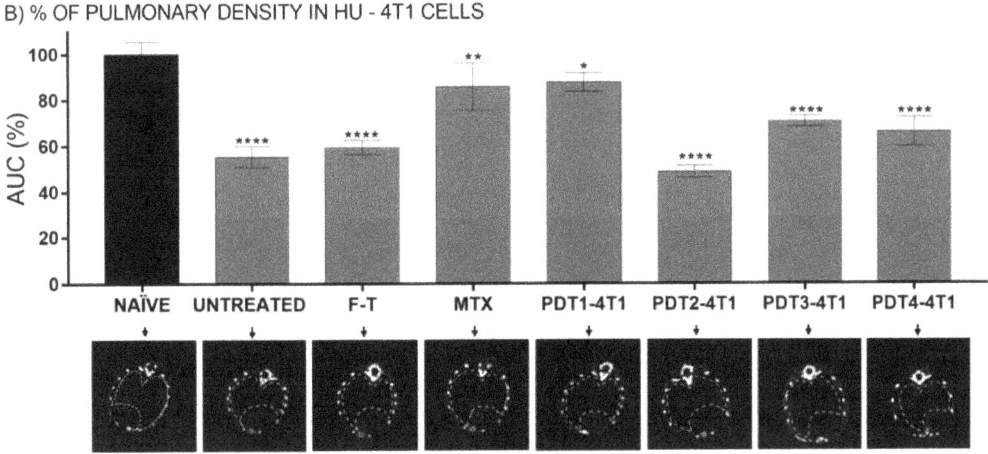

Figure 4. In vivo CT quantification of lung density area using HU values. (**A**) Lung density of animals subjected to vaccination with CT26 cells pretreated with PBS (untreated); F-T; MTX and different PDT protocols. (**B**) Lung density of animals subjected to vaccination with 4T1 cells pretreated with PBS (untreated); F-T; MTX and different PDT protocols. Group of healthy animals (naïve) were used as the control for evaluation of lung density (100%-black bars). Lung 2D representative CT-transverse of the animals references the groups: CT26 cells—PDT protocols: PDT1 = 12.2 nM and 25 J/cm^2; PDT2 = 31.5 nM and 25 J/cm^2; PDT3 = 12.2 nM and 67 J/cm^2; and PDT4 = 31.5 nM and 67 J/cm^2. 4T1 cells—PDT protocols: PDT1 = 9.01 nM and 25 J/cm^2; PDT2 = 19.4 nM and 25 J/cm^2; PDT3 = 9.01 nM and 67 J/cm^2; and PDT4 = 19.4 nM and 67 J/cm^2. MTX (mitoxantrone); F-T (three cycles of freeze-thawing); PBS: phosphate buffered saline. Superscript *, ** and **** means $p < 0.05$, $p < 0.01$ and $p < 0.001$, respectively. For all data, $n = 6$ mice, mean $\pm$ SEM.

4. Conclusions

The present study suggests that the concentration of photosensitizer and the energy dose are important parameters regarding the ability of PDT to induce ICD in CT26 and 4T1 cells. The application of a milder PDT-AlPc-NE rendered the cells more immunogenic than the more intense PDT protocols. Further studies must address how variations on PDT parameters can affect the in vivo activation of the immune system.

Supplementary Materials: The following are available online at https://www.mdpi.com/article/10 .3390/pharmaceutics14010196/s1, Figure S1: Figure S1. Photodynamic effect of AlPc-NE in CT26 cells. The black line represents the CT26 cells exposed to AlPc-NE and maintained in the dark (non-irradiated). The gray line represents the viability of CT26 cells exposed to AlPc-NE for 15 min, washed, left in the dark for different times: (A) 0; (B) 15; (C) 30; (D) 45; (E) 105; and (F) 225 min, respectively. After the incubation, the cells were irradiated for 10 min with a light-emitting diode (LED, λ 660 nm, final energy density of 25 J/cm^2). IC$_{50}$: inhibitory concentrations 50%; IC$_{90}$: inhibitory concentrations 90%. Data are presented as mean $\pm$ SEM for triplicates. Figure S2: Cell viability as a function of the energy density applied (LED, λ 660 nm). The cells were exposed to AlPc IC$_{50}$ and irradiated at the specific incubation-to-irradiation times (LED, λ 660 nm) as follows: (A) 0; (B) 15; (C) 30; (D) 45; (E) 105; and (F) 225 min, respectively. Subscript $^{\#}$ $p < 0.01$ vs control (100%). Data are presented as mean $\pm$ SEM for triplicates. Figure S3. Exposure of calreticulin by CT26 and 4T1 cells exposed to different treatments in vitro. The results for CT26 cells are shown in (A–I) (left panel), and for 4T1 cells in (J–R) (right panel). (A) and (J): Triton X-100 permeabilized cells (TX100); (B) and (K): unpermeabilized control cells; (C) and (L): MTX-treated cells (1.5 µM); (D) and (M): cells subjected to F-T; (E) and (N): PDT1; (F) and (O): PDT2; (G) and (P): PDT3; (H) and (Q): PDT4; (I) and (R): All results for the DAPI marking profile for blue nucleus with green CRT are shown. Data plotted as $\pm$ SEM for triplicates. Figure S4. Exposure of HSP70 by CT26 and 4T1 cells exposed to different treatments in vitro. The results for CT26 cells are shown in (A–I) (left panel), and for 4T1 cells in (J–R) (right panel). (A) and (J): Triton X-100 permeabilized cells (TX100); (B) and (K): unpermeabilized control cells; (C) and (L): MTX-treated cells (1.5 µM); (D) and (M): cells subjected to F-T; (E) and (N): PDT1; (F) and (O): PDT2; (G) and (P): PDT3; (H) and (Q): PDT4; (I) and (R): All results for the DAPI marking profile for blue nucleus with green HSP70 are shown. Data plotted as $\pm$ SEM for triplicates. Figure S5. Exposure of HSP90 by CT26 and 4T1 cells exposed to different treatments in vitro. The results for CT26 cells are shown in (A–I) (left panel), and for 4T1 cells in (J–R) (right panel). (A) and (J): Triton X-100 permeabilized cells (TX100); (B) and (K): unpermeabilized control cells; (C) and (L): MTX-treated cells (1.5 µM); (D) and (M): cells subjected to F-T; (E) and (N): PDT1; (F) and (O): PDT2; (G) and (P): PDT3; (H) and (Q): PDT4; (I) and (R): All results for the DAPI marking profile for blue nucleus with green HSP90 are shown. Data plotted as $\pm$ SEM for triplicates. Figure S6: In vivo computed tomography quantification of the frequency of voxel as a function of HU in the lung. Lung density of animals subjected to vaccination with CT26 cells pretreated: (A) NAÏVE ANIMALS and (A.1) results showed as AUC; (B) PBS (untreated) and (B.1) results showed as AUC AUC; (C) F-T and (C.1) results showed as AUC; (D) MTX and (D.1) results showed as AUC; (E to H) CT26 cells – PDT protocols: PDT1 = 12.2 nM and 25 J/cm^2; PDT2 = 31.5 nM and 25 J/cm^2; PDT3 = 12.2 nM and 67 J/cm^2; and PDT4 = 31.5 nM and 67 J/cm^2 and E.1 to H.1) results showed as AUC, respectively. The assays were performed with a difference of one week (assay 1 to assay 2). For all data, $n = 6$ mice, mean $\pm$ SEM. Figure S7. In vivo computed tomography quantification of the frequency of voxel as a function of HU in the lung. Lung density of animals subjected to vaccination with 4T1 cells pretreated: (A) NAÏVE ANIMALS and (A.1) results showed as AUC; (B) PBS (untreated) and (B.1) results showed as AUC AUC; (C) F-T and (C.1) results showed as AUC; (D) MTX and (D.1) results showed as AUC; (E to H) 4T1 cells – PDT protocols: PDT1 = 9.01 nM and 25 J/cm^2; PDT2 = 19.4 nM and 25 J/cm^2; PDT3 = 9,01 nM and 67 J/cm^2; and PDT4 = 19.4 nM and 67 J/cm^2 and E.1 to H.1) results showed as AUC, respectively. The assays were performed with a difference of one week (assay 1 to assay 2). For all data, $n = 6$ mice, mean $\pm$ SEM.

Author Contributions: Conceptualization, M.C.R. and L.A.M.; Methodology, M.C.R., W.T.d.S.J., T.M., C.L.C.V., J.V.d.O., R.G., T.J.A.P., J.A.V.M., J.P.F.L. and L.A.M.; Formal Analysis, M.C.R., J.P.F.L., R.B.A. and L.A.M.; Investigation, M.C.R., W.T.d.S.J., T.M., C.L.C.V., J.V.d.O., R.G., T.J.A.P., J.A.V.M., J.P.F.L. and L.A.M.; Resources, R.B.A. and L.A.M.; Data Curation, M.C.R. and L.A.M.; Writing—Original Draft Preparation, M.C.R.; Writing—Review and Editing, M.C.R., L.A.M. and J.P.F.L.; Project Administration, L.A.M.; Funding Acquisition, R.B.A. and L.A.M. All authors have read and agreed to the published version of the manuscript.

Funding: This work was funded by the Brazilian government agencies FAP-DF/Brazil (0193.001020/ 2015, and 0193.001626/2017), PRONEX FAP-DF, CNPq (0447628/2014-3, 193.001.200/2016) and Coordenação de Aperfeiçoamento de Pessoal de Nível Superior—Brasil (CAPES)—Finance Code 001.

Institutional Review Board Statement: The study was conducted according to the guidelines of the Declaration of Helsinki, and approved by the Institutional Ethics Committee of the Institute of Biology, University of Brasilia (UnB/Doc n. 5529/2015, approved on 3 March 2015).

Informed Consent Statement: Not applicable.

Data Availability Statement: Data is contained within the article or supplementary material.

Conflicts of Interest: The authors declare no conflict of interest.

References

1. Rodrigues, M.C.; Vieira, L.G.; Horst, F.H.; de Araújo, E.C.; Ganassin, R.; Merker, C.; Muehlmann, L.A. Photodynamic therapy mediated by aluminium-phthalocyanine nanoemulsion eliminates primary tumors and pulmonary metastases in a murine 4T1 breast adenocarcinoma model. *J. Photochem. Photobiol. B Biol.* **2020**, *204*, 111808. [CrossRef] [PubMed]
2. Rodrigues, M.C. Photodynamic therapy based on arrabidaea chica (Crajiru) extract nanoemulsion: In vitro activity against monolayers and spheroids of human mammary adenocarcinoma MCF-7 cells. *J. Nanomed. Nanotechnol.* **2015**, *6*, 1–6. [CrossRef]
3. Zhang, J.; Jiang, C.; Figueiró Longo, J.P.; Azevedo, R.B.; Zhang, H.; Muehlmann, L.A. An updated overview on the development of new photosensitizers for anticancer photodynamic therapy. *Acta Pharm. Sin. B* **2018**, *8*, 137–146. [CrossRef] [PubMed]
4. Monge-Fuentes, V.; Muehlmann, L.A.; Longo, J.P.F.; Silva, J.R.; Fascineli, M.L.; Azevedo, R.B.; Amorim, R.F.B. Photodynamic therapy mediated by acai oil (*Euterpe oleracea Martius*) in nanoemulsion: A potential treatment for melanoma. *J. Photochem. Photobiol. B. Biol.* **2017**, *166*, 301–310. [CrossRef] [PubMed]
5. Garg, A.D.; Agostinis, P. ER stress, autophagy and immunogenic cell death in photodynamic therapy-induced anti-cancer immune responses. *Photochem. Photobiol. Sci. Off. J. Eur. Photochem. Assoc. Eur. Soc. Photobiol.* **2014**, *13*, 474–487. [CrossRef]
6. Longo, J.P.F.; Muehlmann, L.A.; Miranda-Vilela, A.L.; Portilho, F.A.; de Souza, L.R.; Silva, J.R.; Azevedo, R.B. Prevention of distant lung metastasis after photodynamic therapy application in a breast cancer tumor model. *J. Biomed. Nanotechnol.* **2016**, *12*, 689–699. [CrossRef]
7. Morais, J.A.V.; Almeida, L.R.; Rodrigues, M.C.; Azevedo, R.B.; Muehlmann, L.A. The induction of immunogenic cell death by photodynamic therapy in B16F10 cells in vitro is effected by the concentration of the photosensitizer. *Photodiagn. Photodyn. Ther.* **2021**, *2021 35*, 102392. [CrossRef]
8. Tanaka, M.; Kataoka, H.; Yano, S.; Sawada, T.; Akashi, H.; Inoue, M.; Joh, T. Immunogenic cell death due to a new photodynamic therapy (PDT) with glycoconjugated chlorin (G-chlorin). *Oncotarget* **2016**, *7*, 47242. [CrossRef]
9. Thorsson, V.; Gibbs, D.L.; Brown, S.D.; Wolf, D.; Bortone, D.S.; Ou Yang, T.H.; Shmulevich, l.; Eddy, J.A.; Ziv, E.; Gao, G.F.; et al. The immune landscape of cancer. *Immunity* **2018**, *4*, 812–830. [CrossRef]
10. Korbelik, M.; BanÃ¡th, J.; Saw, K.M.; Zhang, W.; Čiplys, E. Calreticulin as cancer treatment adjuvant: Combination with photodynamic therapy and photodynamic therapy-generated vaccines. *Front. Oncol.* **2015**, *5*, 15. [CrossRef] [PubMed]
11. Muehlmann, A.L.; Rodrigues, C.M.; Longo, P.J.; Garcia, P.M.; Py-Daniel, R.K.; Veloso, B.A.; Azevedo, B.R. Aluminium-phthalocyanine chloride nanoemulsions for anticancer photodynamic therapy: Development and in vitro activity against monolayers and spheroids of human mammary adenocarcinoma MCF-7 cells. *J. Nanobiotechnol.* **2015**, *13*, 36. [CrossRef]
12. Kasibhatla, S.; Amarante-Mendes, G.P.; Finucane, D.; Brunner, T.; Bossy-Wetzel, E.; Green, D.R. Acridine orange/ethidium bromide (AO/EB) saining to detect apoptosis. *CSH Protoc.* **2006**. [CrossRef]
13. Garg, A.D.; Vandenberk, L.; Koks, C.; Verschuere, T.; Boon, L.; van Gool, S.W.; Agostinis, P. Dendritic cell vaccines based on immunogenic cell death elicit danger signals and T cell-driven rejection of high-grade glioma. *Sci. Trans. Med.* **2016**, *8*, ra27–ra328. [CrossRef]
14. Panzarini, E.; Inguscio, V.; Fimia, G.M.; Dini, L. Rose bengal acetate photodynamic therapy (RBAc-PDT) induces exposure and release of damage-associated molecular patterns (DAMPs) in human HeLa cells. *PLoS ONE* **2014**, *9*, e105778. [CrossRef]
15. Cheng, Y.; Chang, Y.; Feng, Y.; Liu, N.; Sun, X.; Feng, Y.; Zhang, H. Simulated sunlight-mediated photodynamic therapy for melanoma skin cancer by titanium-dioxide-nanoparticle–gold-nanocluster–graphene heterogeneous nanocomposites. *Small* **2017**, *13*, 1603935. [CrossRef]
16. Garg, A.D.; Nowis, D.; Golab, J.; Vandenabeele, P.; Krysko, D.V.; Agostinis, P. Immunogenic cell death, DAMPs and anticancer therapeutics: An emerging amalgamation. *Biochim. Biophys. Acta Rev. Cancer* **2010**, *1805*, 53–71. [CrossRef]
17. Li, W.; Yang, J.; Luo, L.; Jiang, M.; Qin, B.; Yin, H.; You, J. Targeting photodynamic and photothermal therapy to the endoplasmic reticulum enhances immunogenic cancer cell death. *Nat. Commun.* **2019**, *10*, 3349. [CrossRef]
18. Donohoe, C.; Senge, M.O.; Arnaut, L.G.; Gomes-da-Silva, L.C. Cell death in photodynamic therapy: From oxidative stress to anti-tumor immunity. *Biochim. Biophys. Acta (BBA) Rev. Cancer* **2019**, *1872*, 188308. [CrossRef]
19. Garg, A.D.; Nowis, D.; Golab, J.; Agostinis, P. Photodynamic therapy: Illuminating the road from cell death towards anti-tumour immunity. *Apoptosis* **2010**, *15*, 1050–1071. [CrossRef]
20. Rodríguez, M.E.; Cogno, I.S.; Milla Sanabria, L.S.; Morán, Y.S.; Rivarola, V.A. Heat shock proteins in the context of photodynamic therapy: Autophagy, apoptosis and immunogenic cell death. *Photochem. Photobiol. Sci. Off. J. Eur. Photochem. Assoc. Eur. Soc. Photobiol.* **2016**, *15*, 1090–1102. [CrossRef]

21. Garg, A.D.; Krysko, D.V.; Vandenabeele, P.; Agostinis, P. Hypericin-based photodynamic therapy induces surface exposure of damage-associated molecular patterns like HSP70 and calreticulin. *Cancer Immunol. Immunother.* **2012**, *61*, 215–221. [CrossRef]
22. Doix, B.; Trempolec, N.; Riant, O.; Feron, O. Low photosensitizer dose and early radiotherapy enhance antitumor immune response of photodynamic therapy-based dendritic cell vaccination. *Front. Oncol.* **2019**, *9*, 811. [CrossRef]
23. Kawczyk-Krupka, A.; Czuba, Z.; Latos, W.; Wasilewska, K.; Verwanger, T.; Krammer, B.; Sieroń, A. Influence of ALA-mediated photodynamic therapy on secretion of interleukins 6, 8 and 10 by colon cancer cells in vitro. *Photodiagn. Photodyn. Ther.* **2018**, *22*, 137–139. [CrossRef]
24. Coleman, L.G.; Maile, R.; Jones, S.W.; Cairns, B.A.; Crews, F.T. HMGB1/IL-1β complexes in plasma microvesicles modulate immune responses to burn injury. *PLoS ONE* **2018**, *13*, e0195335. [CrossRef]
25. Andersson, U.; Yang, H.; Harris, H. High-mobility group box 1 protein (HMGB1) operates as an alarmin outside as well as inside cells. *Semin. Immunol.* **2018**, *38*, 40–48. [CrossRef]
26. Gerö, D.; Szoleczky, P.; Módis, K.; Pribis, J.P.; Al-Abed, Y.; Yang, H.; Szabo, C. Identification of Pharmacological Modulators of HMGB1-Induced Inflammatory Response by Cell-Based Screening. *PLoS ONE* **2013**, *8*, e65994. [CrossRef]
27. Yang, H.; Wang, H.; Chavan, S.S.; Andersson, U. High mobility group box protein 1 (HMGB1): The prototypical endogenous danger molecule. *Mol. Med.* **2015**, *21*, S6–S12. [CrossRef]
28. Vénéreau, E.; Ceriotti, C.; Bianchi, M.E. DAMPs from cell death to new life. *Front. Immunol.* **2015**, *6*, 1. [CrossRef]
29. Korbelik, M.; Hamblin, M.R. The impact of macrophage-cancer cell interaction on the efficacy of photodynamic therapy. *Photochem. Photobiol. Sci. Off. J. Eur. Photochem. Assoc. Eur. Soc. Photobiol.* **2015**, *14*, 1403–1409. [CrossRef]
30. Bruchard, M.; Mignot, G.; Derangère, V.; Chalmin, F.; Chevriaux, A.; Végran, F.; Ghiringhelli, F. Chemotherapy-triggered cathepsin B release in myeloid-derived suppressor cells activates the Nlrp3 inflammasome and promotes tumor growth. *Nat. Med.* **2013**, *19*, 57–64. [CrossRef]
31. Galluzzi, L.; Vitale, I.; Warren, S.; Adjemian, S.; Agostinis, P.; Martinez, A.B.; Marincola, F.M. Consensus guidelines for the definition, detection and interpretation of immunogenic cell death. *J. Immunother. Cancer* **2020**, *8*, e000337. [CrossRef]
32. Safavi, A.; Kefayat, A.; Ghahremani, F.; Mahdevar, E.; Moshtaghian, J. Immunization using male germ cells and gametes as rich sources of cancer/testis antigens for inhibition of 4T1 breast tumors' growth and metastasis in BALB/c mice. *Int. Immunopharmacol.* **2019**, *74*, 105719. [CrossRef]
33. Pulaski, B.A.; Ostrand-Rosenberg, S. Mouse 4T1 breast tumor model. *Curr. Protoc. Immunol.* **2001**, *20*, 16. [CrossRef]
34. Brattain, M.; Strobel-Stevens, J.; Fine, D.; Webb, M.; Sarrif, A. Establishment of mouse colonic carcinoma cell lines with different metastatic properties. *Cancer Res.* **1980**, *40*, 2142–2146.
35. Abe, H.; Wada, H.; Baghdadi, M.; Nakanishi, S.; Usui, Y.; Tsuchikawa, T.; Seino, K.I. Identification of a highly immunogenic mouse breast cancer sub cell line, 4T1-S. *Hum. Cell* **2016**, *29*, 58–66. [CrossRef]
36. Kim, K.; Skora, A.D.; Li, Z.; Liu, Q.; Tam, A.J.; Blosser, R.L.; Zhou, S. Eradication of metastatic mouse cancers resistant to immune checkpoint blockade by suppression of myeloid-derived cells. In Proceedings of the National Academy of Sciences of the United States of America, Baltimore, MD, USA, 12 August 2014; Volume 111, pp. 11774–11779. [CrossRef]
37. Sanovic, R.; Verwanger, T.; Hartl, A.; Krammer, B. Low dose hypericin-PDT induces complete tumor regression in BALB/c mice bearing CT26 colon carcinoma. *Photodiagn. Photodyn. Ther.* **2011**, *8*, 291–296. [CrossRef]
38. Rocha, L.B.; Gomes-Da-Silva, L.C.; Dąbrowski, J.M.; Arnaut, L.G. Elimination of primary tumours and control of metastasis with rationally designed bacteriochlorin photodynamic therapy regimens. *Eur. J. Cancer* **2015**, *51*, 1822–1830. [CrossRef]

 *pharmaceutics*

Article

Immunomodulatory Nanoparticles Mitigate Macrophage Inflammation via Inhibition of PAMP Interactions and Lactate-Mediated Functional Reprogramming of NF-κB and p38 MAPK

Jackline Joy Martín Lasola [1], Andrea L. Cottingham [2], Brianna L. Scotland [2], Nhu Truong [2], Charles C. Hong [3], Paul Shapiro [2,4] and Ryan M. Pearson [2,4,*]

1   Department of Microbiology and Immunology, University of Maryland School of Medicine, 685 W. Baltimore Street, Baltimore, MD 21201, USA; jacklinejoy.lasola@som.umaryland.edu
2   Department of Pharmaceutical Sciences, University of Maryland School of Pharmacy, 20 N. Pine Street, Baltimore, MD 21201, USA; acottingham@umaryland.edu (A.L.C.); bscotland@umaryland.edu (B.L.S.); ntruong@umaryland.edu (N.T.); pshapiro@rx.umaryland.edu (P.S.)
3   Division of Cardiovascular Medicine, Department of Medicine, University of Maryland School of Medicine, 110 S. Paca Street, Baltimore, MD 21201, USA; charles.hong@som.umaryland.edu
4   Marlene and Stewart Greenebaum Comprehensive Cancer Center, University of Maryland School of Medicine, 22 S. Greene Street, Baltimore, MD 21201, USA
*   Correspondence: rpearson@rx.umaryland.edu; +1-410-706-3257

Citation: Lasola, J.J.M.; Cottingham, A.L.; Scotland, B.L.; Truong, N.; Hong, C.C.; Shapiro, P.; Pearson, R.M. Immunomodulatory Nanoparticles Mitigate Macrophage Inflammation via Inhibition of PAMP Interactions and Lactate-Mediated Functional Reprogramming of NF-κB and p38 MAPK. *Pharmaceutics* **2021**, *13*, 1841. https://doi.org/10.3390/pharmaceutics13111841

Academic Editors: João Paulo Longo and Luís Alexandre Muehlmann

Received: 3 September 2021
Accepted: 28 October 2021
Published: 2 November 2021

Publisher's Note: MDPI stays neutral with regard to jurisdictional claims in published maps and institutional affiliations.

Abstract: Inflammation is a key homeostatic process involved in the body's response to a multitude of disease states including infection, autoimmune disorders, cancer, and other chronic conditions. When the initiating event is poorly controlled, severe inflammation and globally dysregulated immune responses can occur. To address the lack of therapies that efficaciously address the multiple aspects of the dysregulated immune response, we developed cargo-less immunomodulatory nanoparticles (iNPs) comprised of poly(lactic acid) (PLA) with either poly(vinyl alcohol) (PVA) or poly(ethylene-alt-maleic acid) (PEMA) as stabilizing surfactants and investigated the mechanisms by which they exert their inherent anti-inflammatory effects. We identified that iNPs leverage a multimodal mechanism of action by physically interfering with the interactions between pathogen-associated molecular patterns (PAMPs) and bone marrow-derived macrophages (BMMΦs). Additionally, we showed that iNPs mitigate proinflammatory cytokine secretions induced by LPS via a time- and composition-dependent abrogation of NF-κB p65 and p38 MAPK activation. Lastly, inhibition studies were performed to establish the role of a pH-sensing G-protein-coupled receptor, GPR68, on contributing to the activity of iNPs. These data provide evidence for the multimodal mechanism of action of iNPs and establish their potential use as a novel therapeutic for the treatment of severe inflammation.

Keywords: inflammation; innate immunity; macrophages; Toll-like receptors; TLR; NF-κB; p38 MAPK; sepsis; poly(lactic acid); PLA; lactate; nanoparticles; microparticles

1. Introduction

Severe inflammation is a complex and global multi-step physiological process implicated in the development of a systemic dysregulated immune environment. Using sepsis as an example of severe inflammation, epidemiological data suggests that one-in-five of all global deaths is due to sepsis or sepsis-related causes [1]. However, the standard of care for sepsis has failed to move far beyond antibiotics and supportive care, thus leaving much room for the development of new treatment strategies to improve outcomes. To date, over 100 clinical trials have been conducted for potential therapies, but curative strategies remain elusive [2,3]. Previous attempts to address and manage severe inflammation and sepsis have focused on the development of single molecular agents targeted against specific molecules or aspects of molecular pathways implicated in the development of the severe inflammatory response. Despite these methods often demonstrating outstanding preclinical success, translating these results to viable therapeutics for critically ill patients has been unrealized [4]. It has been hypothesized that these attempts have failed because of the profound clinical heterogeneity of sepsis, the lack of fundamental understandings of the different endotypes of sepsis, and treatments that have been targeted towards only a single molecular pathway, leaving redundant pathways associated with immune activation and a multifaceted immune dysfunction unaddressed [5,6]. Therefore, a significant need exists to develop multimodal therapeutics to address the complexity of immune responses present in severe inflammation and sepsis.

Modulation of the innate immune system using nanoparticles serves as the basis for many new and promising therapies for some of the most prevalent and/or severe diseases [7–9]. We recently reviewed the various strategies of nanoparticle-mediated immunomodulation for the treatment of severe inflammation and sepsis [10]. Three mechanisms were proposed by which nanoparticles can be utilized to offset the negative immune mediators of severe inflammation: (1) sequestration of activating pathogen-associated molecular patterns (PAMPs) or proinflammatory cytokines; (2) functional reprogramming of inflammatory immune cell phenotypes; and (3) redirection of inflammatory immune cell trafficking from sites of inflammation.

Our group [11,12] and others [13] have developed cargo-less immunomodulatory nanoparticles (iNPs) that lacked incorporation of small molecules, proteins, or other immunomodulating agents and showed that the physicochemical properties of the nanoparticles were major contributors to the observed therapeutic effects. In our previous studies, antigen presenting cells treated with cargo-less poly(lactic-co-glycolic acid) (PLGA)- and poly(D,L-lactic acid) (PLA)-based iNPs prepared with highly negative zeta potentials could mitigate proinflammatory cytokine secretions such as IL-6 and TNF-α when stimulated with extracellular and intracellular PAMPs, namely Toll-like receptor 4 (TLR4)-targeted lipopolysaccharide (LPS) and TLR9-targeted unmethylated CpG oligodeoxynucleotides (CpG ODN). Furthermore, their immunomodulatory properties translated into a survival benefit in lethal murine LPS-induced endotoxemia models [11]. Initial analysis hinted at a potential role for modulation of NF-κB, IRF1, and STAT1; however, the mechanisms by which iNPs elicit their favorable therapeutic effects remains poorly understood. For these nanoparticle-based strategies to move forward, a greater understanding of the biological effects of these materials and mechanisms by which they exert their immunomodulatory effects is warranted.

Nanoparticles are complex systems and can function through multiple mechanisms where each component involved in its production (i.e., stabilizing surfactant and polymer composition) can potentially alter cellular and inflammatory mediator interactions including rate of uptake, trafficking, rate of degradation and degradation products, etc. Stabilizing surfactants such as poly(vinyl alcohol) (PVA) and poly(ethylene-alt-maleic acid) (PEMA) are ideal for testing the impact of surface characteristics on nano-bio interactions given the variability in zeta potentials and surface chemistry while allowing for control of iNP size. PLA is ideal for understanding the role of the polymer composition and further use in nanoparticle development due to its Food and Drug Administration (FDA) approved status for internal use in humans. Its degradation occurs via autocatalytic cleavage of the ester bonds through hydrolysis into oligomers and monomers of lactic acid, which are substrates of the Krebs cycle [14]. For this reason, minimal toxicity is usually observed due to its biodegradable and biocompatible properties. Although not toxic, there has been a growing appreciation in immunology of the effects of metabolic byproducts in driving observed immune phenotypes [15–19]. Specifically, lactate has been implicated in modifying inflammatory macrophage responses, although controversy remains as to how lactate acts to do this and whether its role is protective or detrimental [20,21]. Additionally, although PLA is a widely used biomaterial in nanoparticle formulation, its effects following degradation are not well characterized in comparison to other commonly used polymeric materials.

In this study, we assess the physical and biological mechanisms that affect iNP-mediated modulation of macrophage activation by TLR agonists. We hypothesize that the anti-inflammatory effects of iNPs are multimodal, such that the choice of surfactants elicits differences in the nano-bio interactions, while the choice of nanoparticle composition and its degradation products abrogate the activation of proinflammatory cell signaling pathways. Two formulations of iNPs were prepared using PLA with either PVA or PEMA as surfactants to evaluate the role of surface chemistry and charge on inducing anti-inflammatory immune responses. We first evaluated the ability for iNPs to directly interact with PAMPs and the impact of iNP-cell interactions on PAMP-cellular interactions. Next, we assessed the time course-dependent effects of PLA-based iNPs on modulation of NF-κB and p38 mitogen-activated protein kinase (MAPK) signaling. The composition-dependent effects of iNPs on NF-κB and p38 MAPK signaling were subsequently investigated by comparing PLA-based iNPs with commonly utilized commercially available nanoparticles. Lastly, we established a potential role for the pH-sensing G protein-coupled receptor (GPR) 68 on the anti-inflammatory activity of iNPs. Taken together, our study provides evidence for the multimodal mechanisms by which iNPs exert their inherent anti-inflammatory immunomodulatory effects. This work serves as a foundation for further investigation of the inherent immunomodulatory properties of biomaterials and how their specific design features can be tuned to elicit predictable immunological responses through novel strategies and systematic testing with the potential of opening new avenues of research to treat a variety of immune-mediated diseases.

2. Materials and Methods

2.1. Materials

Acid-terminated PLA of low inherent viscosity in hexafluoro-2-propanol ~0.21 dL/g (approx. 11,700 g/mol) was purchased from Lactel Absorbable Polymers (Birmingham, AL, USA). PEMA (MW 400,000 g/mol) was purchased from Polysciences, Inc. (Warrington, PA, USA). PVA (MW 30,000–70,000 g/mol) was obtained from Sigma-Aldrich (St. Louis, MO, USA). Polystyrene (PS) and poly(methyl methacrylate) (PMMA) particles were purchased from Phosphorex (Hopkinton, MA, USA).

ODN 1668 and ODN 1668 FITC (referred to collectively as CpG ODN) were obtained from Invivogen (San Diego, CA, USA); lipopolysaccharide (LPS) and FITC-conjugated LPS from Escherichia coli serotype O111:B4 were obtained from Sigma-Aldrich (St. Louis, MO, USA).

2X SDS-PAGE sample buffer was produced using 4% SDS; 5.7 M β-mercaptoethanol; 0.2 M Tris-HCl, pH 6.8; 20% glycerol and 5 mM EDTA. RIPA Buffer was purchased from Sigma-Aldrich (St. Louis, MO, USA) and both Halt® Protease Inhibitor Cocktail (100X) and Invitrogen NuPAGE 4–12% Bis-Tris Gel were purchased from Thermo Fisher Scientific (Waltham, MA). Doramapimod (also known as BIRB 796) was purchased from Selleck Chemicals (Houston, TX, USA). Ogremorphin (OGM) was graciously provided by Charles C. Hong [22]. 3-hydroxybutyric acid (3-OBA) was purchased from Sigma-Aldrich.

FITC anti-mouse CD14 mAb (Clone Sa14-2) and PE anti-mouse CD284 (TLR4) mAb (Clone SA15-21) were purchased from BioLegend (San Diego, CA, USA). Phospho-NF-κB p65 (Ser536) (93H1) rabbit mAb, NF-κB p65 (D14E12) XP rabbit mAb, phospho-IκB (Ser32) (14D4) rabbit mAb, IκB (44D4) rabbit mAb, phospho-p38 (Thr180/Tyr182) rabbit Ab, phospho-ERK1/2 (Thr202/Tyr204) (197G2) rabbit mAb, total ERK2 rabbit Ab, phospho-SAPK/JNK (Thr183/Tyr185) rabbit Ab, total SAPK/JNK rabbit Ab, phospho-MKK3 (Ser189)/MKK6 (Ser207) rabbit Ab, MKK6 rabbit Ab, phospho-TAK1 (Thr184/187) rabbit Ab, total TAK1 (D94D7) rabbit mAb, total IRAK4 rabbit Ab, and β-Actin (13E5) rabbit mAb were all purchased from Cell Signaling Technology (Danvers, MA). p38α (C20) rabbit mAb was purchased from Santa Cruz Biotechnology (Dallas, TX, USA). Anti-rabbit IgG (H+L), peroxidase labeled secondary Ab was purchased from Sera Care (Milford, MA, USA).

RAW 264.7 cells were purchased from ATCC (Manassas, VA, USA) and cultured in DMEM (Life Technologies, Carlsbad, CA, USA), penicillin (100 units/mL), streptomycin (100 µg/mL), and 10% heat-inactivated fetal bovine serum (FBS) (VWR, Radnor, PA, USA) at 37 °C and 5% CO_2.

2.2. iNP Preparation and Characterization

PLA iNPs were prepared using the oil-in-water (o/w) emulsion-solvent evaporation (SE) technique following a similar method as described [10]. Briefly, 200 mg of PLA was dissolved in ethyl acetate at a concentration of 80 mg/mL, 20 mL of 1% PEMA was added then sonicated at 100% amplitude for 30 s using a Cole-Parmer 500-Watt Ultrasonic Homogenizer to make PLA-PEMA. For PLA-PVA, 200 mg of PLA was dissolved in ethyl acetate at a concentration of 300 mg/mL. To this, 5 mL of 2% PVA was added and sonicated at 40% amplitude for 30 s using the same homogenizer. The resulting o/w emulsion was then poured into 100 mL of magnetically stirred 0.5% PEMA (or 0.5% PVA) overnight to remove ethyl acetate. iNPs were then collected by centrifugation at 12,000× g for 20 min at 4 °C and washed with 40 mL of MilliQ water. The centrifugation and washing steps were repeated two more times. A mixture of sucrose and mannitol were added to the particle suspension as cryoprotectants to achieve a final concentration of 4% and 3% w/v, respectively. The nanoparticles were then frozen at −80 °C and lyophilized for at least 48 h prior to use.

The size and zeta potential of all the particles were determined by dynamic light scattering (DLS) using a Malvern Zetasizer ZSP. Cy5.5-labeled PLA particles were prepared by incorporating 1% w/w of PLA-Cy5.5 into particles as previously described [23].

2.3. Particle-TLR Agonist Association Studies

PLA iNPs (concentrations of iNPs as described in the results) were incubated with 1 µg/mL ODN 1668 FITC or 1 µg/mL FITC LPS in sterile DPBS containing 10% heat inactivated fetal bovine serum (FBS). These samples were incubated for 1 h at 37 °C at 5% CO_2 and vortexed every 10 min. Following incubation, the solutions were centrifuged for 5 min at 12,000× g to pellet iNPs then the supernatant was transferred to black 96-well plates to measure fluorescence at 525 nm with the Molecular Devices (San Jose, CA, USA) SpectraMax iD3 Microplate Reader.

2.4. Mice

Female C57BL/6J (five to seven weeks old) were purchased from The Jackson Laboratories (Bar Harbor, ME, USA). The mice were housed under specific pathogen-free conditions in a facility at the University of Maryland, Baltimore Veterinary Resources. All mouse procedures and experiments were compliant to the protocols of the University of Maryland, Baltimore Institutional Animal Care and Use Committee (IACUC) and approved under IACUC protocol 0818014.

2.5. Isolation and Generation of Bone Marrow-Derived Macrophages (BMMΦs) and Dendritic Cells (BMDCs)

BMMΦs [24] and BMDCs [25] were generated from isolated bone marrow as previously described. Briefly, 5–12-week C57BL/6J female mice were euthanized and the femurs and tibias isolated and flushed with BMMΦ media [RPMI 1640 supplemented with L-glutamine (Life Technologies, Carlsbad, CA), penicillin (100 units/mL), streptomycin (100 µg/mL), 10% heat-inactivated fetal bovine serum (FBS) (VWR, Radnor, PA, USA), and 20% L929 (ATCC) cell-conditioned media] or BMDC media [RPMI 1640 supplemented with L-glutamine, penicillin (100 units/mL), streptomycin (100 µg/mL), 10% FBS, 50 mM β-mercaptoethanol (Sigma-Aldrich) and 20 ng/mL GM-CSF (Peprotech, Rocky Hill, NJ, USA)] using a 1 mL syringe and a 25-gauge needle. Once isolated, the cells were pipetted and filtered through a 40 µm cell strainer then plated in uncoated 10 cm non-tissue culture treated petri dishes. The cells were incubated at 37 °C at 5% CO_2 and the media was replaced on days 0, 3, 6, and 8. BMMΦs and BMDCs were used for experiments between days 8–10.

2.6. Flow Cytometry

Cell staining was conducted according to BioLegend protocols for flow cytometry. Flow cytometry data were collected using a Becton Dickinson LSR II or Becton Dickinson Canto II flow cytometer. Analysis was performed using FCS Express 7 (De Novo Software, Glendale, CA, USA). FcR blocking was performed with the anti-CD16/32 antibody prior to staining. Viability was assessed with 4′,6-diamidino-2-phenylindole dilactate (DAPI) as an exclusion dye for iNP and TLR agonist studies.

2.7. Particle-Cell Association Studies

BMMΦs and RAW 264.7 cells were seeded in sterile 24-well plates at a concentration of 0.2×10^6 cells/well and then treated with 30 μg/mL of Cy5.5-labeled iNPs (PLA-PEMA and PLA-PVA) for 1 hr. All treated wells were washed twice with PBS to remove excess iNPs and replenished with 500 μL of fresh sterile PBS. For fluorescence microscopy, cells were either visualized immediately or fixed with 4% paraformaldehyde prior to visualization with either an ECHO (San Diego, CA, USA) Revolve benchtop fluorescence microscope or Nikon (Tokyo, Japan) Eclipse Ti-E confocal microscope. For flow cytometry, cells were scraped using a blunt 1000 μL pipette tip followed by collection by centrifugation and stained for viability using DAPI dye. Flow cytometry was used to measure Cy5.5 signal on viable (DAPI$^-$) cells.

2.8. Cytokine and Chemokine Secretion Analysis

To evaluate cytokine and chemokine production, BMMΦs were seeded at 0.2×10^6 cells/well in sterile 24-well plates and incubated with 300 μg/mL of the different iNP formulations at 37 °C and 5% CO_2 for three hours. Excess iNPs were removed by washing twice with PBS followed by replacing with complete medium containing 100 ng/mL LPS or 200 ng/mL CpG ODN. After 48 h, cell culture supernatants were collected and analyzed using enzyme-linked immunosorbent assays (ELISA) (BioLegend) to measure murine interleukin-6 (IL-6) and tumor necrosis factor alpha (TNF-α) or Luminex (Austin, TX, USA) Multi-Analyte Profiling technology (xMAP) to assess multiple cytokines and chemokines as described in the text.

2.9. Immunoblotting for Transcriptional Activity

To observe the effects of iNPs on transcriptional activity, BMMΦ were seeded at 1.0×10^6 cells/well in sterile 6-well plates and incubated with 300 μg/mL of the different iNP formulations at 37 °C and 5% CO_2 for three hours. Excess iNPs were removed by washing twice with PBS followed by replacing with BMMΦ media. Cells were then challenged with 100 ng/mL LPS for 0.5, 1, or 4 h where indicated, at 37 °C and 5% CO_2 before washing twice with PBS and harvested using 300 μL RIPA buffer containing 1% Halt$^®$ Protease Inhibitor.

Wells treated with BIRB 796 were made to a concentration of 5 μM and incubated for 15 min prior to LPS induction. BIRB 796 was not washed from the wells. Wells treated with OGM were made to a concentration 10 μM and incubated for 4 h at 37 °C and 5% CO_2, washed twice with PBS to remove excess OGM, replaced with BMMΦ media, followed by iNP treatment as described above. Wells exposed to UV light were exposed in the cell culture hood for 15 min with the plate lids removed. After exposure, media was exchanged and cells were returned to the incubator and harvested at 0.5, 1, or 4 h after exposure.

Protein lysates were generated using 50/50 sample to 2× SDS-PAGE sample buffer. Proteins were then separated by SDS-PAGE and immunoblotted using the antibodies listed above. ECL was used for detection.

2.10. Statistical Analyses

Statistical analyses were performed using Prism 9 (GraphPad, San Diego, CA, USA). Results are reported as mean ± standard deviation (SD). A Student's *t*-test was used to determine the significance of parametric data between groups as labeled. $p \leq 0.05$ is the cutoff for statistical significance and is denoted throughout the text with *. Additional asterisks are used as applicable to denote the following: ** for $p \leq 0.01$, *** for $p \leq 0.001$, and **** for $p \leq 0.0001$. Comparisons that were not statistically significant were denoted with ns ($p > 0.05$).

3. Results

3.1. Fabrication, Characterization, and Stability Assessment of Poly(Lactic Acid) iNPs

iNPs were prepared using PLA by the single emulsion-solvent evaporation method (Figure 1A) with two surfactants—PVA or PEMA. The iNPs produced were similar in size with diameters between 400–600 nm (Figure 1B) with low polydispersity indices (PDI) (Figure 1C). In contrast to size, the zeta potentials of iNPs were significantly different, where PLA-PVA were approximately −17 mV and PLA-PEMA were approximately −40 mV (Figure 1D). We performed additional studies aimed to determine the stability of iNPs following reconstitution in deionized water over 8 h under various storage temperatures [26]. Both PLA-PVA and PLA-PEMA showed less than 10% change in size (Figure 1E). Similarly, the zeta potential of iNPs remained stable with less than 10% variability over 8 h (Figure 1F). Both iNP formulations displayed similar stabilities independent of reconstitution and storage at room temperature or refrigeration.

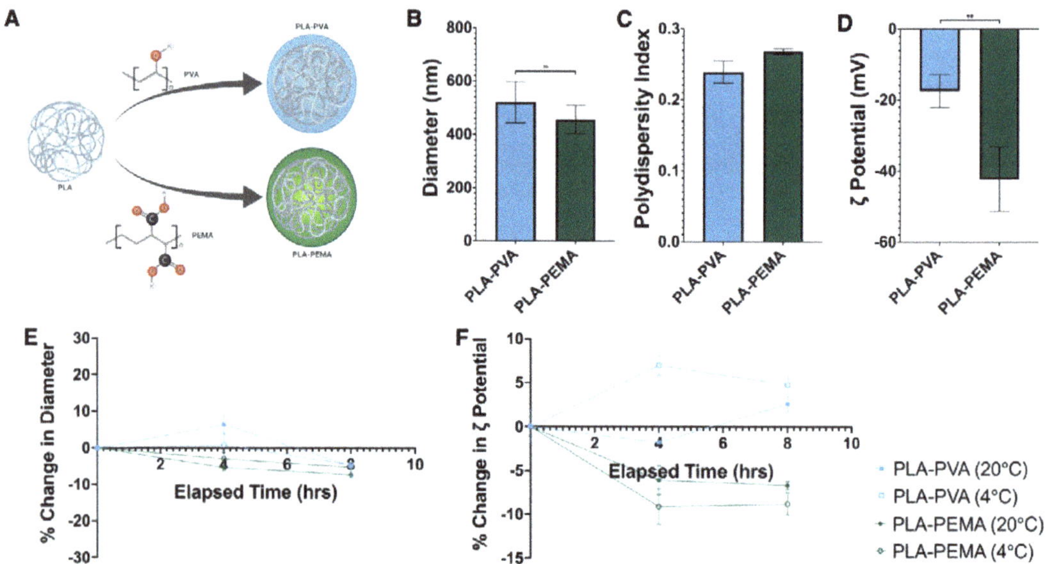

Figure 1. Physicochemical characterization of the synthesized iNPs. (**A**) Schema of the particle formulations utilized for this study. (**B**) Particle diameters were optimized to be in the range of 400–600 nm with (**C**) polydispersity indices in the range of 0.150–0.250. (**D**) Particles were also standardized across surface charge as represented by ζ potential. Additionally, particle stability following reconstitution in distilled water was determined at room temperature (20 °C) and refrigeration (4 °C) over a course of 8 h to confirm stability of particle size (**E**) and zeta potential (**F**). Schematic in (**A**) created with BioRender. Statistical differences between groups were determined by performing Student's *t*-test. Error bars represent SD. ** for $p \leq 0.01$ and ns = not significantly different ($p > 0.05$).

3.2. PLA iNPs Do Not Sequester PAMPs

One possible mechanism for iNP-mediated anti-inflammatory activity is through functioning as a sink to directly bind PAMPs to sequester them away from TLRs expressed on immune cells [10]. To evaluate the possibility of direct interactions between PAMPs and iNPs (Figure 2A), we incubated PLA-PVA or PLA-PEMA with fluorescein (FITC)-labeled LPS or CpG ODN. Following incubation, the samples were centrifuged to pellet the iNPs and the fluorescence intensity of the supernatant was measured. We tested direct iNP interactions with FITC-LPS and FITC-CpG ODN in PBS containing 10% FBS (Figure 2B,C, respectively). Compared to the FITC-LPS or FITC CpG ODN controls (dashed lines), no concentration-dependent reduction in FITC signal was observed for either iNP tested and the FITC signal variation was less than 20% from the control in all cases. These studies established that iNP sequestration of PAMPs is not a major mechanism by which iNPs elicit their inherent anti-inflammatory effects, warranting further investigation to understand if the protective mechanism is driven directly by iNP interaction with the immune cells of interest.

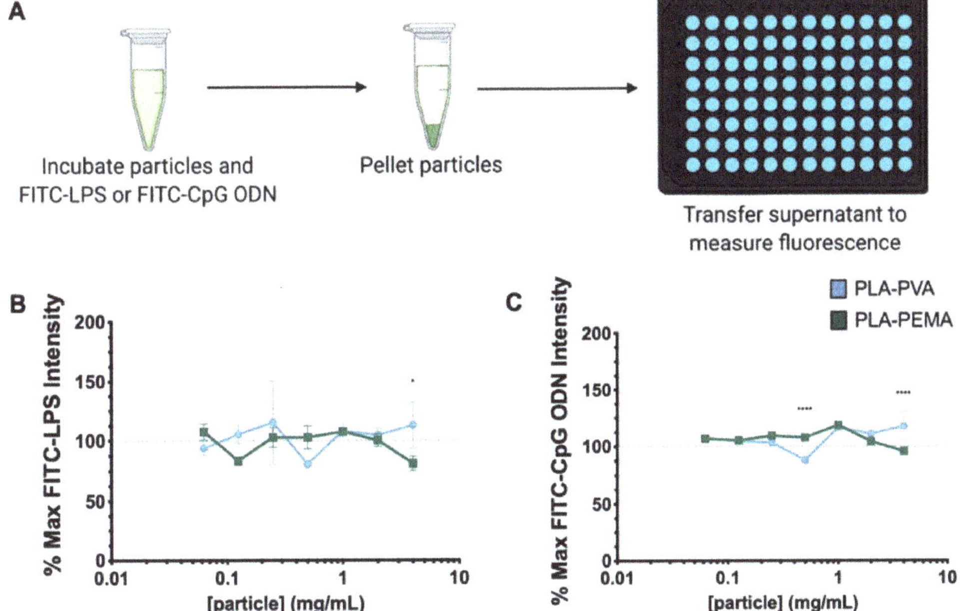

Figure 2. Under typical in vitro serum conditions, both particle types fail to sequester FITC-tagged TLR agonists. (**A**) Particles and FITC-conjugated TLR agonists were co-incubated at 37 °C and 5% CO_2 for 1 h and then pelleted to determine direct interactions between particles and TLR agonists. When co-incubated with PBS containing 10% FBS, both (**B**) FITC-LPS and (**C**) FITC-CpG ODN fail to interact with particles alone as signified by the dashed line representing 100% FITC signal of FITC-LPS (**B**) or FITC-CpG ODN (**C**) alone. Schematic in (**A**) created with BioRender. Statistical differences between groups were determined by performing Student's *t*-test. Error bars represent SD. * for $p \leq 0.05$ and **** for mboxemphp ≤ 0.0001.

3.3. BMMΦs Associate with and Internalize PLA-PEMA More Extensively Than PLA-PVA

As iNPs do not directly interact with PAMPs, we aimed to further understand the differences in cellular interactions and uptake between various iNPs. To assess iNP-cell interactions, we prepared Cy5.5-conjugated versions of iNPs with similar physicochemical characteristics as unlabeled PLA-PEMA and PLA-PVA (Supplemental Figure S1). Fluorescence microscopy showed that BMMΦs displayed a higher propensity to associate with

PLA-PEMA compared to PLA-PVA (Figure 3A), which was also seen using RAW 264.7 cells (Supplemental Figure S2). Flow cytometry was further used to quantitatively measure cell uptake of particles and confirmed that PLA-PEMA associated more rapidly with BMMΦs than PLA-PVA. Within 1 h of iNP incubation with BMMΦs, approximately 75% of BMMΦs were PLA-PEMA-Cy5.5$^+$ while only 30% of BMMΦs were PLA-PVA-Cy5.5$^+$ (Figure 3B). Since the formulations differed mainly in the surfactant choice and the resultant zeta potential of the iNP, these results suggest that the choice of the negatively charged PEMA drives the propensity of BMMΦs to preferentially interact with iNPs compared to those prepared using PVA.

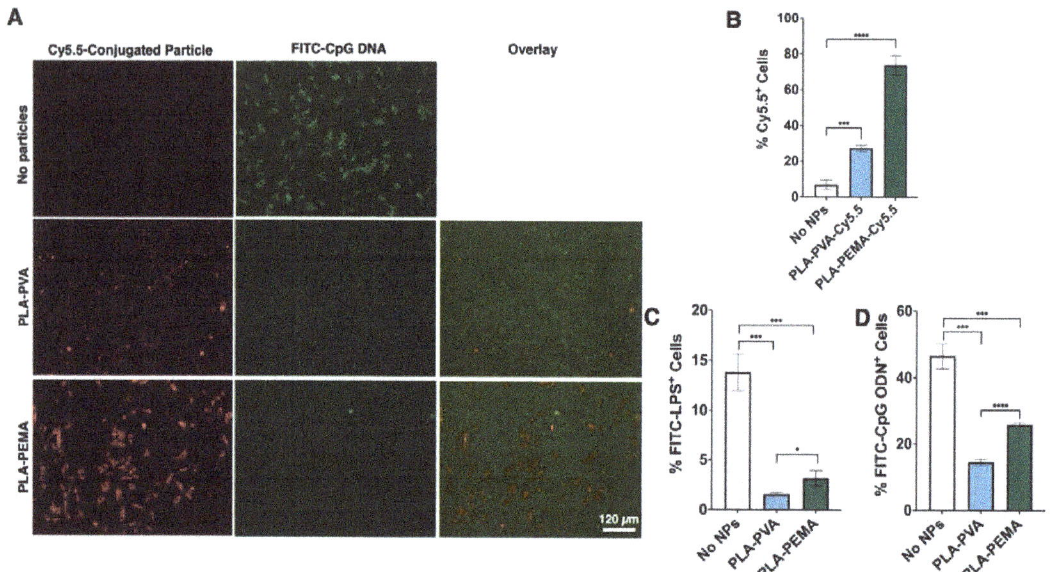

Figure 3. PLA-PVA and PLA-PEMA particle formulations show similar behavior when interacting with TLR agonists and cells with the exception being the increased propensity of PLA-PEMA particle uptake by cells. (**A**) Fluorescence microscopy establishes that PLA-PEMA-Cy5.5 interact to a greater extent with BMMΦs than PLA-PVA-Cy5.5. (**B**) Quantification with flow cytometry after 1-hr co-incubation at 37 °C and 5% CO_2 confirms cells associate to a greater extent with PLA-PEMA-Cy5.5 particles (30 μg/mL) than PLA-PVA-Cy5.5 particles (30 μg/mL). Both PLA-PVA (300 μg/mL) and PLA-PEMA (300 μg/mL) treatments result in dramatic decreases in BMMΦ cellular uptake of (**C**) FITC-LPS (100 ng/mL) and (**D**) FITC-CpG ODN (100 ng/mL) following 18-hr incubation at 37 °C and 5% CO_2. Statistical differences between groups were determined by performing a Student's *t*-test. Error bars represent SD. * for $p \leq 0.05$, *** for $p \leq 0.001$, and **** for $p \leq 0.0001$.

3.4. PLA Particles Hinder LPS and CpG ODN Interaction with BMMΦs

As described above, iNPs do not sequester LPS or CpG ODN (Figure 2), but interact differentially with BMMΦs (Figure 3A,B). Collectively, this suggests that the immunomodulatory activity of iNPs is dependent on their interactions with BMMΦs. To assess this, we first treated BMMΦs with either PLA-PEMA or PLA-PVA followed by incubation with either FITC-LPS (Figure 3C,D) or FITC-CpG ODN (Figure 3A–D). Qualitatively it was observed that despite the greater interaction of PLA-PEMA with BMMΦs, both iNP formulations decreased the association of FITC-CpG ODN with the BMMΦs (Figure 3A). For both LPS and CpG ODN, flow cytometry shows quantitatively that iNP pre-treatment significantly decreased the overall interaction of BMMΦs with LPS and CpG ODN (Figure 3C,D, respectively). These decreases in PAMP interactions with the cells occurs regardless of iNP type, suggesting that iNP-mediated interruption of BMMΦ and LPS or CpG ODN interactions is a process independent of iNP uptake and surfactant composition. Flow cytometry studies (Supplemental Figure S3) revealed reduced CD14 and TLR4 surface molecule expression in response to iNP treatment, suggesting that iNP-mediated disruption of BMMΦ-PAMP interactions may be influenced by the reductions in CD14 and TLR4 surface expression.

3.5. PLA-PEMA and PLA-PVA Inhibit NF-κB Activation, But Do So at Different Rates

PAMPs engage their respective TLRs and initiate complex signaling cascades that eventually lead to the production of cytokines and chemokines and expression of costimulatory molecules that eventually lead to widespread inflammation [27]. Because these signaling cascades are dependent upon activation of key transcriptional nodes, we next investigated the impact of LPS-mediated TLR4 stimulation via activation of the NF-κB p65 transcription factor and p38 MAPK (Figure 4A). In both signaling pathways, phosphorylation of p65 and p38 signifies engagement and activation of the upstream TLRs. To investigate the effects of iNPs, we first incubated the BMMΦs with iNPs prior to LPS stimulation for 0.5, 1, and 4 h to assess the activation of these key signaling pathways. Figure 4B shows that both PLA-PEMA and PLA-PVA decrease the phosphorylation of p65 compared to no particle treatment. Importantly, the decrease in phosphorylation in the case of PLA-PEMA treatment occurs earlier than that seen in PLA-PVA, suggesting that the more extensive uptake of PLA-PEMA compared to PLA-PVA (Figure 3A,B) plays a role in mediating this protective effect against activation of proinflammatory signaling cascades. Along with this, incubation with either PLA-PEMA or PLA-PVA alone results in no alteration in phosphorylation of either p65 nor a decrease in the total amount of the protein (Supplemental Figure S4). We next probed for MAPKs (Figure 4C). MAPKs are key in that they are activated by different stimuli (including LPS), yet p38, ERK1/2, and SAPK/JNK all have the capacity to phosphorylate transcription factors that form the AP-1 complex, a key regulator of the transcription of inflammatory cytokines [28,29]. We can see that phosphorylation of p38 is decreased secondary to LPS stimulation when treated with iNPs and that this result is opposite to that seen with phospho-ERK1/2 and phospho-JNK. Interestingly, when we evaluate the effects of iNP treatment on MAPK activation alone, we see that iNPs stimulate phosphorylation of ERK1/2, an effect not seen with the other probed MAPKs (Supplemental Figure S4). Finally, an investigation of upstream signaling components shows no decrease in phosphorylation of MKK3, MKK6, and TAK1, nor total levels of IRAK4 suggesting that the iNP-mediated effects downstream of LPS stimulation are limited to NF-κB p65 and p38 MAPK (Figure 4D). These data suggest that the iNP-based modifications to the BMMΦs are inherent to their capacity to respond to an inflammatory trigger rather than some basal change to the BMMΦs.

Furthermore, to establish that these transcription changes result in functional changes to the BMMΦs, we used Luminex to establish that these changes in transcription factor activation also resulted in a decrease in cytokine secretions. Indeed, we confirmed this across a multitude of signaling pathways including NF-κB-dependent IL-6 (Figure 5A), IRF3-dependent IFNβ (Figure 5B) [30], and the transcriptionally complex IL-10 (Figure 5C) [31]. Similar experiments conducted with murine macrophage-derived RAW 264.7 cells confirmed iNP-dependent decreases in IL-6 with iNPs following LPS stimulation (Supplemental Figure S5). Interestingly, PLA-PVA treatment resulted in an increase in TNF-α secretion with iNP treatment, while the opposite effect was observed with PLA-PEMA. Additionally, another potential consequence of iNP treatment is the induction of cell death driving the decrease in transcriptional activation and proinflammatory cytokines. We used flow cytometry and cell exclusion dye to establish that our iNPs do not induce cell death and can increase cell survival in the setting of LPS stimulation of BMMΦs (Figure 5D). When taken together, iNPs can drive changes in the function of BMMΦs through reprogramming of transcriptional activation. This leads to decreased proinflammatory cytokine secretions and also aids in extending survival of this cell population.

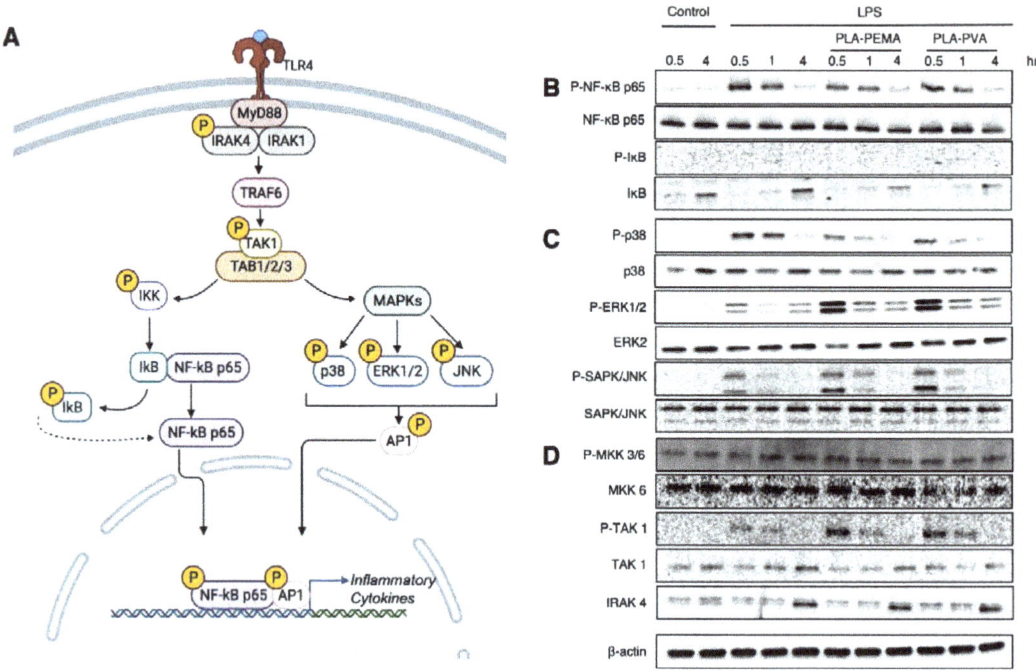

Figure 4. Both iNPs suppress NF-κB p65 and p38 MAPK activation, although at different rates. (**A**) TLR4 activation upon LPS binding triggers a complex signaling cascade where two key nodes of activity involve activation of NF-κB p65 and p38 MAPK to initiate downstream inflammatory cytokine expression. Following PLA-PVA (300 μg/mL) and PLA-PEMA (300 μg/mL) treatment, cells were then stimulated with LPS (100 ng/mL) for 0.5, 1, or 4 h. (**B**) Samples were then immunoblotted for key NF-κB proteins: phospho-NF-κB p65 (Ser536), total NF-κB p65, phospho-IκB (Ser32), and total IκB. (**C**) These samples were also immunoblotted for a variety of MAPKs: phsopho-p38 (Thr180/Tyr182), total p38, phospho-ERK1/2 (Thr202/Tyr204), total ERK2, phospho-SAPK/JNK (Thr183/Tyr185), and total JNK. (**D**) Finally, to address potential impacts of iNPs on upstream signaling proteins, these samples were immunoblotted for phospho-MKK3 (Ser189/)/MKK6 (Ser207), total MKK6, phospho-TAK1 (Thr184/187), total TAK1, IRAK4, and β-Actin. Schematic in (**A**) created with BioRender.

3.6. The PLA Polymer Composition of iNPs Drives the Suppression of NF-κB Signaling

Given that both of our iNP formulations produce an anti-inflammatory immunomodulatory effect, but PLA-PEMA does so more effectively, we next focused our efforts on understanding how these iNPs work by comparing our PLA-PEMA (denoted as PLA in Figures 6 and 7) to commercially available nanoparticle formulations composed of different polymer materials [polystyrene-COOH and poly(methyl methacrylate), herein referred to as PS and PMMA] (Figure 6A). Although both PS and PMMA are non-biodegradable, PS is of particular interest in that it has previously been investigated to be immunomodulatory in studies of inflammatory monocytes via a separate splenic sequestration mechanism [12]. To control for some of the physicochemical properties described as being key to this study (Figure 1), we ensured that the diameter and PDI of the commercial nanoparticles were within range of our iNPs (Supplemental Figure S6). Additionally, because we hypothesize that the lactic acid in our PLA-based iNPs plays a role in mitigating proinflammatory signaling, we also used soluble lactic acid (sLA) as a control to compare its activity to the iNPs. With LPS stimulation (Figure 6B), similarly to sLA, PLA particles suppress NF-κB p65 phosphorylation while PS and PMMA formulations did not, which confirms that the immunomodulatory activity of iNPs was dependent upon the polymers. Of note, when we look at IκB degradation as a marker of NF-κB activation, we see that PLA particles show similar protein levels to PS and PMMA, all of which were lower than for LPS alone and LPS plus sLA. This suggests a PLA-mediated NF-κB suppression unique to the p65 subunit. We then compared the effects of the polymer on mitigating proinflammatory cytokine secretions in response to LPS (Supplementary Figure S7). Again, the PLA-based iNPs successfully suppressed inflammatory cytokine secretion while PS and PMMA showed little immunomodulatory activity as expected based on the NF-κB results.

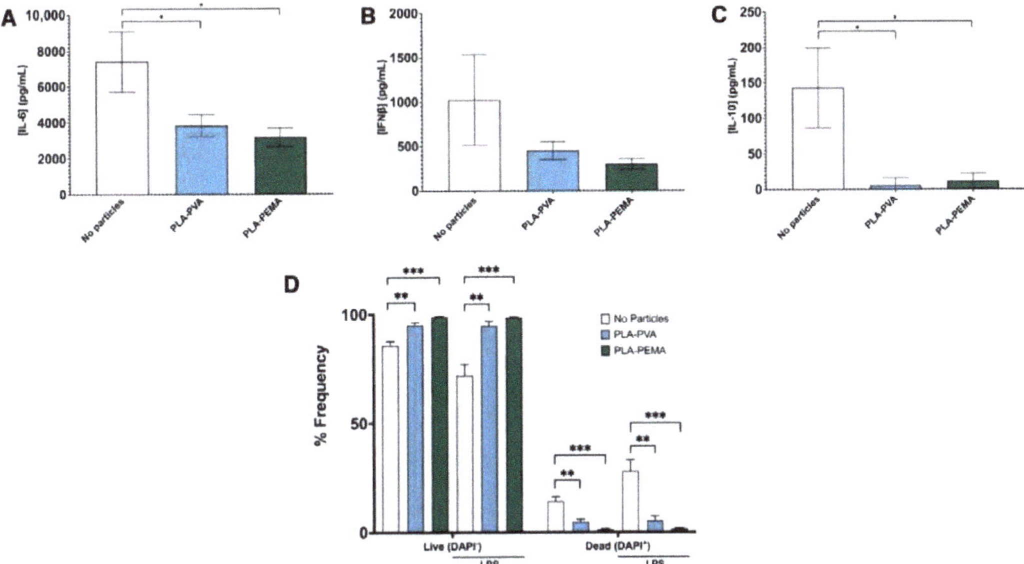

Figure 5. iNP suppression of transcription activation of genes for inflammatory cytokines result in correlated decreases in cytokine production with an increase in BMMΦ survival. Supernatants from cells incubated with LPS for 48 h following iNP treatment were also collected to assess secretion of (**A**) IL-6, (**B**) IFNβ, and (**C**) IL-10, key cytokines produced downstream of TLR4 engagement. (**D**) Additionally, these cells were assessed for viability. Both iNPs, especially PLA-PEMA, result in increased survival based on flow cytometry with DAPI exclusion dye. Statistical differences between groups were determined by performing Student's *t*-test. Error bars represent SD. * for $p \leq 0.05$, ** for $p \leq 0.01$, and *** for $p \leq 0.001$.

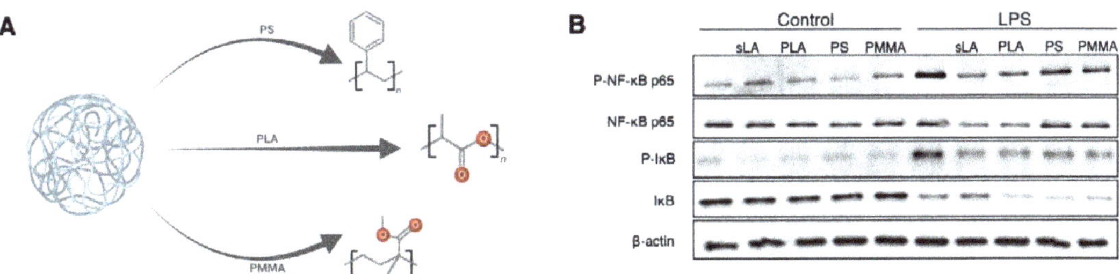

Figure 6. Suppression of NF-κB p65 is dependent upon lactic acid polymer of the particles. (**A**) BMMΦ cells treated with soluble lactic acid (sLA), or nanoparticles composed of PLA-PEMA (PLA), polystyrene-COOH (PS), or poly(methyl methacrylate) (PMMA), for 3 h followed by 1 h LPS stimulation to show differential activation of p65 and greater baseline degradation of IκB compared to sLA and LPS alone following particle incubation and 1h LPS stimulation. (**B**) Samples were immunoblotted for phospho-NF-kB p65 (Ser536), total NF-kB p65, phospho-IκB (Ser32), total IκB, and β-actin. Schematic in (**A**) created with BioRender.

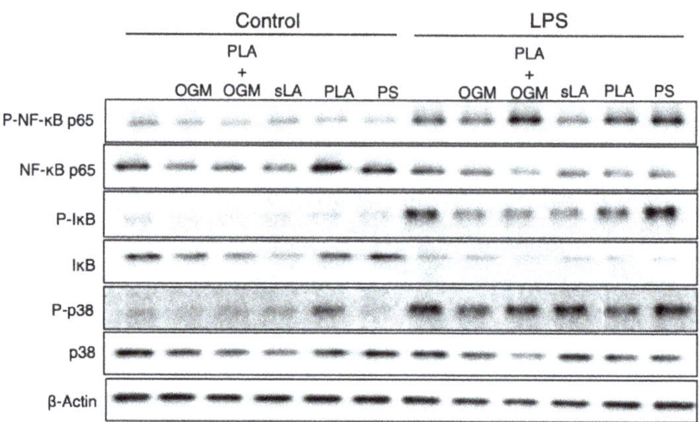

Figure 7. GPR68 inhibition with OGM reverses lactate-mediated suppression of NF-κB p65 activation. BMMΦ cells were treated like previously with soluble lactic acid (sLA) or nanoparticles composed of PLA-PEMA (PLA) or polystyrene-COOH (PS). The addition of OGM, a GPR68 inhibitor, reverses the inhibition of p65 activation seen with PLA following particle incubation and 1-hr LPS stimulation. Samples were immunoblotted for phospho-NF-κB p65 (Ser536), total NF-κB p65, phospho-IκB (Ser32), total IκB, phospho-p38 (Thr180/Tyr182), total p38, and β-actin.

3.7. The Protective Function of the PLA-Based iNPs Depends upon GPR68 Signaling

The inhibition of NF-κB p65 phosphorylation is dependent upon the PLA polymer of our iNPs. We next assessed the mechanism by which the lactic acid from the iNPs elicits its inhibition of inflammatory signaling pathways. Lactic acid is actively removed from the intracellular space [19], therefore we sought to identify the receptor through which the particle-mediated acidity is sensed. Previous work has shown that the GPR68 regulates intestinal inflammation and is a cellular pH sensor [32,33]. We hypothesized that the potential mechanism by which PLA-based iNPs work to inhibit LPS-induced inflammation is through the pH-sensing GPR68. In order to test this, we used OGM [22], a novel inhibitor of GPR68, to block the GPR68-mediated inhibition of inflammation (Figure 7). As expected, PLA iNPs alone showed less NF-κB p65 and p38 MAPK phosphorylation following LPS stimulation than LPS alone or OGM with LPS treatment alone. When we combined both PLA iNPs and OGM with LPS stimulation, not only did OGM increase the level of NF-κB p65 and p38 phosphorylation compared to just PLA iNPs, but it did so to a greater extent than the LPS only control. We further confirmed that the GPR68 inhibitor OGM or GPR81 inhibitor 3-OBA [34] (control) reversed the ability of PLA iNPs to mitigate proinflammatory cytokine secretions (Supplemental Figure S8). These results confirm that inhibition of inflammation is not only mediated by PLA-associated acidification of the microenvironment (negated with OGM) but is specific to sensing of the lactic acid byproduct (inhibited by 3-OBA) of PLA degradation.

4. Discussion

Developing improved treatments for severe inflammation and sepsis is a burgeoning area where nanotechnology-based approaches hold significant promise. Current strategies under development have focused on single-target small molecules and biologics where the failure of these therapeutics in clinical trials suggests a need for strategies with broad activity against proinflammatory immune responses [3,5].

iNPs invoke multiple physical and biological mechanisms to accomplish their protective effects (Figure 8). As shown, both types of iNPs lack an ability to directly bind PAMPs including LPS and CpG ODN (Figure 2) but they do alter the ability of BMMΦs to interact with both PAMPs (Figure 3). With this change in BMMΦ-PAMP interaction, it is important to note that although a physical mechanism inhibiting BMMΦ-PAMP interactions is occurring, we cannot yet formally conclude whether cell surface receptor downregulation is the sole response leading to this change (Supplemental Figure S3) or if iNPs serve to directly prevent the interaction of PAMPs with TLRs. Additionally, PLA (but not PLGA) nanoparticles have been shown to downregulate cell surface expression of CD80, CD86, and MHC class II [11,13], suggesting a mechanism by which there is global downregulation of a multitude of key immune cell surface receptors unique to PLA-based nanoparticles. Indeed, engagement of iNPs may trigger endocytosis of these receptors [35], thus making BMMΦs "blind" to PAMP stimulation and perhaps arrested from engaging T cells through the T cell receptor complex [36].

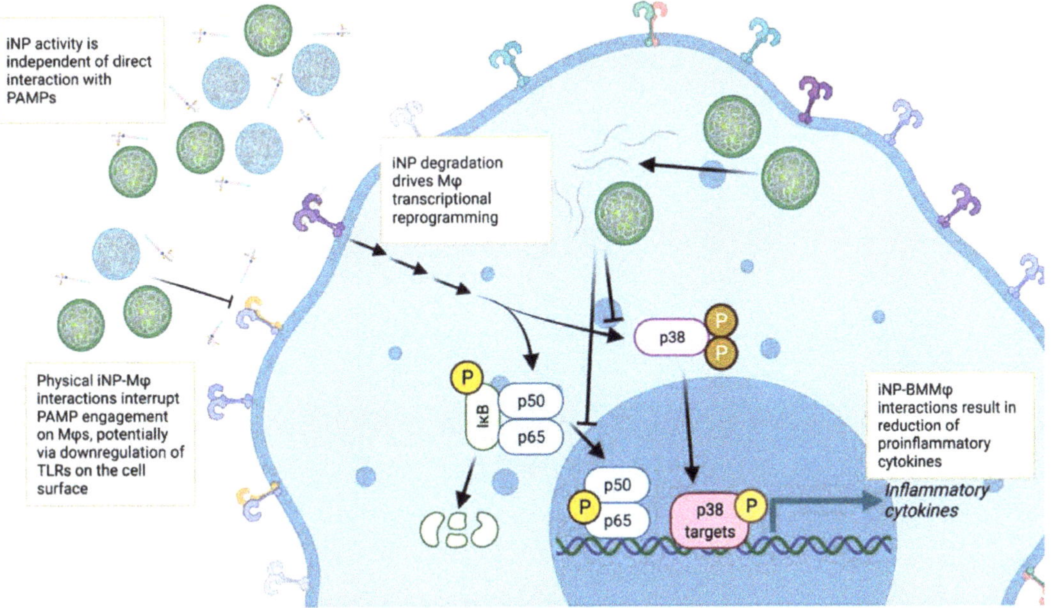

Figure 8. iNPs invoke multiple physical and biological mechanisms to elicit a protective effect in BMMΦs. iNPs interact with BMMΦs to interrupt the engagement of PAMPs on TLRs at both the cell surface and endosomal surface, thus limiting the activation of TLR signaling networks. However, if LPS engages TLR4, the iNPs function via a secondary mechanism whereby their degradation triggers BMMΦ transcriptional reprogramming in response to LPS. This reduces the overall activation and production of inflammation mediators resulting in an overall protective effect to PAMP challenge. Schematic created with BioRender.

Of note, the uptake of iNPs and subsequent cellular transcriptional changes (Figure 4) appear to be independent of this iNP-mediated disruption of BMMΦ-PAMP interaction (Figure 3). Of particular interest is that these iNPs successfully mediate this disruption at two different cellular compartments given that LPS initially binds to TLR4 at the cell surface and TLR4 is then rapidly endocytosed for further LPS binding at the endosomal surface [37,38]. As a further validation of this multi-compartmental activity of iNPs compared to the stimulation of TLR4 at the cell surface prior to endocytosis, TLR9 is endosomal when it is stimulated by CpG ODN [39–41]. This suggests that the iNPs serve to disrupt multiple PAMP recognition pathways at different, distinct locations within the cell that lead to proinflammatory cytokine secretion [11], and further emphasizes their potential to serve as a broad-based therapeutic for inflammation. One curiosity that was encountered when evaluating NF-κB- and MAPK-mediated inflammatory signaling is that of all the analytes probed, iNPs drove downregulation of cytokine secretions independent of the formulation; however, TNF-α secretion was increased with PLA-PVA. TNF-α is produced downstream of NF-κB and MAPK activation, but it also has the additional characteristic of further MAPK activation downstream of engagement of its receptor TNFR1 [42]. Additionally, TNF-α exists preformed as pro-TNF-α at the cell membrane until cleavage to the activated form [43], suggesting the possibility that the PLA-PVA iNPs are less effective at inhibiting this cleavage activity.

This proposed mechanism whereby iNPs suppress BMMΦ-PAMP compliments a similar strategy to one employed by Thamphiwatana et al. [44], where macrophage-like NPs served as a sponge for LPS and proinflammatory cytokines. Rather than induce a competition for LPS binding, our experiments show that our iNPs prevented BMMΦ-LPS and BMMΦ-CpG ODN interactions. In combination, these nanoparticle strategies could be combined to further reduce the overall interactions between BMMΦs and stimulating PAMPs. Alone, our iNPs eliminate the need for any cellular material to generate macrophage-like nanoparticles and simplify the synthesis process for the platform since it only requires off-the-shelf chemical components. This potentially avoids regulatory roadblocks in the future with any putative anti-inflammatory therapeutic containing biological components. Additionally, through usage of strictly polymer-based nanoparticles without the need of chemotherapeutic or biologic payloads, we have shown the inherent immunomodulatory capabilities of iNPs that also lend themselves to further modification to suit the needs of other potential therapeutic applications.

As noted, this physical inhibitory iNP activity is assisted by the additional action of reprogramming the functional phenotype of these BMMΦs (Figures 4 and 5). Through alteration of BMMΦ effector activity secondary to LPS challenge, these iNPs take advantage of the inherent plasticity of BMMΦs to modify their activity at the location of PAMP insult. This strategy is of additional benefit in that it serves as a redundant second mechanism at play to synergize with the initial inhibition of BMMΦ-PAMP interactions (Figure 8). Reports of similar nanoparticle-driven innate cell reprogramming has been shown in models of spinal cord injury [45], experimental autoimmune encephalitis [46], and allergic airway inflammation [47]. The culmination of these studies aids in the idea that the iNP-mediated effects on immunomodulation alter the inherent responses of the BMMΦs independent of potential sequestration mechanisms. Given this change in the effector phenotype of the BMMΦs, it remains to be fully elucidated how exactly iNPs elicit these functional responses. Recent work in bone marrow-derived dendritic cells (BMDCs) with PLGA- and PLA-based particles argues that the released lactate from the degradation of these particles lock dendritic cells in an immature phenotype [13,36]. This further suggests that these presumed inert polymeric materials have inherent biological activities that has thus far been under-appreciated, especially the ability of these biomaterials to functionally reprogram the in-situ activity of a variety of immune cells when challenged by known activators of innate immune cells.

Interestingly, these earlier studies and the work described herein highlight the need for increased understanding of the crosstalk between nanoparticle degradation products and the burgeoning field of immunometabolism. PLA is first biodegraded via non-enzymatic random hydrolytic ester cleavage to form oligomers and monomers of lactic acid via surface and bulk erosion [48]. These oligomers and monomers are then free to interact with cells to interact with a variety of cellular processes including the Krebs cycle [14] and, more importantly for our interests, inflammatory pathways. Although degradation of synthetic polymers is better established via passive hydrolysis rather than enzymatic reactions [49], reports in the literature note the existence of fungal [50] and bacterial [50,51] enzymes that can degrade PLA into its monomeric form suggesting a role for infection to drive PLA degradation [52]. Additionally, macrophages and other innate cells secrete an array of enzymes such as lactate dehydrogenase and its coenzyme NADH-reductase during inflammation that can catalyze the degradation of PLA in the setting of PLA implants [53].

When we consider the converse—the role of lactate in modifying the inflammatory response—we see that lactate has been established to play a role in dampening the proinflammatory response within macrophages. An early study compared the role of lactic acid and hydrochloric acid at inducing different inflammatory patterns in RAW 264.7 stimulated with LPS. In this work they showed that when cells were titrated to more acidic environments such as pH 6.5, HCl treatment essentially drove a proinflammatory response with LPS stimulation as measured by evolution of NO, IL-6-to-IL-10, and NF-κB DNA binding. In contrast, lactic acid treatment (controlled for pH) effectively inhibited LPS-induced NO, IL-6, IL-10, and NF-κB DNA binding [54]. This work is key because it establishes that the acidity of the environment alone does not alone drive the anti-inflammatory effects that we have also observed, but rather that lactate serves as a unique molecule driving the suppression of inflammatory responses in macrophages. Further work built upon this to establish a key role for GPR81, a cell-surface receptor for lactate, in mediating lactate suppression of proinflammatory responses in the GI tract using animal models for dextran sulfate-sodium-induced colitis [55] and acute hepatitis and pancreatitis [56]. Interestingly, in other inflammatory models utilizing macrophages from non-GI sources, the role of GPR81 in lactate-mediated responses remains controversial [20,21] suggesting the potential of other pH-sensing receptors, such as GPR68, to play a complementary role [57].

5. Conclusions

Taken together, this work establishes that iNPs take advantage of multiple mechanisms to mitigate severe inflammatory responses and suggests a novel multimodal approach to improve prospects for patients with sepsis and other inflammation-mediated diseases. Polymer-based nanoparticles show promise in serving as drug carriers for controlled delivery of active chemotherapeutic agents; however, the inherent immunomodulatory nature of the materials themselves remains not well characterized. We have described the nano-bio interactions for PLA-based iNPs with varying surface charge and applied these formulations to modulating BMMΦ activity in response to diverse inflammatory agents. We showed that iNPs modify proinflammatory cytokine secretions and also establish that the mechanisms by which this occurs are broad and rely on both physical interactions and reprogramming of BMMΦs. Physical interaction of the BMMΦs with iNPs limit uptake of LPS and CpG ODN interaction. Furthermore, iNPs elicit intrinsic changes in the BMMΦs through metabolic alterations such that NF-κB and p38 MAPK activity is downregulated in response to LPS stimulation. Future studies aim to address applications of iNPs to improve clinical outcomes in murine models of severe inflammation and sepsis and to further characterize nano-bio interactions of iNPs with other key players of the innate immune response, particularly those regulating immunometabolism.

Supplementary Materials: The following are available online at https://www.mdpi.com/article/10.3390/pharmaceutics13111841/s1, Figure S1: Physicochemical characterization of the iNPs with their fluorophore-conjugated formulations, Figure S2: Confocal microscopy showing extent of iNP uptake in RAW 264.7 cells and subsequent colocalization with FITC-CpG ODN, Figure S3: Flow cytometry reveals iNPs trigger downregulation of CD14 and TLR4 surface expression in bone marrow-derived dendritic cells, Figure S4: Control immunoblots showing p38 inhibition with BIRB 796 and lack of NF-κB p65 and p38 MAPK activation with iNPs alone, Figure S5: LPS-stimulated RAW 264.7 IL-6 and TNF-α secretion using the same experimental design as in Figure 4, Figure S6: Physicochemical characterization of commercially available particles, Figure S7: LPS-stimulated RAW 264.7 IL-6 and TNF-α secretion using the same experimental design as described previously with the inclusion of commercially available particles as described in Figure 5, Figure S8: LPS-stimulated RAW 264.7 IL-6 and TNF-α secretion using the same experimental design as described previously with the inclusion of GPR68 and GPR81 inhibitors described in Figure 6.

Author Contributions: Conceptualization, J.J.M.L. and R.M.P.; methodology, J.J.M.L., P.S. and R.M.P.; investigation, J.J.M.L., A.L.C., B.L.S. and N.T.; formal analysis, J.J.M.L. and A.L.C.; resources, C.C.H., P.S. and R.M.P.; writing—original draft preparation, J.J.M.L.; writing—review and editing, P.S., R.M.P. and C.C.H.; visualization, J.J.M.L., A.L.C. and R.M.P.; funding acquisition, R.M.P. All authors have read and agreed to the published version of the manuscript.

Funding: This research was supported by startup funds from the University of Maryland School of Pharmacy, the New Investigator Award from the American Association of Colleges of Pharmacy (AACP), and the National Institute of General Medical Sciences of the National Institutes of Health under Award Number R35GM142752 awarded to R.M.P. Additional support was provided by the NIH/NIAID Signaling Pathways in Innate Immunity Training Program (NIH T32AI095190) and NIH/NHLBI Interdisciplinary Training Program in Cardiovascular Disease (NIH T32HL007698) to support J.J.M.L. The authors declare no competing conflicts of interest.

Institutional Review Board Statement: The study was conducted according to the guidelines of the University of Maryland, Baltimore Office of Animal Welfare Assurance #D16-00125 (A3200-01). The experiments described were approved by the University of Maryland, Baltimore Institutional Animal Care and Use Committee (IACUC) protocol #0721010.

Informed Consent Statement: Not applicable.

Data Availability Statement: The data presented in this study are available upon request from the corresponding author, R.M.P., upon request.

Acknowledgments: The authors wish to acknowledge the support of the University of Maryland School of Medicine Center for Innovative Biomedical Resources, Flow Cytometry Core—Baltimore, Maryland for technical assistance and experimental design support for this study.

Conflicts of Interest: The authors declare no conflict of interest.

References

1. Rudd, K.E.; Johnson, S.C.; Agesa, K.M.; Shackelford, K.A.; Tsoi, D.; Kievlan, D.R.; Colombara, D.V.; Ikuta, K.S.; Kissoon, N.; Finfer, S.; et al. Global, regional, and national sepsis incidence and mortality, 1990–2017: Analysis for the Global Burden of Disease Study. *Lancet* **2020**, *395*, 200–211. [CrossRef]
2. Delano, M.J.; Ward, P.A. Sepsis-induced immune dysfunction: Can immune therapies reduce mortality? *J. Clin. Investig.* **2016**, *126*, 23–31. [CrossRef]
3. Marshall, J.C. Why have clinical trials in sepsis failed? *Trends Mol. Med.* **2014**, *20*, 195–203. [CrossRef] [PubMed]
4. Fink, M.P. Animal models of sepsis. *Virulence* **2014**, *5*, 143–153. [CrossRef] [PubMed]
5. Remick, D.G. Cytokine therapeutics for the treatment of sepsis: Why has nothing worked? *Curr. Pharm. Des.* **2003**, *9*, 75–82. [CrossRef]
6. Leligdowicz, A.; Matthay, M.A. Heterogeneity in sepsis: New biological evidence with clinical applications. *Crit. Care* **2019**, *23*, 80. [CrossRef]
7. Gomes, A.C.; Mohsen, M.; Bachmann, M.F. Harnessing Nanoparticles for Immunomodulation and Vaccines. *Vaccines* **2017**, *5*, 6. [CrossRef]
8. Feng, X.; Xu, W.; Li, Z.; Song, W.; Ding, J.; Chen, X. Immunomodulatory Nanosystems. *Adv. Sci.* **2019**, *6*, 1900101. [CrossRef] [PubMed]
9. Pearson, R.M.; Casey, L.M.; Hughes, K.R.; Miller, S.D.; Shea, L.D. In vivo reprogramming of immune cells: Technologies for induction of antigen-specific tolerance. *Adv. Drug Deliv. Rev.* **2017**, *114*, 240–255. [CrossRef] [PubMed]
10. Lasola, J.J.M.; Kamdem, H.; McDaniel, M.W.; Pearson, R.M. Biomaterial-Driven Immunomodulation: Cell Biology-Based Strategies to Mitigate Severe Inflammation and Sepsis. *Front. Immunol.* **2020**, *11*, 1726. [CrossRef] [PubMed]
11. Casey, L.M.; Kakade, S.; Decker, J.T.; Rose, J.A.; Deans, K.; Shea, L.D.; Pearson, R.M. Cargo-less nanoparticles program innate immune cell responses to toll-like receptor activation. *Biomaterials* **2019**, *218*, 119333. [CrossRef] [PubMed]
12. Getts, D.R.; Terry, R.L.; Getts, M.T.; Deffrasnes, C.; Muller, M.; van Vreden, C.; Ashhurst, T.M.; Chami, B.; McCarthy, D.; Wu, H.; et al. Therapeutic inflammatory monocyte modulation using immune-modifying microparticles. *Sci. Transl. Med.* **2014**, *6*, 219ra217. [CrossRef] [PubMed]
13. Allen, R.P.; Bolandparvaz, A.; Ma, J.A.; Manickam, V.A.; Lewis, J.S. Latent, Immunosuppressive Nature of Poly(lactic-co-glycolic acid) Microparticles. *ACS Biomater. Sci. Eng.* **2018**, *4*, 900–918. [CrossRef] [PubMed]
14. Pearson, R.M.; Podojil, J.R.; Shea, L.D.; King, N.J.C.; Miller, S.D.; Getts, D.R. Overcoming challenges in treating autoimmuntity: Development of tolerogenic immune-modifying nanoparticles. *Nanomed. Nanotechnol. Biol. Med.* **2018**, *18*, 282–291. [CrossRef] [PubMed]
15. Ivashkiv, L.B. The hypoxia-lactate axis tempers inflammation. *Nat. Rev. Immunol.* **2020**, *20*, 85–86. [CrossRef]
16. Zhang, W.; Wang, G.; Xu, Z.G.; Tu, H.; Hu, F.; Dai, J.; Chang, Y.; Chen, Y.; Lu, Y.; Zeng, H.; et al. Lactate Is a Natural Suppressor of RLR Signaling by Targeting MAVS. *Cell* **2019**, *178*, 176–189 e115. [CrossRef]
17. Colegio, O.R.; Chu, N.Q.; Szabo, A.L.; Chu, T.; Rhebergen, A.M.; Jairam, V.; Cyrus, N.; Brokowski, C.E.; Eisenbarth, S.C.; Phillips, G.M.; et al. Functional polarization of tumour-associated macrophages by tumour-derived lactic acid. *Nature* **2014**, *513*, 559–563. [CrossRef]

18. Zhang, D.; Tang, Z.; Huang, H.; Zhou, G.; Cui, C.; Weng, Y.; Liu, W.; Kim, S.; Lee, S.; Perez-Neut, M.; et al. Metabolic regulation of gene expression by histone lactylation. *Nature* **2019**, *574*, 575–580. [CrossRef]

19. Certo, M.; Tsai, C.H.; Pucino, V.; Ho, P.C.; Mauro, C. Lactate modulation of immune responses in inflammatory versus tumour microenvironments. *Nat. Rev. Immunol.* **2021**, *21*, 151–161. [CrossRef]

20. Yang, K.; Xu, J.; Fan, M.; Tu, F.; Wang, X.; Ha, T.; Williams, D.L.; Li, C. Lactate Suppresses Macrophage Pro-Inflammatory Response to LPS Stimulation by Inhibition of YAP and NF-kappaB Activation via GPR81-Mediated Signaling. *Front. Immunol.* **2020**, *11*, 587913. [CrossRef]

21. Errea, A.; Cayet, D.; Marchetti, P.; Tang, C.; Kluza, J.; Offermanns, S.; Sirard, J.C.; Rumbo, M. Lactate Inhibits the Pro-Inflammatory Response and Metabolic Reprogramming in Murine Macrophages in a GPR81-Independent Manner. *PLoS ONE* **2016**, *11*, e0163694. [CrossRef]

22. Williams, C.H.; Neitzel, L.R.; Karki, P.; Keyser, B.D.; Thayer, T.E.; Wells, Q.S.; Perry, J.A.; Birukova, A.A.; Birukov, K.G.; Hong, C.C. Coupling Metastasis to pH-Sensing GPR68 Using a Novel Small Molecule Inhibitor. *BioRxiv* **2021**, 612549. [CrossRef]

23. McCarthy, D.P.; Yap, J.W.; Harp, C.T.; Song, W.K.; Chen, J.; Pearson, R.M.; Miller, S.D.; Shea, L.D. An antigen-encapsulating nanoparticle platform for TH1/17 immune tolerance therapy. *Nanomed. Nanotechnol. Biol. Med.* **2017**, *13*, 191–200. [CrossRef] [PubMed]

24. Weischenfeldt, J.; Porse, B. Bone Marrow-Derived Macrophages (BMM): Isolation and Applications. *Cold Spring Harb. Protoc.* **2008**, *2008*, 5080. [CrossRef]

25. Lutz, M.B.; Kukutsch, N.; Ogilvie, A.L.; Rossner, S.; Koch, F.; Romani, N.; Schuler, G. An advanced culture method for generating large quantities of highly pure dendritic cells from mouse bone marrow. *J. Immunol. Methods* **1999**, *223*, 77–92. [CrossRef]

26. Freitag, T.L.; Podojil, J.R.; Pearson, R.M.; Fokta, F.J.; Sahl, C.; Messing, M.; Andersson, L.C.; Leskinen, K.; Saavalainen, P.; Hoover, L.I.; et al. Gliadin Nanoparticles Induce Immune Tolerance to Gliadin in Mouse Models of Celiac Disease. *Gastroenterology* **2020**, *158*, 1667–1681.e12. [CrossRef]

27. Akira, S.; Takeda, K. Toll-like receptor signalling. *Nat. Rev. Immunol.* **2004**, *4*, 499–511. [CrossRef] [PubMed]

28. Hambleton, J.; McMahon, M.; DeFranco, A.L. Activation of Raf-1 and mitogen-activated protein kinase in murine macrophages partially mimics lipopolysaccharide-induced signaling events. *J. Exp. Med.* **1995**, *182*, 147–154. [CrossRef] [PubMed]

29. Hambleton, J.; Weinstein, S.L.; Lem, L.; DeFranco, A.L. Activation of c-Jun N-terminal kinase in bacterial lipopolysaccharide-stimulated macrophages. *Proc. Natl. Acad. Sci. USA* **1996**, *93*, 2774. [CrossRef]

30. Platanitis, E.; Decker, T. Regulatory Networks Involving STATs, IRFs, and NFkappaB in Inflammation. *Front. Immunol.* **2018**, *9*, 2542. [CrossRef]

31. Iyer, S.S.; Cheng, G. Role of interleukin 10 transcriptional regulation in inflammation and autoimmune disease. *Crit. Rev. Immunol.* **2012**, *32*, 23–63. [CrossRef] [PubMed]

32. de Valliere, C.; Cosin-Roger, J.; Simmen, S.; Atrott, K.; Melhem, H.; Zeitz, J.; Madanchi, M.; Tcymbarevich, I.; Fried, M.; Kullak-Ublick, G.A.; et al. Hypoxia Positively Regulates the Expression of pH-Sensing G-Protein-Coupled Receptor OGR1 (GPR68). *Cell. Mol. Gastroenterol. Hepatol.* **2016**, *2*, 796–810. [CrossRef]

33. de Valliere, C.; Wang, Y.; Eloranta, J.J.; Vidal, S.; Clay, I.; Spalinger, M.R.; Tcymbarevich, I.; Terhalle, A.; Ludwig, M.G.; Suply, T.; et al. G Protein-coupled pH-sensing Receptor OGR1 Is a Regulator of Intestinal Inflammation. *Inflamm. Bowel Dis.* **2015**, *21*, 1269–1281. [CrossRef]

34. Shen, Z.; Jiang, L.; Yuan, Y.; Deng, T.; Zheng, Y.R.; Zhao, Y.Y.; Li, W.L.; Wu, J.Y.; Gao, J.Q.; Hu, W.W.; et al. Inhibition of G protein-coupled receptor 81 (GPR81) protects against ischemic brain injury. *CNS Neurosci. Ther.* **2015**, *21*, 271–279. [CrossRef] [PubMed]

35. Ciesielska, A.; Matyjek, M.; Kwiatkowska, K. TLR4 and CD14 trafficking and its influence on LPS-induced pro-inflammatory signaling. *Cell. Mol. Life Sci.* **2021**, *78*, 1233–1261. [CrossRef]

36. Sangsuwan, R.; Thuamsang, B.; Pacifici, N.; Allen, R.; Han, H.; Miakicheva, S.; Lewis, J.S. Lactate Exposure Promotes Immuno-suppressive Phenotypes in Innate Immune Cells. *Cell. Mol. Bioeng.* **2020**, *13*, 541–557. [CrossRef] [PubMed]

37. Husebye, H.; Halaas, O.; Stenmark, H.; Tunheim, G.; Sandanger, O.; Bogen, B.; Brech, A.; Latz, E.; Espevik, T. Endocytic pathways regulate Toll-like receptor 4 signaling and link innate and adaptive immunity. *EMBO J.* **2006**, *25*, 683–692. [CrossRef]

38. Kagan, J.C.; Su, T.; Horng, T.; Chow, A.; Akira, S.; Medzhitov, R. TRAM couples endocytosis of Toll-like receptor 4 to the induction of interferon-beta. *Nat. Immunol.* **2008**, *9*, 361–368. [CrossRef]

39. Leifer, C.A.; Kennedy, M.N.; Mazzoni, A.; Lee, C.; Kruhlak, M.J.; Segal, D.M. TLR9 is localized in the endoplasmic reticulum prior to stimulation. *J. Immunol.* **2004**, *173*, 1179–1183. [CrossRef]

40. Leifer, C.A.; Brooks, J.C.; Hoelzer, K.; Lopez, J.; Kennedy, M.N.; Mazzoni, A.; Segal, D.M. Cytoplasmic targeting motifs control localization of toll-like receptor 9. *J. Biol. Chem.* **2006**, *281*, 35585–35592. [CrossRef]

41. Chockalingam, A.; Brooks, J.C.; Cameron, J.L.; Blum, L.K.; Leifer, C.A. TLR9 traffics through the Golgi complex to localize to endolysosomes and respond to CpG DNA. *Immunol. Cell Biol.* **2009**, *87*, 209–217. [CrossRef] [PubMed]

42. Sabio, G.; Davis, R.J. TNF and MAP kinase signalling pathways. *Semin. Immunol.* **2014**, *26*, 237–245. [CrossRef] [PubMed]

43. Gaestel, M.; Kotlyarov, A.; Kracht, M. Targeting innate immunity protein kinase signalling in inflammation. *Nat. Rev. Drug Discov.* **2009**, *8*, 480–499. [CrossRef]

44. Thamphiwatana, S.; Angsantikul, P.; Escajadillo, T.; Zhang, Q.; Olson, J.; Luk, B.T.; Zhang, S.; Fang, R.H.; Gao, W.; Nizet, V.; et al. Macrophage-like nanoparticles concurrently absorbing endotoxins and proinflammatory cytokines for sepsis management. *Proc. Natl. Acad. Sci. USA* **2017**, *114*, 11488–11493. [CrossRef] [PubMed]

45. Park, J.; Zhang, Y.; Saito, E.; Gurczynski, S.J.; Moore, B.B.; Cummings, B.J.; Anderson, A.J.; Shea, L.D. Intravascular innate immune cells reprogrammed via intravenous nanoparticles to promote functional recovery after spinal cord injury. *Proc. Natl. Acad. Sci. USA* **2019**, *116*, 14947–14954. [CrossRef] [PubMed]

46. Saito, E.; Kuo, R.; Pearson, R.M.; Gohel, N.; Cheung, B.; King, N.J.C.; Miller, S.D.; Shea, L.D. Designing drug-free biodegradable nanoparticles to modulate inflammatory monocytes and neutrophils for ameliorating inflammation. *J. Control. Release* **2019**, *300*, 185–196. [CrossRef]

47. Smarr, C.B.; Yap, W.T.; Neef, T.P.; Pearson, R.M.; Hunter, Z.N.; Ifergan, I.; Getts, D.R.; Bryce, P.J.; Shea, L.D.; Miller, S.D. Biodegradable antigen-associated PLG nanoparticles tolerize Th2-mediated allergic airway inflammation pre- and postsensitization. *Proc. Natl. Acad. Sci. USA* **2016**, *113*, 5059–5064. [CrossRef]

48. Pitt, C.G.; Gratzl, M.M.; Kimmel, G.L.; Surles, J.; Schindler, A. Aliphatic polyesters II. The degradation of poly (DL-lactide), poly (epsilon-caprolactone), and their copolymers in vivo. *Biomaterials* **1981**, *2*, 215–220. [CrossRef]

49. Gopferich, A. Mechanisms of polymer degradation and erosion. *Biomaterials* **1996**, *17*, 103–114. [CrossRef]

50. Williams, D.F. Enzymic Hydrolysis of Polylactic Acid. *Eng. Med.* **1981**, *10*, 5–7. [CrossRef]

51. Nakamura, K.; Tomita, T.; Abe, N.; Kamio, Y. Purification and characterization of an extracellular poly(L-lactic acid) depolymerase from a soil isolate, Amycolatopsis sp. strain K104-1. *Appl. Environ. Microbiol.* **2001**, *67*, 345–353. [CrossRef] [PubMed]

52. da Silva, D.; Kaduri, M.; Poley, M.; Adir, O.; Krinsky, N.; Shainsky-Roitman, J.; Schroeder, A. Biocompatibility, biodegradation and excretion of polylactic acid (PLA) in medical implants and theranostic systems. *Chem. Eng. J.* **2018**, *340*, 9–14. [CrossRef] [PubMed]

53. Schakenraad, J.M.; Hardonk, M.J.; Feijen, J.; Molenaar, I.; Nieuwenhuis, P. Enzymatic activity toward poly(L-lactic acid) implants. *J. Biomed. Mater. Res.* **1990**, *24*, 529–545. [CrossRef] [PubMed]

54. Kellum, J.A.; Song, M.; Li, J. Lactic and hydrochloric acids induce different patterns of inflammatory response in LPS-stimulated RAW 264.7 cells. *Am. J. Physiol. Integr. Comp. Physiol.* **2004**, *286*, R686–R692. [CrossRef]

55. Ranganathan, P.; Shanmugam, A.; Swafford, D.; Suryawanshi, A.; Bhattacharjee, P.; Hussein, M.S.; Koni, P.A.; Prasad, P.D.; Kurago, Z.B.; Thangaraju, M.; et al. GPR81, a Cell-Surface Receptor for Lactate, Regulates Intestinal Homeostasis and Protects Mice from Experimental Colitis. *J. Immunol.* **2018**, *200*, 1781–1789. [CrossRef]

56. Hoque, R.; Farooq, A.; Ghani, A.; Gorelick, F.; Mehal, W.Z. Lactate reduces liver and pancreatic injury in Toll-like receptor- and inflammasome-mediated inflammation via GPR81-mediated suppression of innate immunity. *Gastroenterology* **2014**, *146*, 1763–1774. [CrossRef] [PubMed]

57. Wiley, S.Z.; Sriram, K.; Salmeron, C.; Insel, P.A. GPR68: An Emerging Drug Target in Cancer. *Int. J. Mol. Sci.* **2019**, *20*, 559. [CrossRef] [PubMed]

Review

Superparamagnetic Iron Oxide Nanoparticles for Immunotherapy of Cancers through Macrophages and Magnetic Hyperthermia

Alexandre M. M. Dias [1,†], Alan Courteau [1,2,†], Pierre-Simon Bellaye [1,3], Evelyne Kohli [3,4], Alexandra Oudot [1], Pierre-Emmanuel Doulain [5], Camille Petitot [1], Paul-Michael Walker [1,2,4], Richard Decréau [6] and Bertrand Collin [1,6,*]

[1] Centre George-François Leclerc, Service de Médecine Nucléaire, Plateforme d'Imagerie et de Radiothérapie Précliniques, 1 rue du Professeur Marion, 21079 Dijon, France
[2] ImViA Laboratory, EA 7535, University of Burgundy, 21000 Dijon, France
[3] UMR INSERM/uB/AGROSUP 1231, Labex LipSTIC, Faculty of Health Sciences, Université de Bourgogne Franche-Comté, 21079 Dijon, France
[4] University Hospital Centre François Mitterrand, 21000 Dijon, France
[5] Synthesis of Nanohybrids (SON) SAS, 21000 Dijon, France
[6] Institut de Chimie Moléculaire de l'Université de Bourgogne, UMR CNRS/uB 6302, Université de Bourgogne Franche-Comté, 21079 Dijon, France
* Correspondence: bertrand.collin@u-bourgogne.fr; Tel.: +33-3-4534-8071
† These authors contributed equally to this work.

Abstract: Cancer immunotherapy has tremendous promise, but it has yet to be clinically applied in a wider variety of tumor situations. Many therapeutic combinations are envisaged to improve their effectiveness. In this way, strategies capable of inducing immunogenic cell death (e.g., doxorubicin, radiotherapy, hyperthermia) and the reprogramming of the immunosuppressive tumor microenvironment (TME) (e.g., M2-to-M1-like macrophages repolarization of tumor-associated macrophages (TAMs)) are particularly appealing to enhance the efficacy of approved immunotherapies (e.g., immune checkpoint inhibitors, ICIs). Due to their modular construction and versatility, iron oxide-based nanomedicines such as superparamagnetic iron oxide nanoparticles (SPIONs) can combine these different approaches in a single agent. SPIONs have already shown their safety and biocompatibility and possess both drug-delivery (e.g., chemotherapy, ICIs) and magnetic capabilities (e.g., magnetic hyperthermia (MHT), magnetic resonance imaging). In this review, we will discuss the multiple applications of SPIONs in cancer immunotherapy, focusing on their theranostic properties to target TAMs and to generate MHT. The first section of this review will briefly describe immune targets for NPs. The following sections will deal with the overall properties of SPIONs (including MHT). The last section is dedicated to the SPION-induced immune response through its effects on TAMs and MHT.

Keywords: cancer; immunotherapy; superparamagnetic iron oxide; nanoparticles; macrophages; magnetic hyperthermia; theranostics

Citation: Dias, A.M.M.; Courteau, A.; Bellaye, P.-S.; Kohli, E.; Oudot, A.; Doulain, P.-E.; Petitot, C.; Walker, P.-M.; Decréau, R.; Collin, B. Superparamagnetic Iron Oxide Nanoparticles for Immunotherapy of Cancers through Macrophages and Magnetic Hyperthermia. *Pharmaceutics* **2022**, *14*, 2388. https://doi.org/10.3390/pharmaceutics14112388

Academic Editor: Xiangyang Shi

Received: 22 July 2022
Accepted: 25 October 2022
Published: 5 November 2022

Publisher's Note: MDPI stays neutral with regard to jurisdictional claims in published maps and institutional affiliations.

1. Introduction

Cancer ranks as a leading cause of death and an important barrier to increasing life expectancy in every country of the world [1]. Cancer is the first or second leading cause of death before the age of 70 years in a vast majority of countries [2], underlining the urgent need to address unmet needs in oncology. According to the type and stage of cancer, various approaches can be employed. While surgery is usually the first line of treatment, other strategies based on chemotherapy and radiotherapy can also be performed. Even if all these strategies can be combined, the desired success rate in cancer treatment has not yet been achieved, especially due to the iatrogenic disorders they induce. As a consequence, many therapies have been developed to specifically and safely target cancers.

Among these targeted strategies, cancer immunotherapies have now revolutionized the field of oncology by prolonging the survival of more and more patients suffering from aggressive and fatal cancers [3]. In immunotherapy, the agents are designed to induce an immune response against cancer cells and can be used in combination, strengthening their central role as a first-line therapy for many cancers in the future. Immunotherapies can be divided into several classes: (i) immune checkpoint inhibitors (ICIs) [4]; (ii) chimeric antigen receptor (CAR) cell therapies (e.g., CAR Natural Killers (CAR NK) [5]; CAR Macrophages (CAR M) [6] and CAR T-Cells [7]; (iii) cytokines-based immunotherapy [8]; (iv) agonistic antibodies against costimulatory receptors [9]; (v) cancer vaccines [10] and (vi) bispecific antibody therapy.

Another very exciting field to specifically and safely target tumors for diagnosis and therapy relies on the use of nanoparticles (NPs), also called nanomedicines. Their size typically ranges between 1 and 100 nm, they can be made from different materials and have various physicochemical properties (e.g., size, shape, surface features, magnetism, etc.). According to their chemical composition, NPs can be classified into organic (e.g., liposomes, polymeric micelles, dendrimers, etc.), inorganic (e.g., super paramagnetic iron oxide NPs (SPIONs), gold nanorods, carbon nanotubes, etc.), or hybrid (e.g., lipid-polymer NPs, organic-inorganic NPs, etc.) NPs [11]. In addition to their intrinsic properties due to the material they are made of, NPs can be modified with a lot of targeting ligands, affecting their biological behavior accordingly.

Even if it is always discussed, it is commonly accepted that NPs target tumors via two main mechanisms. The first one is passive targeting (enhanced permeability and retention (EPR)). There are a few points about the EPR effect that should be made clear. Despite the fact that the EPR effect is frequently described as a process that enables the delivery and retention of drugs at cancerous sites thanks to structural and architectural abnormalities (such as abnormal fenestrations and structural disorganization), the truth is that this increase in permeability and retention is not yet fully understood and may have other explanations. For instance, it is currently understood that this effect, is also influenced by the impairment of lymphatic drainage and permeability-enhancing factors, including nitric oxide, bradykinin, or vascular endothelial growth factors [12]. Moreover, additional phenomena, such as vascular transcytosis-based nutritional pathways (mediated by caveolae, clathrin-coated pits, and macropinocytotic vesicles), may potentially play a role in NP uptake and, subsequently, the EPR effect, especially for NPs with a size between 50 and 100–150 nm [13]. A second transcytosis pathway, known as the vesiculo-vascular organelle (VVO), has also been identified in normal endothelial cells and may potentially contribute significantly to the EPR effect. This system is made up of a vast network of grouped and connected cytoplasmic vesicles and vacuoles. Therefore, more investigation is required to understand exactly the biophysical and metabolic mechanisms that result in the extravasation of NPs into the tumor and, ultimately, the EPR effect [12].

The second mechanism by which NPs target tumors is the active targeting through an ad hoc surface functionalization (e.g., targeting peptide) of the NPs [14]. Through these mechanisms of targeting, NPs are well-known for their capabilities to release encapsulated or conjugated bioactive agents within tumors. NPs make it possible to improve the bioavailability of drugs, to combine therapeutic agents with imaging (i.e., nanotheranostics) techniques, or to boost antitumor effects [15]. Over the last 20 years, around 80 nanomedicine products have been approved by the Food and Drug Administration (FDA) and the European Medicines Agency (EMA) in various indications including cancers [16]. Unfortunately, in spite of considerable technological success, nanomedicines have demonstrated modest effects on survival and in some examples, less than other approved therapies [17].

Since the majority of patients do not benefit from the currently available immunotherapies, and they can experience severe adverse events, immunotherapeutic nanomedicines might enhance efficacy while mitigating certain life-threatening toxicities [18]. Moreover, aiming for pharmacological synergy, it is also possible to design new combinations associating classical immunotherapies and nanomedicines to overcome their respective weaknesses [19]. In this respect, iron oxide nanoparticles (IONPs) such as SPIONs may be an appealing class as they combine many features that allow targeting of the immune system and tumors for theranostic purposes [20]. SPIONs are typically made up of magnetite (Fe_3O_4) or maghemite (γ-Fe_2O_3) with a core radius ranging from 5 to 15 nm and a hydrodynamic radius (i.e., core with shell and water coat) ranging from 20 to 150 nm [21]. These SPIONs have already been demonstrated to act as advanced platforms for drug delivery and contrast agents in magnetic resonance imaging (MRI) and magnetic hyperthermia (MHT) [22]. Very recently, the theranostic potential of IONPs in cancer immunotherapy has been reported, emphasizing their ability to perform tumor imaging for early assessment of the efficacy of immunotherapy and their capability to alter macrophage polarization [20]. Moreover, more and more studies have demonstrated that SPIONs exhibit the intrinsic capability to stimulate systemic antitumor immune responses through MHT, paving the way for new immunotherapeutic strategies [19].

In this review, we will discuss the multiple applications of SPIONs in cancer immunotherapy, focusing on their intrinsic theranostic properties to target tumor-associated macrophages (TAMs) and to generate MHT in light of their effects on anticancer immunity. The first section of this review will briefly describe immune targets for NPs. The following sections will deal with the overall properties of SPIONs, including the development of MHT. Next, we will see how SPIONs can induce an immune response through the targeting of TME, with a more in-depth focus on TAMs and MHT.

2. Immunity, Cancer, and SPION Nanoparticles

2.1. Immune Targets in Cancer

2.1.1. Immune Cells and Tumor Microenvironment at a Glance

It is now well-established that immune evasion is a hallmark of cancer [23]. This concept is related to cancer immunoediting, comprising three processes: elimination, equilibrium, and escape [24]. Immune cells are notably present within the TME, a complex network made up of numerous cellular (e.g., vascular, stroma cells) and non-cellular (e.g., extracellular matrix, ECM) components, other than tumor cells. These immune cells (from both innate and adaptive immunity) can either promote or prevent tumor growth. Tumor-promoting immune cells include regulatory T cells (Tregs), myeloid-derived suppressor cells (MDSCs), and tumor-associated macrophages (TAMs). Conversely, tumor-preventing immune cells include CD4+ T helper cells (TH), CD8+ cytotoxic T lymphocytes (CTLs), and natural killer (NK) cells [25]. For a greater insight into the functions of the various immune cells, readers are directed to a comprehensive review conducted by Doshi and Asrani [26]. Since TME is a very immunosuppressive milieu, it seems particularly relevant to pharmacologically activate immune cells (within lymphoid organs or the TME itself) or to target immunosuppressive cells or the combination of these approaches [25]. Interestingly, these strategies can be carried out to potentially target all immune "compartments" (i.e., TME, circulation, and myeloid/lymphoid tissues) since the immunological imbalance in cancer goes beyond the primary tumor [27].

2.1.2. Innate and Adaptive Immunity in Cancer

Both innate and adaptive immunity are crucial components in cancer development and progression and the overall immune response relies on the interplay between them. Innate immunity involves various types of myeloid cells: dendritic cells (DCs), monocytes, macrophages, polymorphonuclear leukocytes (PMNs), mast cells, NKs, and natural killer T (NKT) cells [26]. The innate immune system can directly inhibit tumor progression by engaging tumoricidal activity with NKs (recognition of tumor-derived antigens), granulocytes, and macrophages through antibody-dependent cellular cytotoxicity (ADCC) or antibody-dependent cellular phagocytosis (ADCP) [28]. Conversely, the innate immune system can also contribute to immunosuppression. A major example is represented by pro-tumorigenic M2-TAMs, which express multiple immunosuppressive (e.g., prostaglandin

E2, IL10) and tumor-promoting factors leading to suppressed anti-tumor responses [29]. In addition to innate immunity, we find cells from the adaptive immune system, i.e., T-cells and B-cells, whose aim is to eradicate cancer or to inhibit their proliferation through cellular and humoral immunity, respectively. This anticancer response relies notably on the cancer immunity cycle (CIC), a process which can be divided into seven stages starting with cancer antigen release (step 1) and finishing with killing cancer cells (step 7: immunogenic cell death, ICD) through CTLs [30]. Thus, CIC can be self-propagating, leading to an accumulation of immune-stimulatory factors that in principle should amplify and broaden T-cell responses. CIC is also characterized by immune regulatory feedback mechanisms capable of stopping or lowering the immune response. Physiologically, immune tolerance regulating immune responses and preventing tissue damage is mediated by immune check-points which are negative regulators of T-cell activation. They refer to immunosuppressive molecules which can be highly expressed in cancer, mediating tumor immune evasion. The main immune checkpoints are cytotoxic T lymphocyte antigen 4 (CTLA-4, expressed on the activated CD8+ and CD4+ T cells), programmed cell death protein 1 (PD-1, expressed in myeloid, B- and activated T cells), and programmed cell death ligand 1 (PD-L1, myeloid and cancer cells). These immune checkpoints have given rise to one of the most important immunotherapies based on their inhibition: the ICIs [31]. LAG3, TIGIT, and TIM3 are other checkpoint signaling molecules, among many others, extensively studied to understand their role in T-cell functions and their potential as new immunotherapies for cancer [32]. It is important to point out that innate and adaptive immunity are tightly connected, notably due to the involvement of antigen-presenting cells (APCs: DCs, macrophages) and complement proteins, resulting in the activation of a T-cell response and immunological memory [26]. This underlines the great interest in using therapeutic strategies capable of targeting both innate and adaptive immunity through, for example, "all-in-one" modalities such as nanomedicines [33], radiotherapy [34], hyperthermia [35], or various synergistic combinations [36,37].

2.2. Opportunities for Targeting Immune System with SPION Nanoparticles

2.2.1. Rational for Targeting Immune System with SPION Nanoparticles

As previously mentioned, in spite of breakthrough advances due to immunotherapy for cancer treatment, there is an urgent need to overcome some major limitations. Several reasons can be mentioned to explain some issues related to these treatments [31]. First, there is an intrinsic variability between patients' immune systems, especially in a context in which they may be immunocompromised by treatments (radiotherapy, chemotherapy), leading to low response rates. Then, as with any cancer therapy, resistance development is inevitable and can be classified as extrinsic (i.e., related to the patient's gender, TME, gut microbiota) or intrinsic resistance due to the nature of the tumor itself (i.e., "cold" versus "hot" tumors). Finally, safety concerns have been frequently reported through immune-related adverse events (irAEs). Indeed, boosting the innate and/or adaptive immune system has a unique set of inflammatory side effects, which can be life-threatening. In this context, nanomedicines designed to enhance antitumor immunity in a variety of ways might represent an interesting alternative, combining efficacy and safety, alone or in combination with other anticancer strategies [38]. Even if anticancer nanomedicines may enhance tumor targeting, the therapeutic responses cannot be guaranteed, especially when they are used in monotherapy. Indeed, we can observe relapse resulting from the re-establishment of pro-tumorigenic conditions (e.g., progenitor immune cells, re-activation of cancer stem cells) [27]. This underlines the need to develop more holistic approaches, notably based on the immune system, taking into account growth-promoting phenomena that occur inside and outside of tumor tissue. So, according to their design, nanomedicines could take advantage of their numerous properties (e.g., targeting moieties, drug payloads, intrinsic properties such as magnetism) to specifically target the immune system while being able to evade clearance from the bloodstream and reticuloendothelial system (RES). Taken together, these data emphasize the interest in using nanomedicines alone or in combination to target, engage, and modulate immune cells in the TME, circulation, and immune cell-enriched tissues [27].

2.2.2. Nanoparticles to Target Immune TME (iTME)

Very nice and comprehensive reviews related to this topic have been recently published [31,39]. Overall, a lot of different strategies exist in order to target immunity within the TME, especially through its immunosuppressive properties.

The most studied immunosuppressive strategy relies on the inhibition of immune checkpoints, especially in a clinical context with the use of monoclonal antibody-based ICIs targeting CTLA-4, PD1, and PD-L1 [3]. So far, various NPs (organic/inorganic) have been designed with success to deliver ICIs (e.g., siPD-L1, anti-PDL1, anti-PD1, and anti-CTLA-4) in preclinical models [38]. Many other ways to target the immune checkpoints synergistically with ICIs have been performed in combination with various therapeutic modalities such as photothermal therapy (PTT), photodynamic therapy (PDT), radiodynamic therapy, sonodynamic therapy, genetic manipulations, and stimulatory agonists [31]. Another way to remove immunosuppression of the TME is to target indoleamine 2,3-dioxygenase 1 (IDO1), an enzyme producing immunosuppressive metabolites. This enzyme has been shown to be overexpressed in many cancers and many inhibitors have been designed so far. In this context, a prodrug nanoplatorm (approximately 40 nm) has been designed by integrating a PEGylated IDO1 inhibitor (epacadostat) and a photosensitizer (indocyanine green, ICG). Both in vitro and in vivo (B16-F10 cells), the authors demonstrated good efficacy of this strategy, especially in combination with PD-L1 checkpoint blockade [40]. Another way to remove immunosuppression is to reprogram immunosuppressive cells such as M2-like TAMs (*vide infra*) and MDSCs. Phuengkham et al. have targeted both TAMs and MDSCs. They encapsulated resiquimod (TLR7 and 8 agonist) and doxorubicin (to induce ICD) within crosslinked collagen-hyaluronic acid scaffolds. Interestingly, there was subsequent polarization from M2-like to M1-like TAMs associated with the reprogramming of MDSCs into tumor-killing APCs [41].

Another relevant strategy relies on the activation of DCs with nanomedicines. Since DCs are essential to the initiation of the anti-tumor immune responses through the CIC, stimulating their activation has attracted a lot of attention lately. The first strategy to activate DCs is the agonism of the stimulator of interferon genes (STING), a cytosolic receptor mainly localized in the endoplasmic reticulum. When the innate immune system detects DNA from viruses or tumors, cGMAP (an agonist of STING) is produced and activates cells such as DCs, which in turn release type I Interferon (IFN-I). To mimic this mechanism, Wang-Bishop et al. developed a polymeric NP ("polymersome") encapsulated with cGMAP. These STING-activating NPs were able to induce the expression of IFN-I through DCs stimulation. This was associated with ICD, a remodeling of iTME, and a subsequent inhibition of the tumor growth in neuroblastoma tumor-bearing mice [42]. Another approach to activating DCs relies on cancer vaccines (cellular, protein/peptide, and genetic vaccines), through the use of various NPs. This topic has been recently and extensively reviewed [43].

2.2.3. Nanoparticles to Target Circulating Immune Cells

Before homing on diseased areas, nanomedicines can be recognized and interact with circulating immune cells in the bloodstream. In this case, immune cells loaded with nanomedicines become drug carriers, which significantly extend the circulation time of nanoparticles with broad-spectrum tumor-targeting properties. Interestingly, immune cells can cross many biological barriers and are natural carriers due to their homing characteristics (e.g., inflammatory sites, tumors). To do so, monocytes-macrophages, lymphocytes, and neutrophils might represent the most favorable options. Nevertheless, these cells are quite difficult to extract and are not necessarily optimal for drug delivery (e.g., low drug loading efficiency, changes in cell function after drug loading, and degradation of drugs by cellular enzymes).

According to the loading technique, immune cells are simply divided into two categories: "backpacks" (i.e., onto cells through adsorption, ligand-receptor interaction, and chemical coupling) and "Trojan horses" (i.e., into cells through hypotonic hemolysis, electroporation, membrane encapsulation, and phagocytosis). These different aspects have been extensively reviewed by Zhang et al. [44]. As an example, the Trojan horse strategy has been already used with monocytes as cellular vehicles for the co-transport of oxygen-loaded polymer bubbles/a photosensitizer (chlorin e6) and SPIONs to target hypoxic tumors with photodynamic therapy (PDT). Following activation by an external high-frequency magnetic field (HFMF), the co-entrapped SPIONs induced the thermal ablation of murine prostate (Tramp-C1 tumor-bearing mice) while inducing the release of oxygen available for the PDT effect [45].

2.2.4. Nanoparticles to Target Myeloid and Lymphoid Immune Cell-Enriched Tissues

Nanomedicine is well-known to be uptaken by the RES, and various strategies have been documented to overcome this major drawback [46]. RES is a part of the immune system composed of phagocytic cells found in the spleen, liver, lungs, bone marrow, and lymph nodes. So, it is important to consider that targeting immune cells also implies delivering drugs to these immune-cell-rich organs [27]. Nevertheless, this apparent drawback may be advantageous to target immune cell-enriched tissues for both diagnosis and therapy. Indeed, in various cancers, nanoparticles are administered subcutaneously to target lymph nodes for preoperative imaging and intraoperative detection (radioactivity, fluorescence, magnetism).

As an example, a novel mannose-labeled SPION was recently developed (maghemite iron oxide core) to target lymph node resident macrophages, making it possible to perform lymph node imaging in pigs with a substantial percentage of accumulated iron (83%) [47]. Moreover, it is also possible to target RES with NPs to elicit a personalized anti-cancer response through various lipid NP platforms, allowing the targeted delivery of mRNA or gene editing in a tissue-specific manner [27].

3. SPIONs: An Overview

3.1. Biophysical Properties of Superparamagnetic Materials

In this section, we introduce the notion of magnetism and present the specific advantages of superparamagnetic materials for MRI and hyperthermia treatment.

All matter exhibits magnetic properties [48–50]. However, purely diamagnetic materials made of atoms with filled electron shells exhibiting no magnetic moment must be distinguished from atoms containing unpaired electrons generating magnetism. In other words, the electronic configuration of atoms and the collective behavior of individual atomic magnetic moments in a material allows us to classify materials into different magnetic types, as summarized in Table 1.

3.1.1. Magnetic Behavior of Ferromagnetic and Ferrimagnetic Materials

Ferrimagnetism and ferromagnetism are the magnetism types of interest for medical or industrial applications, thanks to the strong magnetic response they provide. Ferromagnetic and ferrimagnetic materials show a similar temperature dependence of magnetization and the ability, under particular conditions, to exhibit a non-zero net magnetization at zero magnetic fields [48].

The ferromagnetic materials (e.g., Fe, Ni, or Co) are characterized by strong (negative) exchange interactions which are opposed to the thermal agitation effect [51]. As a consequence, the atomic magnetic moments undergo a parallel self-alignment inducing a spontaneous magnetization even in the absence of an external magnetic field. Ferrimagnets are characterized by an anti-alignment of atomic magnetic moments of non-equal magnitudes. Iron oxides such as magnetite (Fe_3O_4) and ferrites are examples of ferrimagnetic materials [48,52]. The spontaneous magnetization of these materials remains true below the Curie temperature (Tc) for ferromagnets and below the Néel temperature for ferrimagnets [51,53,54]. Above these critical temperatures, the thermal energy overcomes the exchange interactions and the material is then a paramagnet [48].

Table 1. Types of magnetic materials and their response to the external magnetic field.

Type	Spontaneous Magnetization	Description
Diamagnetism	No	Electron magnetic moment compensation. Magnetic interactions within atoms. No exchange magnetic interaction between atoms and molecules. Weakly repelled by magnetic fields.
Paramagnetism	No	Presence of unpaired electrons in the electronic configuration. Weakly attracted by magnetic fields.
Antiferromagnetism	No	Antiparallel ordered magnetic moments. Canting of magnetic moments leading to the appearance of small net magnetization along the direction of the applied magnetic field.
Ferrimagnetism	Yes	Antiparallel unbalanced magnetic moments. Small net magnetic moment at zero applied magnetic field.
Ferromagnetism	Yes	Parallel magnetic moments. Strong net magnetic moment at zero applied magnetic field.

The understanding and observation of the microscopic magnetic structure of ferromagnetic materials began in the middle of the 20th century [50,55,56]. Ferromagnetic materials are organized in small regions within which the magnetic moments are aligned in parallel, whereas the net magnetization of all regions is null in the absence of an external magnetic field. These regions, called magnetic domains, are separated by micrometric magnetic walls in which the orientation disrupts by 90° or 180° [51]. In the absence of a magnetic field, every domain can have a specific orientation [57,58]. The domain formation relies on a combination of exchange interactions and other contributions such as magnetostatic energy (the energy inherent to time-independent magnetic fields) allowing minimization of its total magnetic energy [50,51,55].

The external magnetic field progressively forces the domains to align in the direction of the field. As a consequence, the domains with a direction close to the applied magnetic field grow to the detriment of others, until all domains finally align in this direction [51]. At this stage, saturation magnetization is achieved [54]. Ferromagnets and ferrimagnets exhibit a non-linear relation between the applied magnetic field intensity H and the resulting magnetization M. M depends on the history of the applied magnetic field [59], or in other words, the magnetization curve of a material does not follow the same path when applying and removing the external magnetic field. Plotting M versus H leads to a hysteresis loop, reproducible in consecutive H cycles [58]. A part of the magnetic moment alignment remains after the magnetic field is removed. This is expressed by the remanence value M_R [48,54] located at the intersection of the hysteresis curve with the ordinate axis in Figure 1. To nullify magnetization, a reverse magnetic field must be applied, reported by the coercivity coefficient [52].

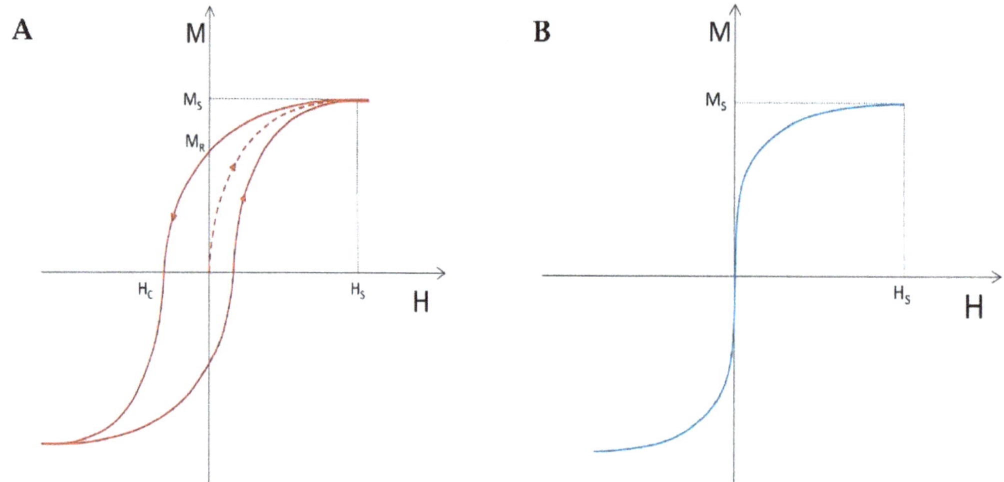

Figure 1. (**A**) Schematic illustration of the typical hysteresis curve of a ferromagnetic material. Starting at field H = 0, M increases towards the saturation magnetization M_S (dotted line) and then decreases following a non-reversible path. M_R represents the remanent magnetization obtained when H reaches zero. H_C represents the coercivity, i.e., the field to apply to nullify the magnetization. The open loop area represents the hysteresis energy losses in the material during the reversal process (heat production). (**B**) Typical magnetization curve of a superparamagnetic material, characterized by the absence of coercivity and hysteresis.

3.1.2. From Ferromagnetic and Ferrimagnetic to Superparamagnetic Behavior

As the size of the particles composing a ferromagnet or ferrimagnet decreases, the amount of energy required to create domain walls in this material increases. Below a critical diameter, the coercivity of the material tends to zero due to the anisotropy energy reduction [51]. When the ferromagnetic or ferrimagnetic particle diameter is small enough (of the order of 100 nm or smaller [57,60,61]), and while the thermal energy overcomes the anisotropy energy, the assembly of individual spin magnetic moments behaves as a single super-spin [62] and the particle exhibits a single magnetic domain structure [52,57]. Although the existence of single-domain ferromagnets was predicted in the 1930s [63], important theoretical advances notably by Néel [64], and new measurement methods were required before applications based on this particular magnetism type emerged [65].

In the absence of an external magnetic field, the single-domain magnetization direction is determined by the magnetocrystalline anisotropy of single-domain nanoparticles, which represents the easy (preferred), intermediate, and hard magnetization directions [51]. In the case when the net magnetization of single-domain particles flips randomly very fast under the influence of thermal fluctuations, the magnetization is nulled [52]. When a magnetic field is applied to them, as shown in Figure 1B, the super-spins of individual particles align in the direction of the magnetic field, and the net magnetization increases rapidly and saturates: a behavior shared with paramagnetism. However, unlike paramagnets, materials of this type have a very high magnetic susceptibility due to the ferromagnetic or ferrimagnetic nature of the super-spin. This remarkable behavior is referred to as superparamagnetism [66]. It must be noted that not all single-domain particles are concerned with superparamagnetism [48].

Superparamagnetic materials exhibit remarkable biophysical properties that have been exploited in the medical field since the late 1970s for diagnostic and therapeutic applications [67–72]. Firstly, their interaction with the protons of water molecules allows them to be used as a contrast agent in MRI. Secondly, when excited by an alternative magnetic field (AMF) at the appropriate frequency and amplitude, they release thermic energy, which led to the development of MHT, suitable for cancer treatment. More generally, they lose their magnetism when the external magnetic field is removed [73]. In addition to their physical properties, the biocompatibility of iron-based particles explains their increasing use in medicine. Indeed, the human body is able to handle, store and eliminate iron, which is used in several physiological processes such as oxygen transport, DNA synthesis, energy production, and metabolism [66].

The biomedical potential of superparamagnetic materials led to the emergence of a new class of biocompatible superparamagnetic agents referred to as SPIONs, or ultra-small SPIONs when their size is up to 50 nm, monocrystalline iron oxide (MION) between 10 nm and 30 nm [74], or sometimes SPIO (for superparamagnetic iron oxide particles) for particles having a diameter greater than 50 nm [75]. For simplification purposes, the appellation of SPION will be used hereafter to designate all magnetic nanometric particles.

Each SPION consists of a core containing water-insoluble magnetite or maghemite crystals made of thousands of paramagnetic Fe ions [75,76] encapsulated in a biodegradable coating, which strongly influences the magnetic properties of the agent. In addition to preventing the aggregation of SPIONs, which increases the risk of vascular embolism, the coating is used to target specific tissues and thus direct its biodistribution. Particles are suspended in a biocompatible fluid before being administered to a subject. The amount of iron ions contained in a single nanoparticle explains the high contrast capabilities of SPION-based agents [61].

3.2. Overview of the Use of SPIONs as MRI Contrast Agents

MRI is a powerful imaging modality for soft tissue imaging as it offers high spatial resolution and tissue discrimination without exposing the subject to ionizing radiations. Hydrogen protons are abundant in our bodies. The proton resonance is obtained by the application of short radio frequency (RF) pulses changing their magnetic moment orientation. After the RF is stopped, relaxation occurs, and magnetic moments realign along their original alignments. The reorientations of spins along the B0 axis, i.e., the spin-lattice relaxation, is characterized by the relaxation time T1, the inverse of which is the relaxation rate R1 = 1/T1 (expressed in s^{-1} or Hz). The disappearance of the magnetization in the transverse plane, named spin-spin relaxation, is characterized by the relaxation time T2 (or relaxation rate R2 = 1/T2). The relaxivity (usually expressed in $s^{-1}mM^{-1}$ or $L.mmol^{-1}.s^{-1}$) expresses the R1 or R2 proton relaxation rate modulation induced by an MRI contrast agent in a biological tissue as a function of its concentration.

SPIONs are commonly employed as the contrast agent to change the tissue relaxation rates of normal or pathological tissues in order to improve the sensitivity and specificity of MRI. SPIONs administered for this purpose should offer high relaxivity and adequate biodistribution without inducing local or systemic toxicity. The effect of SPIONs on T1 and T2 relaxations depends on both the saturation magnetization of the nanoparticles and their interaction with water protons in tissues. The size, shape, and surface coatings of SPIONs strongly modulate their T1 and T2 effects [61]. The physical phenomena resulting in the modification of the spin-lattice and spin-spin relaxation rates in tissues under the influence of SPION nanoparticles are thoroughly detailed elsewhere in the literature [61,75,76]. Briefly, the accumulation of SPIONs in a tissue induces local perturbations of the principal magnetic field of the MRI system, which increase spin dephasing and finally shorten the transverse relaxation time (T2). This arises from the magnetic coupling between protons spin in tissues and the spins of SPIONs. Therefore, the presence of SPIONs causes a negative MRI contrast in tissues.

T2-shortening agents have two major drawbacks: (1) the increased magnetic susceptibility artifacts and (2) the difficult interpretation of low-signal areas which may be confused with bone or vascular structures [77]. This encouraged the development of particles providing contrast in both T1-weighted (T1w) and T2-weighted (T2w) imaging. For instance, gadolinium-labeled magnetite nanoparticles dedicated to positive contrast MR angiography were successfully used in vitro and in vivo by Kellar et al. [78]. More recently, a new class of cubic SPIONs, suitable for use as a dual-mode contrast agent, was presented by Alipour et al. [79]. Although these agents do not yet represent the majority in the literature, their development has been accelerated in recent years, expanding possible diagnostic applications with SPIONs. The typical R2 relaxivity of SPIONs ranges from $100\ \text{s}^{-1}\text{mM}^{-1}$ to a few hundred $\text{s}^{-1}\text{mM}^{-1}$ depending on the characteristics of the particle (composition, coating) and the B0 MRI field [61]. One of the challenges of current studies is to increase the R1 relaxivity) (usually much lower than R2 relaxivity).

The oral administration of SPIONs as a gastrointestinal MRI contrast agent has been considered by several research teams [76]. For instance, Hahn et al. described the improvement of the gastrointestinal tract delineation of MR images provided by a 200 nm SPIO suspended in a low-viscosity food-grade fluid [80]. Their preparation was globally well tolerated by animals and patients. Apart from a few special cases, in the vast majority of studies on the subject since the late 1980s [75], SPIONs are administered intravenously [75]. Unlike low molecular weight water-soluble agents such as gadolinium chelates, SPIONs are usually not transferred to the extracellular-extravascular compartment in healthy subjects and are rapidly eliminated by the RES [67,69,70]. As a consequence, their biodistribution is characterized by a short biological half-life and a significant accumulation in the RES (typically: liver, spleen, bone marrow). As a first approach, SPIONs can thereby be used to enhance malignant lesions within organs of RES [80]. For instance, Weissleder et al. used SPIONs to detect focal splenic tumors with MRI, leading to an important step forward in this domain since the other existing imaging techniques do not provide contrast between such lesions and healthy tissues [67].

On the other hand, the phagocytosis of SPIONs allows the visualization of tissues infiltrated by macrophages during inflammatory processes, which would not be possible with gadolinium-based contrast agents, not internalized by immune cells [81]. Macrophages are the key component of acute inflammation [35]. When an infectious agent is detected, an immune response is set up resulting in vasodilation, higher vascular permeability, and infiltration of free fluid and immune cells (neutrophils and macrophages) in tissues [81,82]. These phenomena are followed by the formation of a fibrotic scar. MRI procedures taking advantage of the phagocytosis of SPIONs for the detection of inflammatory areas and infectious foci, and more generally for the assessment of immune-mediated disorders, were described in the mid-2000s [82,83]. Stoll et al. described the interest in SPIONs in the assessment of central nervous system inflammations [84]. Sillerud et al. detected amyloid-β plaques in a transgenic mouse model of Alzheimer's disease [85]. Ruehm et al. described, in a preclinical assay, the interest in SPIONs as a marker of atherosclerosis (chronic inflammatory response to a vascular wall injury) [86].

Over the last few decades, significant progress has been made in the design of SPIONs, such as the reduction in the average size of these nanoparticles, the improvement of their physico-chemical characteristics, the incorporation of innovative coatings, and especially their surface functionalization. For instance, by decreasing the diameter of their SPIONs to the size range of plasma proteins (i.e., around 10 nm), Weissleder et al. increased their biological half-life and facilitated their transcapillary passage to the interstitium. As a result of these improvements, SPIONs were progressively promoted to the rank of multimodal theranostic nanoprobes with, among others, applications in MHT treatment and immunotherapy. It has been established that MRI examinations performed after the administration of SPIONs offer higher sensitivity and specificity than non-injected MR acquisitions in the diagnosis of lymph node metastasis [87].

SPIONs can also help to distinguish infectious masses from cancerous tumors [81]. Preclinical [88] and clinical studies attested to the benefits of such examinations in axillary node metastases detection in breast cancer patients [87]. Other authors proved the interest in SPIONs for cardiovascular system explorations. Majundar et al. showed the blood-to-background nuclear magnetic resonance (NMR) signal ratio improvement provided by a 72 nm SPION used in rat brain perfusion imaging [72]. Antonelli et al. reported the use of

SPIONs to image atrioventricular fistulas, chronic venous occlusions, and lower extremity arteries [74]. SPIONs internalized by macrophages were also reported as a possible contrast agent of the vascular phase, providing cardiovascular applications such as perfusion and viability imaging [75,78,89]. Moreover, for assessing the inflammatory microenvironment of primary/metastatic tumors and for monitoring the therapeutic response of cancer patients receiving radiotherapy and immunotherapy, non-invasive imaging of TAMs with SPION may offer considerable potential [90].

Limitations to the SPION-enhanced MRI examinations have been mentioned in the literature [81,91]. These imaging procedures are long compared with other imaging modalities, potentially causing a greater motion sensitivity. The concentration of MR probes must reach 0.01 mM to 10 mM for efficient detection [92]. By comparison, in single photon computed emission computed tomography (SPECT) and positron emission tomography (PET) imaging, the tracer can be detected at the picomolar scale [93]. Increased iron levels in the body can lead to tissue damage through oxidative injury. This must be taken into account for repeated examinations or longitudinal studies [81], or if the iron clearance rate of the subject is altered. Particles remaining after a SPION-enhanced acquisition can also cause major susceptibility artifacts and interfere with other MR acquisitions even several months after the injection for liver MRI [81].

3.3. Basic Aspects of Magnetic Hyperthermia (MHT)

3.3.1. Biological Aspects of Hyperthermia

Tumoral tissues potentially contain necrotic, hypoxic, and low pH areas rendering them resistant to chemotherapy and radiotherapy. Moreover, cells in late phase S (Synthesis i.e., DNA replication) are usually more radioresistant than cells in the M phase (Mitosis) but sensitive to heat [94]. In this context, hyperthermia in conjunction with conventional therapies such as chemotherapy drugs or radiotherapy brings a synergistic therapeutic effect [53] and potentially improves tumoral regression [94]. The therapeutic effect of hyperthermia relies on the fact that cancer cells are more sensitive to heat because of their increased metabolic rate [53]. The exposition of cancer cells to a 40–46 °C temperature induces a thermal shock modifying cellular processes, altering the structure and function of proteins and ultimately promoting apoptosis of exposed cells. In addition, heat stress restores blood flow, permeability, pH, and oxygenation of the tumor microenvironment [53,73] and inhibits the repair of ionizing radiation-induced DNA damage [48]. Moreover, MHT may induce effective and genuine immunogenic tumor cell death as recently demonstrated [95]. MHT is the main hyperthermal strategy currently being developed for therapeutic applications. An optimal hyperthermia treatment allows a high heating efficiency, in a short time, and with a minimum concentration to avoid systemic side effects [54].

3.3.2. Heat Production Mechanisms in MHT

MHT relies on the conversion of magnetic energy into thermal energy by the action of an alternative magnetic field with a frequency usually ranging from 100 to 300 kHz and a moderate amplitude [53,54,66]. Four independent mechanisms contribute to heat production: eddy current loss, hysteresis loss, Néel relaxation loss, and Brown relaxation loss [48,53,57]. The relative contributions of these four effects are determined by particle size, magnetic anisotropy, and fluid viscosity. The specific absorption rate (SAR) expressed in W/kg, quantifies the thermal power dissipation. SAR increases in proportion to the thermal energy released in the material [57]. The physical basis of MHT is well described elsewhere [48,54,57,96].

3.4. Design of SPIONs

3.4.1. General Design of Cancer Nanomedicines

Concerning the general design of a given nanomedicine, the function, characteristics, materials, and method of synthesis are the four factors that must be taken into account. Once clearly established, the defined function is then used to specify the essential characteristics of the nanomedicine. Then, the materials used to build the object and its manufacturing method have a significant impact on its ability to exhibit the desired properties [14]. Thus, the physicochemical properties obtained from a rational design will have a major influence on the biodistribution and pharmacokinetics of a nanomedicine. The major physicochemical properties of a given nanomedicine are the size, the shape, the surface charge, and chemistry. The size of nanomedicines is a major feature dictating their overall biodistribution and cell internalization. As previously mentioned, the suitable diameter for nanomedicines targeting cancers ranges from 10 to 100 nm, making it possible to passively target tumors through the EPR effect.

The surface charge, expressed through zeta potential (i.e., the surface charge of an NP in colloidal suspension), is also a major parameter related to nanomedicine biodistribution due to electrostatic interactions between the NPs and various biological molecules. Unlike negatively charged NPs, positively charged NPs are more prone to be uptaken by cells that may induce side effects (non-targeted cells), decrease a drug delivery process to targeted cells, and shorten the circulation time. Zeta potential may also affect the loading capacity of SPIONs according to the charge of the payload drug. The surface chemistry of nanomedicines aims also to significantly improve their targeting properties. This goal can be achieved through various functionalizations with targeting ligands enhancing the affinity of nanomedicines to significantly increase the uptake within the targeted tissues. These functionalizations can also rely on surface modification with various polymers (e.g., polyethylene glycol, PEG, *vide infra*) to protect the nanomedicines from rapid clearance, subsequently making them stealthy. The clearance phenomenon may also be increased through renal elimination. This can be observed with small NPs (approximately 6 nm), positively charged NPs (through the negative glomerular basement membrane), and rod-shaped NPs [14].

3.4.2. Design of SPIONs Suitable for Hyperthermia and Immune System Targeting

SPIONs can be produced through physical, biological, or chemical routes. Nevertheless, chemical syntheses remain the main way to obtain SPIONs. The co-precipitation method has been the starting point for other approaches such as thermal decomposition methods, hydrothermal methods, solvothermal methods, sol-gel methods, micelle methods, and many other methods [97–99].

The method of synthesis drives the size, shape, colloidal stability, and magnetic properties of the SPIONs. For biomedical applications, SPIONs need to be modified to enhance their stability. This goal can be achieved through the grafting of various polymers such as PEG, polyethyleneimine (PEI), polyacrylic acid (PAA), polyvinylpyrrolidone (PVP), dextran, chitosan, and many others [97].

For instance, polymers such as PEG are now well-known to increase the biocompatibility, colloidal dispersion, and stability of SPIONs while conferring them relative stealthiness towards the RES. [100]. Silanes are also common coating polymers of SPIONs, used for example to modify the surface (aminosilane type shell) of NanoTherm® (size about 15 nm). This is the only SPION approved to treat glioblastoma with MHT induced with AMF [97]. Nevertheless, it is important to underline that NanoTherm® has to be injected directly into the tumor.

Many interesting preclinical/clinical studies on magnetic hyperthermia (MHT) with SPIONs have been carried out, but very few have led to advanced clinical phases [101]. In spite of its well-known efficacy on cancer cells, the main drawback of hyperthermia is its lack of selectivity between healthy and tumor tissues. To overcome this issue, SPIONs might be one of the most promising solutions if tumor targeting techniques are used (i.e., intra-tumor implantation, pharmacological targeting, and/or magnetic field application). In this context, several clinical trials have been performed with SPIONs, especially in the field of prostate cancer and glioblastoma [102]. Even if most studies have proven good efficacy, methodology/instrumentation issues impair the broader use of magnetic hyperthermia. Nevertheless, the case of Nanotherm® is particularly interesting and gives rise to much hope in this field. Indeed, it is the only approved nanoparticle for MHT of glioblastoma (CE marking, approval in 2010 as a class III medical device) and has recently (2021) achieved a new accomplishment in prostate cancer, being allowed by the FDA to move towards a pivotal phase 2b clinical trial.

So, we could consider that the year 2021 is likely to be a turning point for MHT with the arrival in early clinical phases of new iron nanoparticles from the NoCanTher project (RCL-01, a 149 nm iron oxide nanoparticles coated with dextran) to treat locally advanced pancreatic ductal adenocarcinoma [103]. Based on their designs, we can consider that these products developed and used in clinical settings belong to the first generation of SPIONs. Indeed, these SPIONs are based on a magnetic core decorated with an organic coating without any targeting moieties. This justifies their implantation in situ with surgical procedures to achieve MHT with ad hoc devices. Moreover, the cancers addressed by these new therapies are well-known to be particularly challenging from a pharmacokinetics point of view (blood-brain barrier, blood-prostate barrier), justifying once again, the intratumor injection. We could consider that the intratumoral delivery of SPIONs for MHT is the counterpart of what is classically done in the framework of brachytherapy to treat many cancers (gynecological, prostate, and skin).

Based on these recent data, we assume that MHT with SPIONs is still in its infancy, paving the way to future smarter approaches. Thus, the modularity and theranostic capabilities of SPIONs should make it possible to design and develop a new generation of tumor-selective drugs until clinical phases. Ideally, this new generation should be suitable for the intravenous route with an optimal tumor uptake guided with MRI, allowing us to perform a safe and efficient MHT procedure.

So, for reasons of therapeutic refinement, SPIONs may also acquire active targeting capabilities through, for example, surface modification with antibodies, targeting peptides, or any other molecules with biological targeting capability [98]. However, it should be remembered that the changes in the surface of SPIONs may modulate the thickness of the overall surface coating, affecting the performances of T2 relaxation (MRI) and MHT [104,105]. Overall, when designing SPIONs for MHT, a balance must be achieved between the size of the magnetic core to maximize heat release (>10 nm) and colloidal stability in biological media required for intravenous injection (ideally <50 nm) [106]. Shape is also a major parameter to take into account when designing SPIONS for MHT purposes. For example, cubic-shaped SPIONs (from 17 to 61 nm) have been found to be more efficient in vitro to induce MHT when compared to spherical ones [107]. This effect was verified in vivo in subcutaneous A431 tumor-bearing mice, showing that cubic-shaped SPIONs coated with a polymer shell were able to induce effective MHT and heat-mediated chemotherapy [108].

Immune cells involved in the immunotherapy mechanism can be targeted with SPIONs. Thus, with a more or less sophisticated design based on the strategies previously evoked, SPIONs can be used for cancer vaccines, the guidance of magnetized cytotoxic cells to tumor sites, drug delivery of immune checkpoint inhibitors, the polarization of macrophages, and to trigger magnetic hyperthermia [109]. Of course, the modular construction of SPIONs and their magnetic properties allow us to consider combinatorial immunotherapies in the same nanomedicine [110]. Below, we will emphasize these approaches with recent studies given as examples of SPIONs designed for immunotherapeutic uses. For cancer vaccines, strategies based on ovalbumin bound to SPIONs (size around 200 nm, zeta potential around −22 mV) have been successfully evaluated as a vaccine delivery platform and immune potentiator, showing the activation of immune cells and cytokine production [111]. SPIONs can also be used as platforms to magnetically guide immune cells such as T cells to a region of interest. To do so, Ortega et al. designed several SPIONs coated with dimercaptosuccinic acid (DMSA), 3-aminopropyl-triethoxysilane (APS), or dextran (6 kDa). The size of these SPIONs ranged from 82 to 120 nm (zeta potential from −34 to +38 mV) and made it possible to activate in vivo the migration of T cells, loaded with SPIONs, through the application of an external magnetic field [112].

Another major way to target immunity with SPIONs is to target immune checkpoints since they are becoming a standard regimen in oncology. Very recently, Kiani et al. designed sophisticated SPIONs (90 nm, zeta potential of 28.6 mV), covered by chitosan, functionalized with TAT peptide (cell-penetrating peptide) and loaded with siRNA to silence two of the most important T-cell immune checkpoints (PD-1 and A2aR) [113]. These SPIONs significantly inhibited tumor growth (in CT26 and 4T1 mouse tumors) associated with an important anti-tumor immune response and survival time. SPIONs can also be designed to induce the repolarization of M2 to M1 (*vide infra*). In this way, Zhang et al. perform a study with differently charged SPIONs in order to see potential preferential differences in polarizing macrophages [114]. They synthesized three differently charged SPIONS (zeta potentials of +44.72 mV, −0.282 mV, and −27.31 mV for sizes about 19.4 nm, 15.9 nm, and 21.3 nm, respectively). Interestingly, they demonstrated that positively charged SPIONs had the highest cellular uptake and higher macrophage polarization effect (i.e., M2-like macrophages toward M1-like macrophages).

The shape of SPIONs is also an important parameter affecting the immunological response. Among the various existing shapes (e.g., spheres, rods, cubes, etc.) that have been designed so far, octapod- and plate-shaped SPIONs showed a higher immunomodulatory potential. The shape also influences the targeting and uptake within immune cells. For example, the internalization of spherical SPIONs is increased when compared to non-spherical ones. Conversely, at similar size and charge, spherical SPIONs are less efficient at diffusing across the vascular wall when compared to rod- or bar-shaped SPIONs [110].

In the context of immunotherapy, SPIONs are particularly suitable platforms for theranostic combinations. In this way, Wang et al. designed spherical SPIONs suitable for MRI, targeting M2-like macrophages and MHT in breast tumor-bearing mice [115]. They obtained a multifunctional SPION (hydrodynamic diameter of 20 nm), with efficient targeting capability, high relaxivity (149 $s^{-1}mM^{-1}$), and satisfactory magnetic hyperthermia performance in vitro. In vivo MRI showed that M2-targeting SPIONs had a good biodistribution within tumors, also indicating the optimal timing for MHT. The MHT procedure induced both a decrease in the population of M2-like TAMs and tumoral volume associated with iTME remodeling (notably through a significant increase in CTLs). To go further, we invite the reader to consult a recent review related to the enhancement of CD8+ T-Cell-Mediated tumor immunotherapy via MHT used alone or in combination [116].

Due to the intrinsic versatility of nanomedicine, the various data in the literature show that there is no real consensus on the design of SPIONs. This suggests that the design of a given nanoparticle must be thought of in terms of its future application allowing us to imagine the most suitable specifications resulting from an optimal design. Figure 2 summarizes the design process of theranostic SPIONs emphasizing MHT and targeting M2-like tumor-associated macrophages. The first of these steps (Figure 2A) is the synthesis of the magnetic core (bare SPIONs), which influences its magnetic properties. Unless there is a magnetic field, magnetization equals 0. The core radius usually ranges from 5 to 15 nm. Many synthesis methods are available and drive the convenience of manufacturing, the control of shape, size, composition, and the polydispersity index (i.e., estimation of the average uniformity of a nanoparticle solution) of SPIONs. The second step is the surface engineering of SPIONs (Figure 2B). SPIONs can be coated with various organic moieties for biocompatibility (e.g., PEG, chitosan), targeting (e.g., mAbs, peptides), and theranostic (e.g., radionuclides, chemotherapeutics) purposes. Targeting molecules (e.g., carbohydrates such as mannose to target M2-like CD206 receptors) can also be bound to the biocompatible moieties. The surface engineering will influence the hydrodynamic size (i.e., core with shell and water coat—typically between 20 and 150 nm), zeta potential (i.e., the electric charge on the surface of a given nanoparticle, crucial for colloidal stability, typical absolute value: |30| mV), cellular uptake, toxicity, and hydrophilicity. Size also influences the EPR effect (i.e., passive targeting of tumors, up to 100–150 nm). Finally, the theranostic capabilities of SPIONs are assessed (Figure 2C). Due to their intrinsic superparamagnetic properties, the application of a magnetic field makes it possible to perform MRI, MPI, and MHT (with AMF) and concentrate SPIONs within tumors. Interestingly, decorated SPIONs can target tumors and their microenvironment (e.g., M2-like macrophages through their CD206 receptor) to either exert their diagnostic (MRI, multimodal imaging such as PET-MRI, MPI) and/or their therapeutic (MHT, drug delivery) properties according to the design. In the context of immunotherapy, SPIONs might be particularly appealing through the combination in the same agent of immunogenic cell death inducers such as MHT and/or other thermal/phototherapies (e.g., photothermal therapy, photodynamic therapy), chemotherapy (e.g., doxorubicin), and radiotherapy in addition to macrophage repolarization from M2 to M1 phenotype. This combination makes it possible to boost both innate and adaptative immunity against tumors through the production of various tumoricidal mediators (cytokines such as IL1, TNF-α, and reactive oxygen species). Overall, in addition to these outstanding theranostic properties, SPIONs possess other many advantages such as long-term chemical stability, biocompatibility, and safety. Nevertheless, especially for MHT, the targeting strategies need to be improved to achieve a high concentration of SPIONs within targeted tissues to significantly reduce non-specific heating and increase efficacy. Moreover, in the context of clinical perspectives, all metallic material within 40 cm of the treatment area must be removed prior to alternating magnetic field exposure [117,118].

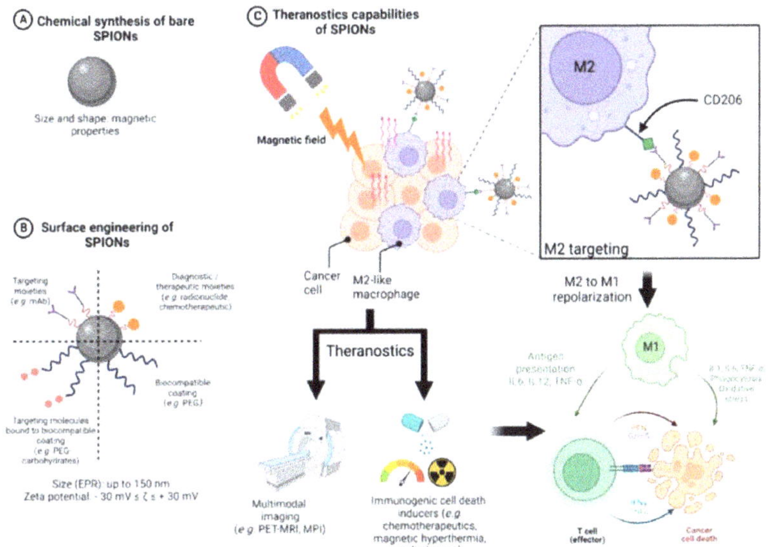

Figure 2. Design and theranostic applications of SPIONs suitable for MHT and macrophage targeting in cancers. Abbreviations. PEG: Poly-Ethylene Glycol; mAbs: monoclonal antibodies; EPR: Enhanced Permeability and Retention; MRI: Magnetic Resonance Imaging; MPI: Magnetic Particle Imaging; PET: Positron Emission Tomography; IL: Interleukin; TNF-α: Tumor Necrosis Factor-α; AMF: Alternating Magnetic Field; PFN: Perforins; GzmB: Granzyme B; IFNγ: Interferony. Created with BioRender.com.

4. Targeting the Immune System with SPIONs

4.1. Magnetic Hyperthermia Based on SPIONs as an Immune Trigger against Tumors

Cancer cells are more sensitive to hyperthermia (elevation of temperature to 40–45 °C) than normal cells [119–121]. This may be because cancer cells have a more accelerated metabolism [122] or because there is poor vascular distribution in cancerous tissue, leading to an accumulation of fever and heat stress [123].

In this sense, several methods of increasing the temperature in order to eradicate tumors have been investigated, such as those based on radiofrequency, microwaves, or ultrasound [124]. It is in this context that SPIONs can be used to generate heat via the use of electromagnetic energy, the so-called MHT [125]. Indeed, as previously seen and thanks to their magnetic properties, when subjected to an AMF, SPIONs are able to produce heat [118]. Furthermore, since SPIONs can be functionalized on their surface with molecules that target cancer cells, it would then be possible to induce localized hyperthermia. This last point is particularly important since a key disadvantage of classical methods of hyperthermia induction is the lack of selectivity [118].

Starting from this premise, only a few clinical trials have been conducted since 2006 to investigate the impact of thermotherapy based on SPIONs on different cancers, mostly glioblastoma and prostate cancer. SPION-based thermotherapy has also been investigated to treat other carcinomas (ovarian, cervical, and rectal) and sarcomas (chondro-, rhabomyo-, and parapharyngeal sarcoma) [126–129]. In general, these studies have shown that it was possible to have an increase in intratumoral temperature thanks to the combination of SPIONs and AMF. For instance, in prostate cancer, maximum temperatures up to 55 °C were reached [127]. Moreover, in glioblastoma, patients' overall survival was improved following MHT treatment [129]. In addition, both of these studies highlighted the fact that only moderate side effects were observed, with no serious complications [128,129].

Recently, a phase 0 clinical trial (NCT02033447) investigating SPIONs-induced MHT with AMF has been completed but, as far as we know, no results have been published so far. Interestingly, another recent phase I (NCT04316091) clinical trial will study MHT in osteosarcoma with SPIONs triggered by spinning magnetic fields (SMF, a new type of magnetic field) in association with neoadjuvant chemotherapy [118]. Despite the fact that the feasibility of SPIONs-induced hyperthermia has been demonstrated at both preclinical and clinical levels, the low number of clinical trials can be partly explained by the fact that this thermotherapy is at the interface of several disciplines (physics, chemistry, biology, medicine, pharmacology) with potential issues to in designing ad hoc SPIONs. Therefore, a better understanding of the mechanism of this therapy in preclinical models, including its action on the immune system, is needed. Indeed, beyond the fact that hyperthermia can directly cause cancer cell death by necrotizing tissues [125], this therapy can also indirectly cause cancer cell death by activating antitumor immunity through ICD [124]. In this sense, Persano et al., in the context of glioblastoma, investigated the impact of magnetic hyperthermia on U87 cells in vitro following an iron oxide nanotube treatment. Interestingly, after thermotherapy, U87 cells displayed a different immunological profile (with an increase in stress-associated signals), making them more likely to be phagocyted by macrophages or killed by NK cells [130].

Other recent studies have demonstrated the impact of SPION-based MHT on the immune system. Carter et al. [125] demonstrated in a subcutaneous syngeneic (GL261 cells, glioblastoma) mouse model (C57BL/6), that magnetic hyperthermia treatment following intratumoral injection of Perimag-COOH SPIONs (dextran-coated, negatively charged and with a hydrodynamic diameter about 130 nm), induced an increase in the proportion of CD8+ T cells within tumors, which is a well-known good prognostic factor [131]. Carter et al. also demonstrated in this mouse model that magnetic hyperthermia treatment was able to reduce tumor growth when compared to control groups [125]. Covarrubias et al. showed in another syngeneic (4T1) mouse model (BALB/c), that IONPs-induced hyperthermia decreased immune cell subpopulations, including those from the innate system (such as neutrophils, dendritic cells, and macrophages) and adaptive system (i.e., CD4+ and CD8+ T cells). Interestingly, subsequent treatment with immune checkpoint inhibitors favored tumor repopulation with the infiltration of innate and adaptive immune cells within tumors [132]. More research is needed to fully assess the effects of SPION-based MHT on the tumor microenvironment. Finally, SPIONs may be useful in treating tumors, in addition to their capacity to cause hyperthermia, by reversing the immunosuppressive tumor microenvironment, which includes, among other things, their influence on macrophage polarization.

4.2. SPIONs and Immunomodulation of the Monocyte-Macrophage Axis

4.2.1. Solid Tumors and TME

Cancers could be divided into two main types, solid and liquid tumors. Both of them are characterized by uncontrolled cell growth. Whereas liquid tumors, also known as blood cancers, can affect blood cells and their precursors [133], solid tumors can occur in many parts of the body and they can be separated into two major groups according to where they originate: carcinoma (epithelial tissue) and sarcoma (connective tissue) [134]. However, in compliance with Global Cancer Statistics 2020, solid tumors alone account for approximately 90% of adult human cancers [1]. Solid tumors are not only composed of cancer cells. Immune cells, such as B and T lymphocytes or macrophages, as well as non-immune cells, including endothelial and stroma cells, are part of a highly complex ecosystem, which directly interacts with cancer cells, called the TME [135]. The TME is also composed of several non-cellular effectors, such as cytokines, chemokines, and the extracellular matrix (ECM) [136]. Moreover, two key hallmarks of the TME include hypoxia, resulting from anarchic neo-angiogenesis and promoting tumor aggressiveness, and immunosuppression, whereby cancer cells manage to escape from immune cells [137,138].

Immunosuppressive effects observed in the TME are sustained by a group of cells, called immunosuppressive cells, such as regulatory T cells, regulatory B cells, MDSCs, and TAMs [139]. TAMs have an important role in cancer progression as they can account for up to 50% of some solid tumors [140]. The vast majority of TAMs exhibit an immunosuppressive and pro-tumoral M2-like phenotype [141]. However, TAMs can also display an M1-like phenotype that could be correlated with tumor regression [142].

4.2.2. Macrophage Polarization

Two major macrophage phenotypes have been described, the classically activated M1 phenotype, characterized by pro-inflammatory properties, and alternatively the activated M2 phenotype, characterized by an anti-inflammatory and a tolerogenic activity [142]. One of the macrophages' fundamental features, besides the fact that they display an important phagocytic activity, is their plasticity. They are the most plastic cells of the entire hematopoietic system [143]. In specific terms, macrophages are able to modify their phenotype according to signals perceived in their environment (cytokines, microbial particles, apoptotic bodies, activated lymphocytes) [144]. One of the current challenges in cancer treatment is to find a way to switch TAMs from an M2-like pro-tumoral into an M1-like anti-tumoral phenotype [145].

Among others, the main *stimuli* of M1 polarization are those triggering a pro-inflamma tory response such as bacterial wall components (Lipopolysaccharide, LPS, and Lipoteichoic acid, LTA), viruses or cytokines (interferon gamma, IFN-γ, and granulocyte-macrophage colony-stimulating factor, GM-CSF). By contrast, the main *stimuli* promoting M2 polarization include interleukins IL-4, IL-13, IL-10, and the cytokine M-CSF (macrophage colony-stimulating factor), which activate a tolerogenic or even anti-inflammatory phenotype [146].

This concept of M1 or M2 phenotype (derived from naïve macrophages or M0) is based on in vitro models (Figure 3) where many polarization markers (Table 2) have been identified [147]. Nevertheless, in vivo, given the complexity of the cellular and cytokine environment (specifically in the TME), M1-like or M2-like macrophage terms are preferentially used [148]. M1 and M2 polarization represent two extremes of the macrophage polarization spectrum [142] between which there are various degrees of polarization towards which macrophages are able to converge according to environmental signals and their concentration [149].

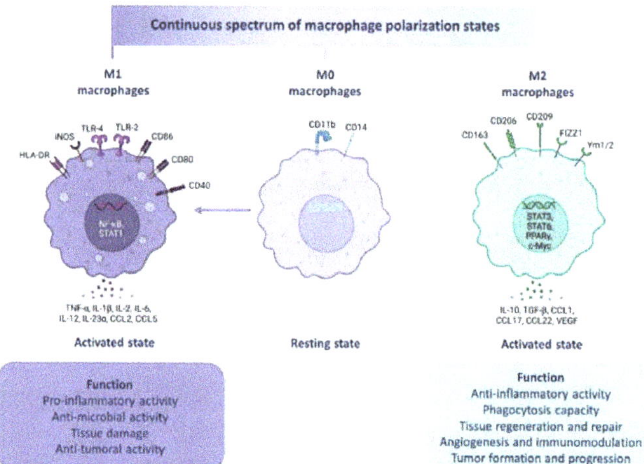

Figure 3. Spectrum of macrophage polarization. Depending on the signals perceived in its milieu (microbial products, damaged cells, cytokines), a naive M0 macrophage can be activated and polarize towards a plethora of different phenotypes. The two extremes of this continuous polarization spectrum are, on the one hand, M1 macrophages, known to have pro-inflammatory activity, and on the other hand, M2 macrophages, known to have anti-inflammatory function. These two extremes were obtained in in vitro models where their polarization markers (such as membrane receptors, transcription factors, cytokines) were identified. However, in vivo, macrophages present in an organism, or in a tumor, will tend towards an M1 or M2 phenotype. Due to the great complexity of the in vivo milieu, these macrophages will never reach the level of polarization that macrophages obtained in vitro. These macrophages in vivo will thus be called M1-like or M2-like [149–152]. Created with BioRender.com.

Moreover, once a macrophage is polarized, this polarization is not definitive. Thus, depending on environmental signals variation, such as a treatment, an M1-like macrophage may switch to an M2-like phenotype or *vice versa*. This phenomenon, based on macrophage plasticity, is known as repolarization [153]. Therefore, treatments promoting the repolarization of TAMs, such as SPIONs, might be a potential therapeutic lead to inhibit cancer development or even contribute to cancer regression in solid tumors.

Table 2. Markers of macrophage polarization.

Molecule Family	Polarization Marker	M0, M1 or M2 Marker	Species	References
Enzyme	Arg1	M2	Murine	[154]
	iNOS	M2	Murine	[154]
Membrane receptors	CD11b	M0	Human/Murine	[155]
	CD14	M0	Human	[155]
	CD40	M1	Human/Murine	[156,157]
	CD80	M1	Human/Murine	[158]
	CD86	M1	Human/Murine	[158]
	CD163	M2	Human/Murine	[155]
	CD206	M2	Human/Murine	[155]
	F4/80	M0	Murine	[155]

Table 2. *Cont.*

	IL-1β	M1	Human/Murine	[26,159]
	IL-2	M1	Human/Murine	[26]
	IL-6	M1	Human/Murine	[154]
	IL-10	M2	Human/Murine	[158]
Cytokines	IL-12	M1	Human/Murine	[26]
	IL-23α	M1	Human/Murine	[26]
	CCL2	M1	Human/Murine	[160]
	TNF-α	M1	Human/Murine	[154]
	TGF-β	M2	Human/Murine	[158]
	VEGF	M2	Human/Murine	[161]

Abbreviations: Arg1: Arginase 1; CCL2: Chemokine (C-C motif) Ligand 2; CD: Cluster of Differentiation; iNOS: inducible Nitric Oxyde Synthase; IL: Interleukins; TGF-β: Transforming Growth Factor beta; Tumor Necrosis Factor alpha; VEGF: Vascular Endothelial Growth Factor.

4.2.3. Macrophage Origin

There are three main cell groups present in peripheral blood, in other words, the blood circulating throughout the body. Erythrocytes and thrombocytes, which are anucleated cells, and leukocytes (or white blood cells), which are nucleated cells that have a role in immunity. Among leukocytes, two subdivisions exist. The first includes granulocytes (including neutrophils, eosinophils, and basophils), which have the particularity of having a multilobed (polynuclear) nucleus, while the second subdivision includes peripheral blood mononuclear cells (PBMCs).

PBMCs include lymphocytes (T, B, and NK), dendritic cells, and monocytes [162]. These circulating monocytes arise from bone marrow, then migrate into tissues through blood and thanks to local signals (essentially cytokines) differentiate into macrophages [163]. These cells, known as tissue-resident macrophages, have a very long lifespan ranging from a few months to years [164]. Tissue-resident macrophages remain in tissues and contribute to their proper functioning (tissue surveillance and clearing) [165].

Furthermore, the differentiation of monocytes into macrophages takes place in two successive steps. First, in a process called maturation, monocytes transform into naïve macrophages (also called M0). Then, in a second step, these cells could be activated and polarized towards a phenotype (M1 or M2) depending on environmental signals [166]. In some organs, such as the gut, the origin and renewal of tissue-resident macrophages rely exclusively upon circulating monocytes [167]. However, the origin and renewal of resident macrophages from other tissues, such as the brain, liver, or lung, is through embryonic precursors produced either by the yolk sac or by the fetal liver. These precursors act as stem cells by ensuring the renewal of these macrophage populations throughout life [168,169].

In tumors, despite there being widespread recognition that TAMs derive predominantly from circulating monocytes, some studies based on murine models of brain, lung, and pancreatic cancers showed that a significant part of TAMs also derived from tissue-resident macrophages [147].

4.2.4. Impact of SPIONs on Monocytes and Macrophages

Since macrophages play an important role in immunosurveillance supported by major phagocytic activity and given their significant presence in tumors, it is essential to assess the impact of SPIONs, as vectors of anti-cancer therapies, on these cells and their precursors (monocytes), whether from safety (cytotoxicity, inflammation) or functional perspectives (polarization, biological responses). In this sense, several recent studies have examined the impact of SPIONs based on in vitro models of monocytes and macrophage cells. All in vitro macrophage models described in this review that were used to study SPIONs are detailed in Figure 4. These macrophage models can also be found in the first two columns of Table 3, a table that depicts the effects of SPIONs on monocyte/macrophage polarization and biological responses.

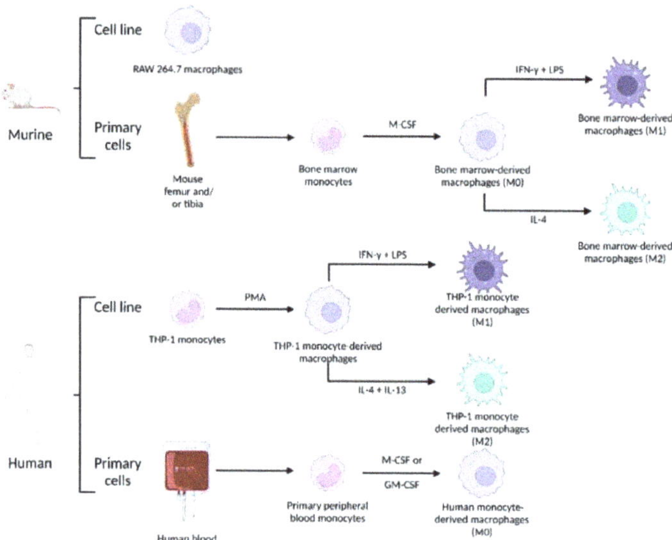

Figure 4. Some of the in vitro models based on murine or human macrophages. Created with BioRender.com.

One of the first parameters to take into account when evaluating the impact of SPIONs on monocytes or macrophages is whether these nanoparticles can undergo rapid uptake. In general, monocytes or macrophages are able to uptake SPIONs relatively rapidly (few hours) [114,170]. Wu et al. demonstrated in primary human monocyte cells that SPIONs can be identified in phagosomes or in cytoplasm [171]. However, there are two noteworthy items regarding the cellular uptake of SPIONs: their size and their charge. Indeed, SPIONs with a size up to 150 nm show a high uptake (Table 3), whereas those with a size above 200 nm showed limited cellular uptake [172]. Zhang et al. [114] demonstrated that the surface charge of SPIONs influenced their uptake rate by murine macrophages. Thus, positively charged SPIONs (+) have a higher rate of uptake than negatively charged SPIONs (−), and negatively charged SPIONs (−) in turn displayed a higher uptake rate than neutral SPIONs (N). Sharkey et al. [173] have also demonstrated that positively charged SPIONs (DEAE-Dextran) provided the best uptake when compared to negatively (CM-Dextran) or neutral (Dextran) ones.

Another important parameter to consider is the impact of SPIONs on cell viability in order to take advantage of the beneficial effects provided by anti-cancer therapies while minimizing the harmful adverse effects potentially induced by SPION vectors, especially since SPIONs cytotoxicity remains unclear [171]. In fact, several variables considerably complicate the evaluation of SPION cytotoxicity. These variables include, for instance, the duration of cell exposure to SPIONs. In order to reduce SPION-related cytotoxicity, Sharkey et al. [173] reduced from 24 h or 48 h to 4 h the incubation time of SPIONs with bone marrow derived-macrophages and no significant decrease in cell viability or increase in cytotoxicity was observed.

Table 3. Impact of SPION treatment on monocytes and macrophages.

Species	Cell Model	SPIONs	Biological Responses	Polarization Markers	References
Murine	RAW 264.7	SPIONs (+): +44.72 mV Size: 19.4 nm	Important uptake Cytotoxicity	Increase in iNOS and TNF-α Decrease in IL-10 and VEGF Increase in CD80 and decrease in CD206 in vivo	[114]
Murine	RAW 264.7	SPIONs (−): −27.31 mV Size: 21.3 nm	Important uptake Cytotoxicity	Increase in iNOS and TNF-α Decrease in IL-10 and VEGF Decrease in CD206 in vivo	[114]
Murine	RAW 264.7	SPIONs (N): −0.282 mV Size: 15.9 nm	Uptake No cytotoxicity	Increase in TNF-α Decrease in IL-10 and VEGF Slight decrease in CD206 in vivo	[114]
Murine	RAW 264.7	PEI-SPIONs (+): from +52.2 to +67.1 mV Size: 139–144 nm	Activation of TLR4, MAPK (p44/p42; p38; JNK) ROS production Modulation of podosome formation	Increase in CD40, CD80, CD86 and IL-12 Increase in IL10	[174]
Murine	Bone Marrow-Derived Macrophages	DEAE-Dextran 1:4 (+): +16.8 mV Size: 68 nm	Low cell viability reduction Low cytotoxicity High iron uptake No impact on phagocytosis	Increase in CD86, IL-1β, IL-12β and TNF-α Decrease in CD206 and Arg1	[173]
Murine	Bone Marrow-Derived Macrophages	CM-Dextran (−): −11.6 mV Size: 34.3 nm	Reduction in cell viability Low cytotoxicity Low iron uptake	NR	[173]
Murine	Bone Marrow-Derived Macrophages	Dextran (N): −3.3 mV Size: 36 nm	Reduction in cell viability Low cytotoxicity Low iron uptake	NR	[173]
Murine	Bone Marrow-Derived Macrophages	Resovist®: Ferucarbotran Carboxydextrane-coated SPIONs Size: 45–60 nm; core size: 5.8 nm	Activation of TLR4	Increase in IL-1β, IL-2, IL-12, CCL2 and TNF-α	[175]
Murine	Bone Marrow-Derived Macrophages	DMSA SPIONs (−): −29.3 mV Size: 65 nm; core size: 10 nm	Fast uptake No reduction in cell viability Activation of MAPK (ERK) and AKT Decrease in transferrin receptor ROS production	Increase in IL-23α and CCL2 No variation in IL-12 Increase in IL-10	[176]

Table 3. *Cont.*

Species	Cell Model	SPIONs	Biological Responses	Polarization Markers	References
Murine	Bone Marrow-Derived Macrophages	APS SPIONs (+): +33.3 Mv Size: 54 nm; core size: 8.3 nm	Fast uptake No reduction in cell viability Activation of MAPK (ERK) and AKT Decrease in transferrin receptor Important ROS production	Increase in IL-23α and CCL2 No variation in IL-12 Increase in IL-10	[176]
Murine	Bone Marrow-Derived Macrophages	AD SPIONs (+): +40.3 nm Size: 150 nm; core size: 6.8 nm	Fast uptake No reduction in cell viability Activation of MAPK (ERK) and AKT Decrease in transferrin receptor Important ROS production	Increase in IL-23α and CCL2 No variation in IL-12 No variation in IL-10	[176]
Human	THP-1 monocytes	Dextran-coated SPIONs Size: 83.5 and 86 nm; core size: 6.48 nm	Fast uptake	No increase in CD14, CD11b or CD86 Increase in IL-1β secretion Slight decrease in IL-10 secretion	[170]
Human	THP-1 Monocyte-derived macrophages	Dextran-coated SPIONs Size: 83.5 and 86 nm; core size: 6.48 nm	Fast uptake	No variation in CD14, CD11b or CD86 No variation in IL-1β No variation in IL-10 secretion	[170]
Human	THP-1 Monocyte-derived macrophages	Resovist®: Ferucarbotran Carboxydextrane-coated SPIONs Size: 45–60 nm; core size: 5.8 nm	Increase in Ferritin	Increase in CD86 and TNF-α on M2 macrophages	[177]
Human	THP-1 Monocyte-derived macrophages	DMSA SPIONs (−): −29.3 mV Size: 65 nm; core size: 10 nm	Fast uptake No reduction in cell viability Activation of MAPK (ERK) and AKT No activation of p38 nor JNK Decrease in transferrin receptor ROS production	Increase in CD86, TGF-β No variation in IL-12, IL-23α nor CCL2Increase in IL-10	[176]
Human	THP-1 Monocyte-derived macrophages	APS SPIONs (+): +33.3 Mv Size: 54 nm; core size: 8.3 nm	Fast uptake No reduction in cell viability Activation of MAPK (ERK) and AKT No activation of p38 nor JNK Decrease in transferrin receptor and FPN-1 ROS production	Increase in CD86, TGF-β No variation in IL-12, IL-23α nor CCL2 Increase in IL-10	[176]

Table 3. *Cont.*

Species	Cell Model	SPIONs	Biological Responses	Polarization Markers	References
Human	THP-1 Monocyte-derived macrophages	AD SPIONs (+): +40.3 nm Size: 150 nm; core size: 6.8 nm	Fast uptake No reduction in cell viability Activation of MAPK (ERK) and AKT No activation of p38 nor JNK Decrease in transferrin receptor and FPN-1 ROS production	No variation in CD86, IL-12, IL-23α nor CCL2	[176]
Human	Primary peripheral blood monocytes	Dextran-coated SPIONs Size: 83.5 and 86 nm; core size: 6.48 nm	Fast uptake	NR	[170]
Human	Primary peripheral blood monocytes	Starch-coated SPIONs (−) Size: 200 nm	Low uptake No toxic effects Disruption of actin skeleton	Decrease in IL-6 No variation in IL-1β No variation in IL-10	[172]
Human	Primary peripheral blood monocytes	Dextran SPIONs (−): −11 mV Size: 62,8 nm	Uptake in phagosomes or cytoplasm No decrease in cell viability nor cytotoxicity Activation of MAPK (ERK; p38; JNK)	Increase in IL-1β and TNF-α	[171]
Human	Human Monocyte-derived macrophages	DEAE-Dextran 1:4 (+): +16.8 mV Size: 68 nm	Important cell viability reduction Cytotoxicity Iron uptake	NR	[173]
Human	Human Monocyte-derived macrophages	Resovist®: Ferucarbotran Carboxydextrane-coated SPIONs Size: 45 and 60 nm; core size: 5.8 nm	Increase in Ferritin	NR	[177]

Abbreviations: AD: AminoDextran; APS: 3-AminoPropyl-triethoxySilane; DEAE: DiEthylAminoEthyl; DMSA: DiMercaptoSuccinic Acid; ERK: Extracellular signal-Regulated Kinase; FPN-1: FerroPortiN-1; JNK: Jun N-terminal Kinase; MAPK: Mitogen-Activated Protein Kinase; NR: Not Reported; N: Neutral; PEI: Polyethylenimine; ROS: Reactive Oxygen Species; TLR-4: Toll-Like Receptor 4. Additional descriptions: Effects described concern in vitro models unless otherwise specified. The mV number concerns the Z potential. Cellular uptake assays (e.g., iron assays) were performed to determine whether SPIONs are internalized in cells, while cytotoxicity assays (e.g., ATP assays) attempted to evaluate the degree of SPION toxicity.

In this particular case, Sharkey et al. aimed at labeling macrophages with SPIONs before injecting them in mice in order to visualize SPIONs-labelled macrophages by MRI. Therefore, reducing incubation time for ex vivo labeling is possible (for imaging purposes) [173]. However, for studies that aim at evaluating treatments with SPIONs (therapy purposes), systemic administration of SPIONs does not allow the control of the incubation time. In order to reduce cytotoxicity, experiments have shown that SPIONs coated with biocompatible polymers such as dextran, polyethylene glycol, or starch were less cytotoxic [178,179]. Most of the SPIONs listed in Table 3 were coated with these molecules. Another means of decreasing cytotoxicity is to choose biocompatible iron oxides cores such as magnetite (Fe_3O_4), maghemite (γ-Fe_2O_3), or hematite (α-Fe_2O_3) [179] and adapt their concentration below 100 mg/mL [170].

There is no doubt that SPIONs exert an important modulation of macrophages' biological responses. Kodali et al. showed in a bone-marrow-derived macrophages model that 1052 genes were differently expressed between macrophages treated with SPIONs and controls [180]. The challenge resides in the understanding of which cell signaling pathways are involved. Several studies clearly demonstrated that SPIONs activate the MAPK signaling pathway through the phosphorylation of the downstream mediator ERK1/2 [171,174,176] (Figure 5, 1). One of the most important signaling pathways implied in cell proliferation is the MAP kinase pathway. This signaling pathway can also be activated in case of stress such as DNA damage or heat shock. In this case, the effects of this pathway will be more oriented toward differentiation or apoptosis rather than cell proliferation [181]. Three studies clearly demonstrated that SPION treatment activates the MAPK signaling pathway [171,174,176]. Indeed, downstream mediator ERK1/2 was phosphorylated in those studies. Interestingly, the activation of other MAPK downstream mediators (other than ERK1/2) has been shown to be highly dependent on the type of SPION coating or cellular model used in these studies. As such, PEI-coated SPIONs activated p38 and JNK downstream mediators in RAW 264.7 macrophages [174] as well as dextran-coated SPIONs in primary peripheral blood monocytes [171]. However, DMSA-, APS-, and AD-coated SPIONs induced no phosphorylation of p38 nor JNK in THP-1 cells (monocyte cell line) [176] (Figure 5, 1). Studies have demonstrated that the action of SPIONs on macrophages is, at least in part, mediated by the family of toll-like receptors (TLRs). TLRs are receptors present on cells of the innate immune system, mainly monocytes and macrophages. There are so far ten TLRs that have been discovered in humans [182]. The ligands recognized by these receptors are very variable, either by their structure (LPS, LTA, peptidoglycans, flagellin, RNA, DNA) or by their origin (derived from bacteria, viruses, parasites, or *fungi*) [183]. It has been demonstrated that there is a crosstalk between MAPK and TLR signaling pathways in THP-1 cells, especially with TLR4 [184], the receptor that binds LPS [185] (Figure 5, 2). Moreover, since PEI was linked to the activation of TLR4 [186], Mulens-Arias et al. have demonstrated that the TLR4 signaling pathway is also activated by PEI-coated SPIONs, at least partly, since an inhibitor of TLR4 (CLI-095, also known as TAK-242) reduced IL-1β and VEGFA mRNA induction upon PEI-coated SPION treatment [174]. Jin et al. likewise demonstrated that TLR4 was involved following a SPION treatment [175]. Finally, another signaling pathway that has been described as being activated by SPIONs is the AKT signaling pathway, which could be activated by metabolic stress, such as ROS production [187], which will be discussed below. Indeed, Rojas et al. showed that DMSA-, APS- and AD-coated SPIONs activated the AKT signaling pathway in murine bone marrow-derived macrophages [176] (Figure 5, 1). One point that remains to be elucidated is whether SPIONs activate these various signaling pathways through their interaction with cell membrane receptors (e.g., TLR4) or classical internalization (e.g., phagocytosis), or both. Other signaling pathways should also be studied in detail, such as G protein-coupled receptors, knowing that there is a link between ROS production and AMPK phosphorylation [187], or cytokines and JAK (janus kinases) protein activation, since their triggering has already been shown in an in vitro model of human endothelial cells following a nanoparticle treatment [188].

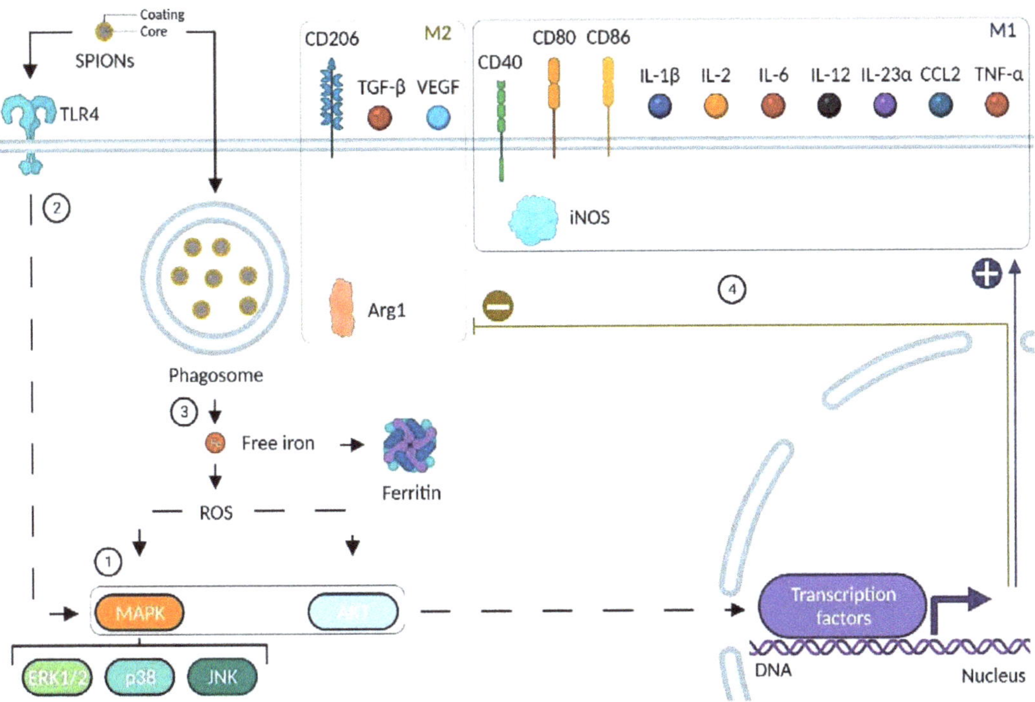

Figure 5. Impact of SPION treatment on polarization markers of macrophages. Created with BioRender.com.

SPIONs have also been described as directly impacting macrophage iron uptake as well as the expression level of iron-related proteins [173,176,177]. SPIONs are incorporated and degraded inside macrophages. Since the core of SPIONs is composed of iron, their degradation results in an increase in intracellular iron concentration. This iron accumulation in macrophages is thought to promote an M1-like phenotype [189]. M1-like macrophages display an iron storage phenotype. Consequently, these cells express higher levels of proteins involved in iron retention such as ferritin (a multimeric protein that is the main iron storage complex in cells [190]) or transferrin receptor 1 also known as CD71 (a transmembrane protein involved in iron uptake thanks to its binding to iron-loaded transferrin [191]). Conversely, M2-like macrophages present an iron export phenotype with an increase in ferroportin (a transmembrane protein involved in iron release [190]). In this context, Laskar et al. have demonstrated that SPIONs increase the expression of ferritin on THP-1 and human monocyte-derived macrophages [177]. Moreover, SPION treatment caused a decrease in transferrin receptors in M2 bone marrow-derived macrophages as well as in THP-1 monocyte-derived macrophages [176]. Moreover, ferroportin-1 expression was also decreased after 48h following AD- and APS-coated SPION treatment in THP-1 monocyte derived-macrophages [176]. Taken together, these results demonstrate that SPIONs will tend to cause iron accumulation in macrophages, a feature mainly observed in M1-like macrophages.

SPIONs degradation by macrophages may result in free iron atoms in the cytoplasm [192]. These atoms can in turn promote reactive oxygen species (ROS) production in a non-enzymatical way (Fenton chemistry, Figure 5, 3) [193]. In macrophages, ROS are associated with a pro-inflammatory M1-like phenotype since their production is used to destroy pathogens by a mechanism known as respiratory or oxidative burst [194] triggering inflammation [195] via the activation of the NF-κB (nuclear factor-κ B) signaling pathway.

PEI-, DMSA-, APS-, and AD-coated SPIONs have been described as inducing ROS production in murine (RAW 264.7 macrophages and bone marrow-derived macrophages) or human (THP-1 monocyte-derived macrophages) macrophages [174,176]. In addition, depending on the type of coating, ROS production levels may vary. AD-coated SPION treatment resulted in more ROS production than SPIONs coated with DMSA in murine macrophages derived from bone marrow [176]. ROS overproduction forms an integral part of oxidative stress that can be deleterious to cells, especially macrophages. The impact of SPIONs on ROS production must be carefully assessed in order to avoid cytotoxic effects linked to oxidative stress and maximize their safety [192].

Other macrophage biological responses have been demonstrated following SPION treatment. DEAE-dextran-coated SPIONs have no impact on macrophage phagocytic activities [173]. This is particularly interesting since one of the main roles of macrophages, phagocytosis, allows them to monitor their microenvironment against possible pathogens and ensure clearance of cellular debris leading to tissue homeostasis [196]. A treatment that would dampen this key feature could therefore prove to be deleterious to the organism. Mulens-Arias et al. demonstrated that PEI-coated SPIONs induced podosome formation in RAW 264.7 macrophages [174], and Gonnissen et al. showed that starch-coated SPIONs led to the disruption of the cytoskeleton in human monocytes [172]. These results all point in the same direction and may suggest that even though SPIONs do not impact the phagocytic activity of macrophages, they could somehow stimulate it indirectly since there is a link between phagocytosis and the formation of podosomes and their transient disruption in human macrophages [197].

Last, but certainly not least, it is also important to highlight that SPIONs exert an impact on macrophage polarization. Different polarization markers (M1 or M2) mainly from three molecule families (enzymes, membrane receptors, and cytokines) vary following SPION treatment (Figure 5, 4).

Firstly, in murine macrophages (RAW 264.7 or bone marrow-derived macrophages), SPIONs increased the expression of iNOS (M1 marker) and decreased the expression of Arg1 (M2 marker) [114,173].

Secondly, the expression of M1-like membrane receptors such as CD40 or CD80 in RAW 264.7 macrophages was increased following SPION treatment [174]. The expression of CD86, another widely used M1 marker, was increased as well when PEI-, DEAE-, Carboxydextrane-, DMSA- and APS-coated SPIONs were used regardless of whether this was assessed on murine or human macrophages [173,174,176,177]. However, no increase in CD86 expression was observed with dextran- or AD-coated SPIONs in human macrophages (THP-1) [170,176]. In addition, a decrease in the expression of M2-like membrane receptor CD206 expression was observed following DEAE-dextran-coated SPION treatment in murine macrophages from bone marrow [173]. An emphasis is needed in one experiment led by Zhang et al. [114]. Murine macrophages (RAW 264.7) were treated with SPIONs and then co-injected with HT1080 cells (fibrosarcoma cell line) in mice. Then, tumors were harvested and analyzed by immunohistochemistry. Compared to the control group (tumor cells co-injected with untreated macrophages), the treated group (tumor cells co-injected with macrophages pre-treated with SPIONs (+)) display an increase in CD80 (M1 marker) and a decrease in CD206 (M2 marker) in vivo. Moreover, SPION (+) pre-treated macrophages were shown to have an important tumor inhibition ability since tumor growth was reduced by at least threefold vs. the control group. These results showed that SPIONs (+) could repolarize macrophages and inhibit tumor growth.

Thirdly, cytokines (interleukins, chemokines, TNF-α, TGF-β, VEGF) expression was evaluated after SPION treatment. In general, SPION treatment had induced an increase in the expression of pro-inflammatory cytokines such as IL-1β, IL-2, IL-6, IL-12, IL-23α, CCL2, and TNF-α in murine or human macrophages [114,170,171,173–177]. Once again, the effects observed may be different for the type of coating and the cell model of use. Thus, in bone marrow-derived macrophages, Carboxydextrane-coated SPIONs induced an increase in IL-12 while DMSA-, APS-, and AD-coated SPIONs did not [175,176]. There is one aspect, however, that must be underlined. Human primary peripheral blood monocytes were treated with dextran-coated SPIONs. This treatment induced the production of pro-inflammatory cytokines such as IL-1β and TNF-α at similar levels to those induced by LPS treatment [171]. This point needs to be further investigated, above all with a view to intravenous treatment with SPIONs. Moreover, the expression of anti-inflammatory cytokines, IL-10 and TGF-β, was also altered by SPION treatment in murine or human macrophages. However, no clear trend was observed. In murine macrophages (RAW 264.7), a decrease and an increase in IL-10 levels has been noted depending on the type of SPIONs being used [114,175]. In human macrophage models, an increase [176], a decrease [170], and no variation [172] in IL-10 levels have been described following SPION treatment. The expression of the other anti-inflammatory cytokine, TGF-β, was increased [176] in THP-1 monocyte-derived macrophages when treated with DMSA- and APS-coated SPIONs. It would be interesting to investigate the variation in this cytokine with different SPIONs in other cell models. Finally, in murine macrophages (RAW 264.7), VEGF expression has been found to decrease following SPION treatment. Since VEGF is considered a marker of M2 polarization involved in angiogenesis [198], it would be worth checking its variation in macrophages from human origin.

To summarize, SPIONs appear to globally induce a trend towards an M1-like pro-inflammatory phenotype (Figure 5) in macrophages (increase in M1- and decrease in M2-associated markers). Despite the fact that cytotoxicity and inflammation related to SPIONs remain issues to be improved, having nanoparticles in the context of cancer biology that would repolarize M2-like macrophages into M1-like macrophages may appear appealing.

In conclusion, given the great heterogeneity of SPIONs (size, surface charge, shape, coating, core composition), it is essential to evaluate the impact of newly synthesized SPIONs on the monocyte-macrophage axis, preferably on primary cell lines as they are closer to physiological conditions and human pathologies [199].

5. Conclusions-Perspectives-Outlook

Cancer immunotherapy has tremendous promise, but it has yet to be clinically applied in a wider variety of tumor situations. The main difficulties are toxicity and therapeutic responsiveness limited to a small subset of patients. The variation in patient response rates reflects the various paths tumors use to regulate the various immune-evasion mechanisms occurring in the tumor microenvironment. As a result, it appears that immunotherapy focused against one particular protumoral mechanism is not effective enough at producing a noticeable therapeutic impact. To ensure the creation of novel, efficient cancer treatments, it is extremely desirable to combine therapy approaches that simultaneously target several cancer immuno-evasion systems, albeit this may result in higher toxicity. In this way, immunogenic cell death (ICD)-based strategies have attracted a lot of scientific attention to address the current constraints in treating solid tumors. Indeed, ICD triggers the immune response against the tumor through the activation of dendritic cells, initiating a cascade process leading to an antigen-specific T-cell response. Even though ICD has the effect of boosting the immune system to eliminate the cancer cells, in many instances, the response is insufficient but has been shown to be significantly improved with immune checkpoint inhibitors. ICD can be induced by some chemotherapies (e.g., doxorubicin, 5-fluorouracil) and external beam therapies such as radiotherapy, photodynamic therapy, and hyperthermia.

On that basis, it seems that nanomedicines can offer the possibility of combining these different approaches in the same drug and thus considerably improve the effectiveness of cancer immunotherapy. Indeed, according to their design and the materials they are made of, NPs can act as drug-delivery vehicles and be sensitive to a physical stimulus for either diagnosis and/or therapy (theranostic potential). As vehicles for the precise delivery of tumor antigens and/or immunostimulatory molecules to specific cells located in lymphoid organs or in the tumor microenvironment, nanoparticle-based delivery systems have recently demonstrated a great potential to improve the effectiveness and safety profile of conventional immunotherapeutics. Among these nanomedicines, magnetic NPs such as SPIONs might have enormous potential for safe, more efficient, and individualized cancer treatment. SPIONs have strong biomedical potential because of their high stability, biocompatibility, and low toxicity. Like most nanomedicines, SPIONs enable localized delivery of payload drugs. They also allow us to perform a rational design of novel combinatorial therapies based on immunotherapeutic treatments. In this way, they can target the adaptive and/or innate immune system through their use with/as immunomodulatory therapies (e.g., M2-like TAMs polarization to M1-like phenotype), therapeutic vaccines, and adoptive cell therapies (e.g., cell tracking of chimeric antigen receptor (CAR) T cells). Moreover, and this is what differentiates them from other NPs, due to their distinct ability to react only to an applied external magnetic field, SPIONs are attracting a lot of attention. Indeed, this property is particularly intriguing for biomedical applications and has allowed the development of novel immunotherapeutic approaches that rely on heating capability (magnetic hyperthermia, thermoresponsive drug release), magnetically controlled navigation (i.e., to guide drugs and cell therapies at the target region under a magnetic field), and image-guided techniques, such as magnetic resonance imaging and magnetic particle imaging, a new SPION-based molecular imaging technique. Moreover, due to their versatility, SPIONs make it possible to perform multimodal imaging such as simultaneous PET-MRI, especially for cell tracking. Combining the two imaging modalities may provide at early time points the fast localization and absolute quantification of radiolabeled SPIONs using PET, while MRI gives high-resolution anatomical background information for long-term NP follow-up. This innovative simultaneous approach allows us to overcome the respective limitations of each modality (i.e., resolution for PET and sensitivity for MRI).

As soon as a nanoparticle must be designed, we have to consider that there must be an ad hoc specification, i.e., making this nanoparticle compatible with its further use as a drug or medical device. For an immunotherapeutic approach, due to the complexity of tumor biology, a disease-driven approach should be proposed for the rational design of SPIONs rather than the traditional formulation-driven approach ("one-size-fits-all"). The specification of tumor-targeted SPIONs with immunotherapeutic capabilities will depend on the application, and it is necessary to take into account their multimodal potential, especially for theranostics:

- Good magnetic properties for imaging (MRI and MPI) and hyperthermia (magnetic core > 10 nm).
- Suitable size for passive tumor targeting through EPR effect (typically below 100 nm).
- Surface chemistry: coating to avoid aggregation, conjugations: with targeting moieties if pharmacological selectivity is desired, bifunctional chelating agents for radiolabeling purposes (nuclear imaging, targeted radionuclide therapy), fluorophores, photosensitizers, etc.
- The shape has to be considered since it is recognized as a parameter affecting the immunological response.
- A requirement for standard and optimized zeta potential values: typically, the higher, the better (good stability with absolute zeta values > |30| mV).

In spite of the strong theranostic potential of SPIONs, the limited quantity of SPION-based nanomedicines in clinical trials and on the market demonstrates a number of challenges to be overcome in order to facilitate their translation from the bench to the bedside. The safety of metallic NPs remains a major concern. To evaluate SPION-based nanomedicine biocompatibility and enhance its therapeutic benefits, a detailed investigation of how it interacts with the host tissues is essential. Previous clinical use of SPION formulations that have received FDA/EMA approval has already shown their acceptable safety and biocompatibility, which is unmatched by other metal-based nanoparticle systems. This offers a benefit in using SPIONs as nanomedicines to boost therapeutic results as improvements in cancer immunotherapy are made. Nevertheless, there are still some major regulatory and industrial hurdles to be overcome prior to reaching the market, due to the complex nature of nanomedicine when compared to conventional pharmaceutical products with a single agent. It is also important to consider the impacts of nanoparticles in general, and metallic NPs such as SPIONs in particular, on the environment, society, and ethics to make them acceptable in a biomedical context.

Overall, the unique properties and versatility of SPIONs pave the way for new approaches in the fields of drug delivery and theranostics for cancer immunotherapy, contributing to the personalization of treatments, especially to manage cancers with high unmet medical needs.

Author Contributions: Conceptualization, A.M.M.D., A.C. and B.C.; writing—original draft preparation, A.M.M.D., A.C. and B.C.; writing—review and editing, A.M.M.D., A.C., P.-S.B., E.K., A.O., P.-M.W. and B.C.; visualization, P.-E.D., C.P. and R.D.; supervision, B.C.; project administration, B.C.; funding acquisition, B.C. and A.O. All authors have read and agreed to the published version of the manuscript.

Funding: This research was funded by a French Government grant managed by the French National Research Agency under the program "Investissements d'Avenir" with reference ANR-10-EQPX-05-01/IMAPPI Equipex. It was also supported by the Institut de Chimie Moléculaire de l'Université de Bourgogne (ICMUB UMR CNRS 6302) and by the Centre George-François Leclerc (Preclinical imaging and radiotherapy platform, nuclear medicine department).

Institutional Review Board Statement: Not applicable.

Informed Consent Statement: Not applicable.

Data Availability Statement: Not applicable.

Acknowledgments: This work was performed within Pharm'image, a regional center of excellence in pharmaco-imaging. Mélanie Guillemin and Marie Monterrat are gratefully and warmly thanked for their technical support. Jérémy Paris, CEO of SON SAS, is also thanked for his scientific advice.

Conflicts of Interest: The authors declare no conflict of interest. The company had no role in the design of the study; in the collection, analyses, or interpretation of data; in the writing of the manuscript, and in the decision to publish the results.

References

1. Sung, H.; Ferlay, J.; Siegel, R.L.; Laversanne, M.; Soerjomataram, I.; Jemal, A.; Bray, F. Global Cancer Statistics 2020: GLOBOCAN Estimates of Incidence and Mortality Worldwide for 36 Cancers in 185 Countries. *CA Cancer J. Clin.* **2021**, *71*, 209–249. [CrossRef] [PubMed]
2. World Health Organization. Available online: http://who.int/data/gho/data/themes/mortality-and-global-health-estimates/ghe-leading-causes-of-death (accessed on 23 May 2022).
3. Waldman, A.D.; Fritz, J.M.; Lenardo, M.J. A guide to cancer immunotherapy: From T cell basic science to clinical practice. *Nat. Rev. Immunol.* **2020**, *20*, 651–668. [CrossRef] [PubMed]
4. Ribas, A.; Wolchok, J.D. Cancer immunotherapy using checkpoint blockade. *Science* **2018**, *359*, 1350–1355. [CrossRef] [PubMed]
5. Xie, G.; Dong, H.; Liang, Y.; Ham, J.D.; Rizwan, R.; Chen, J. CAR-NK cells: A promising cellular immunotherapy for cancer. *EBioMedicine* **2020**, *59*, 102975. [CrossRef] [PubMed]
6. Klichinsky, M.; Ruella, M.; Shestova, O.; Lu, X.M.; Best, A.; Zeeman, M.; Schmierer, M.; Gabrusiewicz, K.; Anderson, N.R.; Petty, N.E.; et al. Human chimeric antigen receptor macrophages for cancer immunotherapy. *Nat. Biotechnol.* **2020**, *38*, 947–953. [CrossRef]
7. Larson, R.C.; Maus, M.V. Recent advances and discoveries in the mechanisms and functions of CAR T cells. *Nat. Rev. Cancer* **2021**, *21*, 145–161. [CrossRef]
8. Atallah-Yunes, S.A.; Robertson, M.J. Cytokine Based Immunotherapy for Cancer and Lymphoma: Biology, Challenges and Future Perspectives. *Front. Immunol.* **2022**, *13*, 872010. [CrossRef]
9. Mayes, P.A.; Hance, K.W.; Hoos, A. The promise and challenges of immune agonist antibody development in cancer. *Nat. Rev. Drug Discov.* **2018**, *17*, 509–527. [CrossRef]
10. Saxena, M.; van der Burg, S.H.; Melief, C.J.M.; Bhardwaj, N. Therapeutic cancer vaccines. *Nat. Rev. Cancer* **2021**, *21*, 360–378. [CrossRef]
11. Yao, Y.; Zhou, Y.; Liu, L.; Xu, Y.; Chen, Q.; Wang, Y.; Wu, S.; Deng, Y.; Zhang, J.; Shao, A. Nanoparticle-Based Drug Delivery in Cancer Therapy and Its Role in Overcoming Drug Resistance. *Front. Mol. Biosci.* **2020**, *7*, 193. [CrossRef]
12. Nel, A.R.E.; Meng, H. New Insights into "Permeability" as in the Enhanced Permeability and Retention Effect of Cancer Nanotherapeutics. *ACS Nano* **2017**, *11*, 9567–9569. [CrossRef]
13. Tuma, P.; Hubbard, A.L. Transcytosis: Crossing cellular barriers. *Physiol. Rev.* **2003**, *83*, 871–932. [CrossRef]

14. Heshmati Aghda, N.; Dabbaghianamiri, M.; Tunnell, J.W.; Betancourt, T. Design of smart nanomedicines for effective cancer treatment. *Int. J. Pharm.* **2022**, *621*, 121791. [CrossRef]

15. Niculescu, A.G.; Grumezescu, A.M. Novel Tumor-Targeting Nanoparticles for Cancer Treatment-A Review. *Int. J. Mol. Sci.* **2022**, *23*, 5253. [CrossRef]

16. Halwani, A.A. Development of Pharmaceutical Nanomedicines: From the Bench to the Market. *Pharmaceutics* **2022**, *14*, 106. [CrossRef]

17. Sun, D.; Zhou, S.; Gao, W. What Went Wrong with Anticancer Nanomedicine Design and How to Make It Right. *ACS Nano* **2020**, *14*, 12281–12290. [CrossRef]

18. Martin, J.D.; Cabral, H.; Stylianopoulos, T.; Jain, R.K. Improving cancer immunotherapy using nanomedicines: Progress, opportunities and challenges. *Nat. Rev. Clin. Oncol.* **2020**, *17*, 251–266. [CrossRef]

19. Soetaert, F.; Korangath, P.; Serantes, D.; Fiering, S.; Ivkov, R. Cancer therapy with iron oxide nanoparticles: Agents of thermal and immune therapies. *Adv. Drug Deliv. Rev.* **2020**, *163–164*, 65–83. [CrossRef]

20. Canese, R.; Vurro, F.; Marzola, P. Iron Oxide Nanoparticles as Theranostic Agents in Cancer Immunotherapy. *Nanomaterials* **2021**, *11*, 1950. [CrossRef]

21. Dulinska-Litewka, J.; Lazarczyk, A.; Halubiec, P.; Szafranski, O.; Karnas, K.; Karewicz, A. Superparamagnetic Iron Oxide Nanoparticles-Current and Prospective Medical Applications. *Materials* **2019**, *12*, 617. [CrossRef]

22. Montiel Schneider, M.G.; Martin, M.J.; Otarola, J.; Vakarelska, E.; Simeonov, V.; Lassalle, V.; Nedyalkova, M. Biomedical Applications of Iron Oxide Nanoparticles: Current Insights Progress and Perspectives. *Pharmaceutics* **2022**, *14*, 204. [CrossRef] [PubMed]

23. Hanahan, D. Hallmarks of Cancer: New Dimensions. *Cancer Discov.* **2022**, *12*, 31–46. [CrossRef] [PubMed]

24. Wilczynski, J.R.; Nowak, M. Cancer Immunoediting: Elimination, Equilibrium, and Immune Escape in Solid Tumors. *Exp. Suppl.* **2022**, *113*, 1–57. [CrossRef] [PubMed]

25. Nam, J.; Son, S.; Park, K.S.; Zou, W.; Shea, L.D.; Moon, J.J. Cancer nanomedicine for combination cancer immunotherapy. *Nat. Rev. Mater.* **2019**, *4*, 398–414. [CrossRef]

26. Doshi, A.S.; Asrani, K.H. Chapter Two-Innate and adaptive immunity in cancer. In *Cancer Immunology and Immunotherapy*; Amiji, M.M., Milane, L.S., Eds.; Academic Press: Cambridge, MA, USA, 2022; pp. 19–61.

27. Sofias, A.M.; Combes, F.; Koschmieder, S.; Storm, G.; Lammers, T. A paradigm shift in cancer nanomedicine: From traditional tumor targeting to leveraging the immune system. *Drug Discov. Today* **2021**, *26*, 1482–1489. [CrossRef]

28. Ginefra, P.; Lorusso, G.; Vannini, N. Innate Immune Cells and Their Contribution to T-Cell-Based Immunotherapy. *Int. J. Mol. Sci.* **2020**, *21*, 4441. [CrossRef]

29. Anfray, C.; Ummarino, A.; Andon, F.T.; Allavena, P. Current Strategies to Target Tumor-Associated-Macrophages to Improve Anti-Tumor Immune Responses. *Cells* **2019**, *9*, 46. [CrossRef]

30. Zuo, S.; Song, J.; Zhang, J.; He, Z.; Sun, B.; Sun, J. Nano-immunotherapy for each stage of cancer cellular immunity: Which, why, and what? *Theranostics* **2021**, *11*, 7471–7487. [CrossRef]

31. Alexander, C.A.; Yang, Y.Y. Harnessing the combined potential of cancer immunotherapy and nanomedicine: A new paradigm in cancer treatment. *Nanomedicine* **2022**, *40*, 102492. [CrossRef]

32. Qin, S.; Xu, L.; Yi, M.; Yu, S.; Wu, K.; Luo, S. Novel immune checkpoint targets: Moving beyond PD-1 and CTLA-4. *Mol. Cancer* **2019**, *18*, 155. [CrossRef]

33. Viana, I.M.O.; Roussel, S.; Defrene, J.; Lima, E.M.; Barabe, F.; Bertrand, N. Innate and adaptive immune responses toward nanomedicines. *Acta Pharm. Sin. B* **2021**, *11*, 852–870. [CrossRef]

34. Yu, S.; Wang, Y.; He, P.; Shao, B.; Liu, F.; Xiang, Z.; Yang, T.; Zeng, Y.; He, T.; Ma, J.; et al. Effective Combinations of Immunotherapy and Radiotherapy for Cancer Treatment. *Front. Oncol.* **2022**, *12*, 809304. [CrossRef]

35. Hurwitz, M.D. Hyperthermia and immunotherapy: Clinical opportunities. *Int. J. Hyperth.* **2019**, *36*, 4–9. [CrossRef]

36. Hartl, C.A.; Bertschi, A.; Puerto, R.B.; Andresen, C.; Cheney, E.M.; Mittendorf, E.A.; Guerriero, J.L.; Goldberg, M.S. Combination therapy targeting both innate and adaptive immunity improves survival in a pre-clinical model of ovarian cancer. *J. Immunother. Cancer* **2019**, *7*, 199. [CrossRef]

37. Zhu, E.F.; Gai, S.A.; Opel, C.F.; Kwan, B.H.; Surana, R.; Mihm, M.C.; Kauke, M.J.; Moynihan, K.D.; Angelini, A.; Williams, R.T.; et al. Synergistic innate and adaptive immune response to combination immunotherapy with anti-tumor antigen antibodies and extended serum half-life IL-2. *Cancer Cell* **2015**, *27*, 489–501. [CrossRef]

38. Gowd, V.; Ahmad, A.; Tarique, M.; Suhail, M.; Zughaibi, T.A.; Tabrez, S.; Khan, R. Advancement of cancer immunotherapy using nanoparticles-based nanomedicine. *Semin. Cancer Biol.* **2022**, *86*, 624–644. [CrossRef]

39. Hu, M.; Huang, L. Strategies targeting tumor immune and stromal microenvironment and their clinical relevance. *Adv Drug Deliv. Rev.* **2022**, *183*, 114137. [CrossRef]

40. Liu, Y.; Lu, Y.; Zhu, X.; Li, C.; Yan, M.; Pan, J.; Ma, G. Tumor microenvironment-responsive prodrug nanoplatform via co-self-assembly of photothermal agent and IDO inhibitor for enhanced tumor penetration and cancer immunotherapy. *Biomaterials* **2020**, *242*, 119933. [CrossRef]

41. Phuengkham, H.; Song, C.; Lim, Y.T. A Designer Scaffold with Immune Nanoconverters for Reverting Immunosuppression and Enhancing Immune Checkpoint Blockade Therapy. *Adv. Mater.* **2019**, *31*, e1903242. [CrossRef]

42. Wang-Bishop, L.; Wehbe, M.; Shae, D.; James, J.; Hacker, B.C.; Garland, K.; Chistov, P.P.; Rafat, M.; Balko, J.M.; Wilson, J.T. Potent STING activation stimulates immunogenic cell death to enhance antitumor immunity in neuroblastoma. *J. Immunother. Cancer* **2020**, *8*. [CrossRef]

43. Viswanath, D.I.; Liu, H.C.; Huston, D.P.; Chua, C.Y.X.; Grattoni, A. Emerging biomaterial-based strategies for personalized therapeutic in situ cancer vaccines. *Biomaterials* **2022**, *280*, 121297. [CrossRef] [PubMed]

44. Zhang, T.; Tai, Z.; Cui, Z.; Chai, R.; Zhu, Q.; Chen, Z. Nano-engineered immune cells as "guided missiles" for cancer therapy. *J. Control Release* **2022**, *341*, 60–79. [CrossRef] [PubMed]

45. Huang, W.C.; Shen, M.Y.; Chen, H.H.; Lin, S.C.; Chiang, W.H.; Wu, P.H.; Chang, C.W.; Chiang, C.S.; Chiu, H.C. Monocytic delivery of therapeutic oxygen bubbles for dual-modality treatment of tumor hypoxia. *J. Control Release* **2015**, *220*, 738–750. [CrossRef] [PubMed]

46. Milligan, J.J.; Saha, S. A Nanoparticle's Journey to the Tumor: Strategies to Overcome First-Pass Metabolism and Their Limitations. *Cancers* **2022**, *14*, 1741. [CrossRef] [PubMed]

47. Krishnan, G.; Cousins, A.; Pham, N.; Milanova, V.; Nelson, M.; Krishnan, S.; Shetty, A.; van den Berg, N.; Rosenthal, E.; Krishnan, S.; et al. Preclinical evaluation of a mannose-labeled magnetic tracer for enhanced sentinel lymph node retention in the head and neck. *Nanomedicine* **2022**, *42*, 102546. [CrossRef]

48. Dennis, C.L.; Ivkov, R. Physics of heat generation using magnetic nanoparticles for hyperthermia. *Int. J. Hyperth.* **2013**, *29*, 715–729. [CrossRef]

49. Sarychev, V.T. Electromagnetic field of a rotating magnetic dipole and electric-charge motion in this field. *Radiophys. Quantum Electron.* **2009**, *52*, 8. [CrossRef]

50. Stöhr, J.; Siegmann, H.C. *Magnetism: From Fundamentals to Nanoscale Dynamics*; Springer: Berlin/Heidelberg, Germany; New York, NY, USA, 2006; p. 820.

51. Spaldin, N.A. *Magnetic materials: Fundamentals and Applications*, 2nd ed.; Cambridge University Press: Cambridge, UK; New York, NY, USA, 2011; p. 274.

52. Nisticò, R. A synthetic guide toward the tailored production of magnetic iron oxide nanoparticles. *Boletín Soc. Española Cerámica Y Vidr.* **2021**, *60*, 29–40. [CrossRef]

53. Rajan, A.; Sahu, N.K. Review on magnetic nanoparticle-mediated hyperthermia for cancer therapy. *J. Nanopart Res.* **2020**, *22*, 319. [CrossRef]

54. Shaterabadi, Z.; Nabiyouni, G.; Soleymani, M. Physics responsible for heating efficiency and self-controlled temperature rise of magnetic nanoparticles in magnetic hyperthermia therapy. *Prog. Biophys. Mol. Biol.* **2018**, *133*, 9–19. [CrossRef]

55. Cullity, B.D.; Graham, C.D. *Introduction to Magnetic Materials*, 2nd ed.; IEEE/Wiley: Hoboken, NJ, USA, 2009; p. 544.

56. Williams, H.J.; Bozorth, R.M.; Shockley, W. Magnetic Domain Patterns on Single Crystals of Silicon Iron. *Phys. Rev.* **1949**, *75*, 155–178. [CrossRef]

57. Laurent, S.; Dutz, S.; Häfeli, U.O.; Mahmoudi, M. Magnetic fluid hyperthermia: Focus on superparamagnetic iron oxide nanoparticles. *Adv. Colloid Interface Sci.* **2011**, *166*, 8–23. [CrossRef]

58. Saslow, W.M. The Magnetism of Magnets. In *Electricity, Magnetism, and Light*; Elsevier: Amsterdam, The Netherlands, 2002; pp. 384–418.

59. Aharoni, A. *Introduction to the Theory of Ferromagnetism*, 2nd ed.; Oxford University Press: Oxford, UK, 2007.

60. Gupta, A.K.; Wells, S. Surface-Modified Superparamagnetic Nanoparticles for Drug Delivery: Preparation, Characterization, and Cytotoxicity Studies. *IEEE Trans. Nanobioscience* **2004**, *3*, 66–73. [CrossRef]

61. Kandasamy, G.; Maity, D. Recent advances in superparamagnetic iron oxide nanoparticles (SPIONs) for in vitro and in vivo cancer nanotheranostics. *Int. J. Pharm.* **2015**, *496*, 191–218. [CrossRef]

62. Khot, V.M.; Salunkhe, A.B.; Ruso, J.M.; Pawar, S.H. Improved magnetic induction heating of nanoferrites for hyperthermia applications: Correlation with colloidal stability and magneto-structural properties. *J. Magn. Magn. Mater.* **2015**, *384*, 335–343. [CrossRef]

63. Frenkel, J.; Doefman, J. Spontaneous and Induced Magnetisation in Ferromagnetic Bodies. *Nature* **1930**, *126*, 274–275. [CrossRef]

64. Néel, L. Théorie du traînage magnétique des substances massives dans le domaine de Rayleigh. *J. Phys. Radium* **1950**, *11*, 49–61. [CrossRef]

65. Bean, C.P.; Livingston, J.D. Superparamagnetism. *J. Appl. Phys.* **1959**, *30*, S120–S129. [CrossRef]

66. Savliwala, S.; Chiu-Lam, A.; Unni, M.; Rivera-Rodriguez, A.; Fuller, E.; Sen, K.; Threadcraft, K.; Rinaldi, C. Chapter 13-Magnetic nanoparticles. In *Nanoparticles for Biomedical Applications*; Elsevier: Amsterdam, The Netherlands, 2020; pp. 195–221.

67. Weissleder, R.; Hahn, P.F.; Stark, D.D.; Elizondo, G.; Saini, S.; Todd, L.E.; Wittenberg, J.; Ferrucci, J.T. Superparamagnetic iron oxide: Enhanced detection of focal splenic tumors with MR imaging. *Radiology* **1988**, *169*, 399–403. [CrossRef]

68. Weissleder, R.; Elizondo, G.; Wittenberg, J.; Rabito, C.A.; Bengele, H.H.; Josephson, L. Ultrasmall superparamagnetic iron oxide: Characterization of a new class of contrast agents for MR imaging. *Radiology* **1990**, *175*, 489–493. [CrossRef]

69. Stark, D.D.; Weissleder, R.; Elizondo, G.; Hahn, P.F.; Saini, S.; Todd, L.E.; Wittenberg, J.; Ferrucci, J.T. Superparamagnetic iron oxide: Clinical application as a contrast agent for MR imaging of the liver. *Radiology* **1988**, *168*, 297–301. [CrossRef] [PubMed]

70. Saini, S.; Stark, D.D.; Hahn, P.F.; Wittenberg, J.; Brady, T.J.; Ferrucci, J.T., Jr. Ferrite particles: A superparamagnetic MR contrast agent for the reticuloendothelial system. *Radiology* **1987**, *162*, 211–216. [CrossRef] [PubMed]

71. Saini, S.; Stark, D.D.; Hahn, P.F.; Bousquet, J.C.; Introcasso, J.; Wittenberg, J.; Brady, T.J.; Ferrucci, J.T., Jr. Ferrite particles: A superparamagnetic MR contrast agent for enhanced detection of liver carcinoma. *Radiology* **1987**, *162*, 217–222. [CrossRef] [PubMed]

72. Majumdar, S.; Zoghbi, S.S.; Gore, J.C. Regional differences in rat brain displayed by fast MRI with superparamagnetic contrast agents. *Magn. Reson. Imaging* **1988**, *6*, 611–615. [CrossRef]

73. Hervault, A.; Thanh, N.T.K. Magnetic nanoparticle-based therapeutic agents for thermo-chemotherapy treatment of cancer. *Nanoscale* **2014**, *6*, 11553–11573. [CrossRef]

74. Antonelli, A.; Magnani, M. SPIO nanoparticles and magnetic erythrocytes as contrast agents for biomedical and diagnostic applications. *J. Magn. Magn. Mater.* **2022**, *541*, 168520. [CrossRef]

75. Bjørnerud, A.; Johansson, L. The utility of superparamagnetic contrast agents in MRI: Theoretical consideration and applications in the cardiovascular system: Superparamagnetic contrast agents. *NMR Biomed.* **2004**, *17*, 465–477. [CrossRef]

76. Koenig, S.H.; Kellar, K.E. Theory of $1/T1$ and $1/T2$ NMRD profiles of solutions of magnetic nanoparticles. *Magn. Reson. Med.* **1995**, *34*, 227–233. [CrossRef]

77. Bae, K.H.; Kim, Y.B.; Lee, Y.; Hwang, J.; Park, H.; Park, T.G. Bioinspired Synthesis and Characterization of Gadolinium-Labeled Magnetite Nanoparticles for Dual Contrast T_1- and T_2-Weighted Magnetic Resonance Imaging. *Bioconjugate Chem.* **2010**, *21*, 505–512. [CrossRef]

78. Kellar, K.E.; Fujii, D.K.; Gunther, W.H.H.; Briley-Saebo, K.; Bjornerud, A.; Spiller, M.; Koenig, S.H. NC100150 injection, a preparation of optimized iron oxide nanoparticles for positive-contrast MR angiography. *J. Magn. Reson. Imaging* **2000**, *11*, 488–494. [CrossRef]

79. Alipour, A.; Soran-Erdem, Z.; Utkur, M.; Sharma, V.K.; Algin, O.; Saritas, E.U.; Demir, H.V. A new class of cubic SPIONs as a dual-mode T1 and T2 contrast agent for MRI. *Magn. Reson. Imaging* **2018**, *49*, 16–24. [CrossRef]

80. Hahn, P.F.; Stark, D.D.; Lewis, J.M.; Saini, S.; Elizondo, G.; Weissleder, R.; Fretz, C.J.; Ferrucci, J.T. First clinical trial of a new superparamagnetic iron oxide for use as an oral gastrointestinal contrast agent in MR imaging. *Radiology* **1990**, *175*, 695–700. [CrossRef]

81. Neuwelt, A.; Sidhu, N.; Hu, C.-A.A.; Mlady, G.; Eberhardt, S.C.; Sillerud, L.O. Iron-Based Superparamagnetic Nanoparticle Contrast Agents for MRI of Infection and Inflammation. *Am. J. Roentgenol.* **2015**, *204*, W302–W313. [CrossRef]

82. Kaim, A.H.; Jundt, G.; Wischer, T.; O'Reilly, T.; Fröhlich, J.; von Schulthess, G.K.; Allegrini, P.R. Functional-Morphologic MR Imaging with Ultrasmall Superparamagnetic Particles of Iron Oxide in Acute and Chronic Soft-Tissue Infection: Study in Rats. *Radiology* **2003**, *227*, 169–174. [CrossRef]

83. Lutz, A.M.; Weishaupt, D.; Persohn, E.; Goepfert, K.; Froehlich, J.; Sasse, B.; Gottschalk, J.; Marincek, B.; Kaim, A.H. Imaging of Macrophages in Soft-Tissue Infection in Rats: Relationship between Ultrasmall Superparamagnetic Iron Oxide Dose and MR Signal Characteristics. *Radiology* **2005**, *234*, 765–775. [CrossRef]

84. Stoll, G.; Bendszus, M. New approaches to neuroimaging of central nervous system inflammation. *Curr. Opin. Neurol.* **2010**, *23*, 282–286. [CrossRef]

85. Sillerud, L.O.; Solberg, N.O.; Chamberlain, R.; Orlando, R.A.; Heidrich, J.E.; Brown, D.C.; Brady, C.I.; Vander Jagt, T.A.; Garwood, M.; Vander Jagt, D.L. SPION-Enhanced Magnetic Resonance Imaging of Alzheimer's Disease Plaques in AβPP/PS-1 Transgenic Mouse Brain. *JAD* **2013**, *34*, 349–365. [CrossRef]

86. Ruehm, S.G.; Corot, C.; Vogt, P.; Kolb, S.; Debatin, J.F. Magnetic Resonance Imaging of Atherosclerotic Plaque With Ultrasmall Superparamagnetic Particles of Iron Oxide in Hyperlipidemic Rabbits. *Circulation* **2001**, *103*, 415–422. [CrossRef]

87. Harnan, S.E.; Cooper, K.L.; Meng, Y.; Ward, S.E.; Fitzgerald, P.; Papaioannou, D.; Ingram, C.; Lorenz, E.; Wilkinson, I.D.; Wyld, L. Magnetic resonance for assessment of axillary lymph node status in early breast cancer: A systematic review and meta-analysis. *Eur. J. Surg. Oncol.* **2011**, *37*, 928–936. [CrossRef]

88. Weissleder, R.; Elizondo, G.; Wittenberg, J.; Lee, A.S.; Josephson, L.; Brady, T.J. Ultrasmall superparamagnetic iron oxide: An intravenous contrast agent for assessing lymph nodes with MR imaging. *Radiology* **1990**, *175*, 494–498. [CrossRef]

89. Chambon, C.; Clement, O.; Le Blanche, A.; Schouman-Claeys, E.; Frija, G. Superparamagnetic iron oxides as positive MR contrast agents: In vitro and in vivo evidence. *Magn. Reson. Imaging* **1993**, *11*, 509–519. [CrossRef]

90. Serkova, N.J. Nanoparticle-Based Magnetic Resonance Imaging on Tumor-Associated Macrophages and Inflammation. *Front. Immunol.* **2017**, *8*, 590. [CrossRef] [PubMed]

91. Storey, P.; Lim, R.P.; Chandarana, H.; Rosenkrantz, A.B.; Kim, D.; Stoffel, D.R.; Lee, V.S. MRI assessment of hepatic iron clearance rates after USPIO administration in healthy adults. *Investig. Radiol.* **2012**, *47*, 717–724. [CrossRef] [PubMed]

92. Sargsyan, S.A.; Thurman, J.M. Molecular imaging of autoimmune diseases and inflammation. *Mol. Imaging* **2012**, *11*, 251–264. [CrossRef] [PubMed]

93. Weissleder, R. *Molecular Imaging: Principles and Practice*; People's Medical Publishing House-USA: Shelton, CT, USA, 2010.

94. Peiravi, M.; Eslami, H.; Ansari, M.; Zare-Zardini, H. Magnetic hyperthermia: Potentials and limitations. *J. Indian Chem. Soc.* **2022**, *99*, 100269. [CrossRef]

95. Yan, B.; Liu, C.; Wang, S.; Li, H.; Jiao, J.; Lee, W.S.V.; Zhang, S.; Hou, Y.; Hou, Y.; Ma, X.; et al. Magnetic hyperthermia induces effective and genuine immunogenic tumor cell death with respect to exogenous heating. *J. Mater. Chem. B* **2022**, *10*, 5364–5374. [CrossRef]

96. Deatsch, A.E.; Evans, B.A. Heating efficiency in magnetic nanoparticle hyperthermia. *J. Magn. Magn. Mater.* **2014**, *354*, 163–172. [CrossRef]

97. Jiao, W.; Zhang, T.; Peng, M.; Yi, J.; He, Y.; Fan, H. Design of Magnetic Nanoplatforms for Cancer Theranostics. *Biosensors* **2022**, *12*, 38. [CrossRef]

98. Palanisamy, S.; Wang, Y.M. Superparamagnetic iron oxide nanoparticulate system: Synthesis, targeting, drug delivery and therapy in cancer. *Dalton Trans.* **2019**, *48*, 9490–9515. [CrossRef]

99. Samrot, A.V.; Sahithya, C.S.; Selvarani, A.J.; Purayil, S.K.; Ponnaiah, P. A review on synthesis, characterization and potential biological applications of superparamagnetic iron oxide nanoparticles. *Curr. Res. Green Sustain. Chem.* **2021**, *4*, 100042. [CrossRef]

100. Berret, J.F.; Graillot, A. Versatile Coating Platform for Metal Oxide Nanoparticles: Applications to Materials and Biological Science. *Langmuir* **2022**, *38*, 5323–5338. [CrossRef]

101. Nuzhina, J.V.; Shtil, A.A.; Prilepskii, A.Y.; Vinogradov, V.V. Preclinical Evaluation and Clinical Translation of Magnetite-Based Nanomedicines. *J. Drug Deliv. Sci. Technol.* **2019**, *54*, 101282. [CrossRef]

102. Wlodarczyk, A.; Gorgon, S.; Radon, A.; Bajdak-Rusinek, K. Magnetite Nanoparticles in Magnetic Hyperthermia and Cancer Therapies: Challenges and Perspectives. *Nanomaterials* **2022**, *12*, 1807. [CrossRef]

103. Rubia-Rodriguez, I.; Santana-Otero, A.; Spassov, S.; Tombacz, E.; Johansson, C.; De La Presa, P.; Teran, F.J.; Del Puerto Morales, M.; Veintemillas-Verdaguer, S.; Thanh, N.T.K.; et al. Whither Magnetic Hyperthermia? A Tentative Roadmap. *Materials* **2021**, *14*, 706. [CrossRef]

104. Tong, S.; Hou, S.; Zheng, Z.; Zhou, J.; Bao, G. Coating optimization of superparamagnetic iron oxide nanoparticles for high T2 relaxivity. *Nano Lett.* **2010**, *10*, 4607–4613. [CrossRef]

105. Liu, X.L.; Fan, H.M.; Yi, J.B.; Yang, Y.; Choo, E.S.G.; Xue, J.M.; Fan, D.D.; Ding, J. Optimization of surface coating on Fe3O4 nanoparticles for high performance magnetic hyperthermia agents. *J. Mater. Chem.* **2012**, *22*, 8235. [CrossRef]

106. Egea-Benavente, D.; Ovejero, J.G.; Morales, M.D.P.; Barber, D.F. Understanding MNPs Behaviour in Response to AMF in Biological Milieus and the Effects at the Cellular Level: Implications for a Rational Design That Drives Magnetic Hyperthermia Therapy toward Clinical Implementation. *Cancers* **2021**, *13*, 4583. [CrossRef]

107. Guardia, P.; Di Corato, R.; Lartigue, L.; Wilhelm, C.; Espinosa, A.; Garcia-Hernandez, M.; Gazeau, F.; Manna, L.; Pellegrino, T. Water-soluble iron oxide nanocubes with high values of specific absorption rate for cancer cell hyperthermia treatment. *ACS Nano* **2012**, *6*, 3080–3091. [CrossRef]

108. Mai, B.T.; Balakrishnan, P.B.; Barthel, M.J.; Piccardi, F.; Niculaes, D.; Marinaro, F.; Fernandes, S.; Curcio, A.; Kakwere, H.; Autret, G.; et al. Thermoresponsive Iron Oxide Nanocubes for an Effective Clinical Translation of Magnetic Hyperthermia and Heat-Mediated Chemotherapy. *ACS Appl. Mater. Interfaces* **2019**, *11*, 5727–5739. [CrossRef]

109. Chung, S.; Revia, R.A.; Zhang, M. Iron oxide nanoparticles for immune cell labeling and cancer immunotherapy. *Nanoscale Horiz.* **2021**, *6*, 696–717. [CrossRef]

110. Persano, S.; Das, P.; Pellegrino, T. Magnetic Nanostructures as Emerging Therapeutic Tools to Boost Anti-Tumour Immunity. *Cancers* **2021**, *13*, 2735. [CrossRef] [PubMed]

111. Zhao, Y.; Zhao, X.; Cheng, Y.; Guo, X.; Yuan, W. Iron Oxide Nanoparticles-Based Vaccine Delivery for Cancer Treatment. *Mol. Pharm.* **2018**, *15*, 1791–1799. [CrossRef] [PubMed]

112. Sanz-Ortega, L.; Rojas, J.M.; Marcos, A.; Portilla, Y.; Stein, J.V.; Barber, D.F. T cells loaded with magnetic nanoparticles are retained in peripheral lymph nodes by the application of a magnetic field. *J. Nanobiotechnology* **2019**, *17*, 14. [CrossRef] [PubMed]

113. Karoon Kiani, F.; Izadi, S.; Ansari Dezfouli, E.; Ebrahimi, F.; Mohammadi, M.; Chalajour, H.; Mortazavi Bulus, M.; Nasr Esfahani, M.; Karpisheh, V.; Mahmoud Salehi Khesht, A.; et al. Simultaneous silencing of the A2aR and PD-1 immune checkpoints by siRNA-loaded nanoparticles enhances the immunotherapeutic potential of dendritic cell vaccine in tumor experimental models. *Life Sci.* **2022**, *288*, 120166. [CrossRef] [PubMed]

114. Zhang, W.; Cao, S.; Liang, S.; Tan, C.H.; Luo, B.; Xu, X.; Saw, P.E. Differently Charged Super-Paramagnetic Iron Oxide Nanoparticles Preferentially Induced M1-Like Phenotype of Macrophages. *Front. Bioeng. Biotechnol.* **2020**, *8*, 537. [CrossRef]

115. Wang, W.; Li, F.; Li, S.; Hu, Y.; Xu, M.; Zhang, Y.; Khan, M.I.; Wang, S.; Wu, M.; Ding, W.; et al. M2 macrophage-targeted iron oxide nanoparticles for magnetic resonance image-guided magnetic hyperthermia therapy. *J. Mater. Sci. Technol.* **2021**, *81*, 77–87. [CrossRef]

116. Zhang, Y.; Gao, X.; Yan, B.; Wen, N.; Lee, W.S.V.; Liang, X.J.; Liu, X. Enhancement of CD8(+) T-Cell-Mediated Tumor Immunotherapy via Magnetic Hyperthermia. *ChemMedChem* **2022**, *17*, e202100656. [CrossRef]

117. Chang, D.; Lim, M.; Goos, J.; Qiao, R.; Ng, Y.Y.; Mansfeld, F.M.; Jackson, M.; Davis, T.P.; Kavallaris, M. Biologically Targeted Magnetic Hyperthermia: Potential and Limitations. *Front. Pharmacol.* **2018**, *9*, 831. [CrossRef]

118. Pucci, C.; Degl'Innocenti, A.; Belenli Gumus, M.; Ciofani, G. Superparamagnetic iron oxide nanoparticles for magnetic hyperthermia: Recent advancements, molecular effects, and future directions in the omics era. *Biomater. Sci.* **2022**, *10*, 2103–2121. [CrossRef]

119. Imashiro, C.; Takeshita, H.; Morikura, T.; Miyata, S.; Takemura, K.; Komotori, J. Development of accurate temperature regulation culture system with metallic culture vessel demonstrates different thermal cytotoxicity in cancer and normal cells. *Sci. Rep.* **2021**, *11*, 21466. [CrossRef]

120. Kanamori, T.; Miyazaki, N.; Aoki, S.; Ito, K.; Hisaka, A.; Hatakeyama, H. Investigation of energy metabolic dynamism in hyperthermia-resistant ovarian and uterine cancer cells under heat stress. *Sci. Rep.* **2021**, *11*, 14726. [CrossRef]

121. van der Zee, J. Heating the patient: A promising approach? *Ann. Oncol.* **2002**, *13*, 1173–1184. [CrossRef]
122. Lee, S.-Y.; Fiorentini, G.; Szasz, A.M.; Szigeti, G.; Szasz, A.; Minnaar, C.A. Quo Vadis Oncological Hyperthermia (2020)? *Front. Oncol.* **2020**, *10*, 1690. [CrossRef]
123. Yagawa, Y.; Tanigawa, K.; Kobayashi, Y.; Yamamoto, M. Cancer immunity and therapy using hyperthermia with immunotherapy, radiotherapy, chemotherapy, and surgery. *J. Cancer Metastasis Treat.* **2017**, *3*, 218–230. [CrossRef]
124. Adnan, A.; Munoz, N.M.; Prakash, P.; Habibollahi, P.; Cressman, E.N.K.; Sheth, R.A. Hyperthermia and Tumor Immunity. *Cancers* **2021**, *13*, 2507. [CrossRef]
125. Carter, T.J.; Agliardi, G.; Lin, F.Y.; Ellis, M.; Jones, C.; Robson, M.; Richard-Londt, A.; Southern, P.; Lythgoe, M.; Zaw Thin, M.; et al. Potential of Magnetic Hyperthermia to Stimulate Localized Immune Activation. *Small* **2021**, *17*, e2005241. [CrossRef]
126. Maier-Hauff, K.; Rothe, R.; Scholz, R.; Gneveckow, U.; Wust, P.; Thiesen, B.; Feussner, A.; von Deimling, A.; Waldoefner, N.; Felix, R.; et al. Intracranial thermotherapy using magnetic nanoparticles combined with external beam radiotherapy: Results of a feasibility study on patients with glioblastoma multiforme. *J. Neurooncol.* **2007**, *81*, 53–60. [CrossRef]
127. Johannsen, M.; Gneveckow, U.; Thiesen, B.; Taymoorian, K.; Cho, C.H.; Waldofner, N.; Scholz, R.; Jordan, A.; Loening, S.A.; Wust, P. Thermotherapy of prostate cancer using magnetic nanoparticles: Feasibility, imaging, and three-dimensional temperature distribution. *Eur. Urol.* **2007**, *52*, 1653–1661. [CrossRef]
128. Wust, P.; Gneveckow, U.; Johannsen, M.; Bohmer, D.; Henkel, T.; Kahmann, F.; Sehouli, J.; Felix, R.; Ricke, J.; Jordan, A. Magnetic nanoparticles for interstitial thermotherapy–feasibility, tolerance and achieved temperatures. *Int. J. Hyperth.* **2006**, *22*, 673–685. [CrossRef]
129. Maier-Hauff, K.; Ulrich, F.; Nestler, D.; Niehoff, H.; Wust, P.; Thiesen, B.; Orawa, H.; Budach, V.; Jordan, A. Efficacy and safety of intratumoral thermotherapy using magnetic iron-oxide nanoparticles combined with external beam radiotherapy on patients with recurrent glioblastoma multiforme. *J. Neurooncol.* **2011**, *103*, 317–324. [CrossRef]
130. Persano, S.; Vicini, F.; Poggi, A.; Fernandez, J.L.C.; Rizzo, G.M.R.; Gavilan, H.; Silvestri, N.; Pellegrino, T. Elucidating the Innate Immunological Effects of Mild Magnetic Hyperthermia on U87 Human Glioblastoma Cells: An In Vitro Study. *Pharmaceutics* **2021**, *13*, 1668. [CrossRef] [PubMed]
131. Li, X.; Gruosso, T.; Zuo, D.; Omeroglu, A.; Meterissian, S.; Guiot, M.C.; Salazar, A.; Park, M.; Levine, H. Infiltration of CD8(+) T cells into tumor cell clusters in triple-negative breast cancer. *Proc. Natl. Acad. Sci. USA* **2019**, *116*, 3678–3687. [CrossRef] [PubMed]
132. Covarrubias, G.; Lorkowski, M.E.; Sims, H.M.; Loutrianakis, G.; Rahmy, A.; Cha, A.; Abenojar, E.; Wickramasinghe, S.; Moon, T.J.; Samia, A.C.S.; et al. Hyperthermia-mediated changes in the tumor immune microenvironment using iron oxide nanoparticles. *Nanoscale Adv.* **2021**, *3*, 5890–5899. [CrossRef] [PubMed]
133. Forte, D.; Barone, M.; Palandri, F.; Catani, L. The "Vesicular Intelligence" Strategy of Blood Cancers. *Genes* **2021**, *12*, 416. [CrossRef] [PubMed]
134. Saggioro, M.; D'Angelo, E.; Bisogno, G.; Agostini, M.; Pozzobon, M. Carcinoma and Sarcoma Microenvironment at a Glance: Where We Are. *Front. Oncol.* **2020**, *10*, 76. [CrossRef] [PubMed]
135. Giraldo, N.A.; Sanchez-Salas, R.; Peske, J.D.; Vano, Y.; Becht, E.; Petitprez, F.; Validire, P.; Ingels, A.; Cathelineau, X.; Fridman, W.H.; et al. The clinical role of the TME in solid cancer. *Br. J. Cancer* **2019**, *120*, 45–53. [CrossRef]
136. Li, X.; Yang, Y.; Huang, Q.; Deng, Y.; Guo, F.; Wang, G.; Liu, M. Crosstalk Between the Tumor Microenvironment and Cancer Cells: A Promising Predictive Biomarker for Immune Checkpoint Inhibitors. *Front. Cell Dev. Biol.* **2021**, *9*, 738373. [CrossRef]
137. Jin, M.Z.; Jin, W.L. The updated landscape of tumor microenvironment and drug repurposing. *Signal Transduct. Target. Ther.* **2020**, *5*, 166. [CrossRef]
138. Liao, D.; Johnson, R.S. Hypoxia: A key regulator of angiogenesis in cancer. *Cancer Metastasis Rev.* **2007**, *26*, 281–290. [CrossRef]
139. Mulens-Arias, V.; Rojas, J.M.; Barber, D.F. The Use of Iron Oxide Nanoparticles to Reprogram Macrophage Responses and the Immunological Tumor Microenvironment. *Front. Immunol.* **2021**, *12*, 693709. [CrossRef]
140. Vitale, I.; Manic, G.; Coussens, L.M.; Kroemer, G.; Galluzzi, L. Macrophages and Metabolism in the Tumor Microenvironment. *Cell Metab.* **2019**, *30*, 36–50. [CrossRef]
141. Komohara, Y.; Fujiwara, Y.; Ohnishi, K.; Takeya, M. Tumor-associated macrophages: Potential therapeutic targets for anti-cancer therapy. *Adv. Drug Deliv. Rev.* **2016**, *99*, 180–185. [CrossRef]
142. De Palma, M.; Lewis, C.E. Macrophage regulation of tumor responses to anticancer therapies. *Cancer Cell* **2013**, *23*, 277–286. [CrossRef]
143. Wynn, T.A.; Chawla, A.; Pollard, J.W. Macrophage biology in development, homeostasis and disease. *Nature* **2013**, *496*, 445–455. [CrossRef]
144. Sica, A.; Mantovani, A. Macrophage plasticity and polarization: In vivo veritas. *J. Clin. Investig.* **2012**, *122*, 787–795. [CrossRef]
145. Pittet, M.J.; Michielin, O.; Migliorini, D. Clinical relevance of tumour-associated macrophages. *Nat. Rev. Clin. Oncol.* **2022**, *19*, 402–421. [CrossRef]
146. Roszer, T. Understanding the Mysterious M2 Macrophage through Activation Markers and Effector Mechanisms. *Mediators Inflamm.* **2015**, *2015*, 816460. [CrossRef]
147. DeNardo, D.G.; Ruffell, B. Macrophages as regulators of tumour immunity and immunotherapy. *Nat. Rev. Immunol.* **2019**, *19*, 369–382. [CrossRef]
148. Mantovani, A.; Marchesi, F.; Malesci, A.; Laghi, L.; Allavena, P. Tumour-associated macrophages as treatment targets in oncology. *Nat. Rev. Clin. Oncol.* **2017**, *14*, 399–416. [CrossRef]

149. Rayahin, J.E.; Gemeinhart, R.A. Activation of Macrophages in Response to Biomaterials. *Results Probl. Cell Differ.* **2017**, *62*, 317–351. [CrossRef]

150. Galdiero, M.R.; Biswas, S.K.; Mantovani, A. Polarized Activation of Macrophages. In *Macrophages: Biology and Role in the Pathology of Diseases*; Biswas, S.K., Mantovani, A., Eds.; Springer: New York, NY, USA, 2014; pp. 37–57.

151. Yu, T.; Gan, S.; Zhu, Q.; Dai, D.; Li, N.; Wang, H.; Chen, X.; Hou, D.; Wang, Y.; Pan, Q.; et al. Modulation of M2 macrophage polarization by the crosstalk between Stat6 and Trim24. *Nat. Commun.* **2019**, *10*, 4353. [CrossRef] [PubMed]

152. Buchacher, T.; Ohradanova-Repic, A.; Stockinger, H.; Fischer, M.B.; Weber, V. M2 Polarization of Human Macrophages Favors Survival of the Intracellular Pathogen Chlamydia pneumoniae. *PLoS ONE* **2015**, *10*, e0143593. [CrossRef] [PubMed]

153. van Dalen, F.J.; van Stevendaal, M.; Fennemann, F.L.; Verdoes, M.; Ilina, O. Molecular Repolarisation of Tumour-Associated Macrophages. *Molecules* **2018**, *24*, 9. [CrossRef] [PubMed]

154. Murray, P.J.; Allen, J.E.; Biswas, S.K.; Fisher, E.A.; Gilroy, D.W.; Goerdt, S.; Gordon, S.; Hamilton, J.A.; Ivashkiv, L.B.; Lawrence, T.; et al. Macrophage activation and polarization: Nomenclature and experimental guidelines. *Immunity* **2014**, *41*, 14–20. [CrossRef] [PubMed]

155. Boutilier, A.J.; Elsawa, S.F. Macrophage Polarization States in the Tumor Microenvironment. *Int. J. Mol. Sci.* **2021**, *22*, 6995. [CrossRef]

156. Jablonski, K.A.; Amici, S.A.; Webb, L.M.; Ruiz-Rosado Jde, D.; Popovich, P.G.; Partida-Sanchez, S.; Guerau-de-Arellano, M. Novel Markers to Delineate Murine M1 and M2 Macrophages. *PLoS ONE* **2015**, *10*, e0145342. [CrossRef]

157. Vogel, D.Y.; Glim, J.E.; Stavenuiter, A.W.; Breur, M.; Heijnen, P.; Amor, S.; Dijkstra, C.D.; Beelen, R.H. Human macrophage polarization in vitro: Maturation and activation methods compared. *Immunobiology* **2014**, *219*, 695–703. [CrossRef]

158. Nascimento Da Conceicao, V.; Sun, Y.; Ramachandran, K.; Chauhan, A.; Raveendran, A.; Venkatesan, M.; DeKumar, B.; Maity, S.; Vishnu, N.; Kotsakis, G.A.; et al. Resolving macrophage polarization through distinct Ca^{2+} entry channel that maintains intracellular signaling and mitochondrial bioenergetics. *iScience* **2021**, *24*, 103339. [CrossRef]

159. Orecchioni, M.; Ghosheh, Y.; Pramod, A.B.; Ley, K. Macrophage Polarization: Different Gene Signatures in M1(LPS+) vs. Classically and M2(LPS-) vs. Alternatively Activated Macrophages. *Front. Immunol.* **2019**, *10*, 1084. [CrossRef]

160. Sierra-Filardi, E.; Nieto, C.; Dominguez-Soto, A.; Barroso, R.; Sanchez-Mateos, P.; Puig-Kroger, A.; Lopez-Bravo, M.; Joven, J.; Ardavin, C.; Rodriguez-Fernandez, J.L.; et al. CCL2 shapes macrophage polarization by GM-CSF and M-CSF: Identification of CCL2/CCR2-dependent gene expression profile. *J. Immunol.* **2014**, *192*, 3858–3867. [CrossRef]

161. Sica, A.; Erreni, M.; Allavena, P.; Porta, C. Macrophage polarization in pathology. *Cell Mol. Life Sci.* **2015**, *72*, 4111–4126. [CrossRef]

162. Sen, P.; Kemppainen, E.; Oresic, M. Perspectives on Systems Modeling of Human Peripheral Blood Mononuclear Cells. *Front Mol. Biosci.* **2017**, *4*, 96. [CrossRef]

163. Italiani, P.; Boraschi, D. Development and Functional Differentiation of Tissue-Resident Versus Monocyte-Derived Macrophages in Inflammatory Reactions. *Results Probl Cell Differ* **2017**, *62*, 23–43. [CrossRef]

164. Isidro, R.A.; Appleyard, C.B. Colonic macrophage polarization in homeostasis, inflammation, and cancer. *Am. J. Physiol. Gastrointest Liver Physiol.* **2016**, *311*, G59–G73. [CrossRef]

165. Murray, P.J.; Wynn, T.A. Protective and pathogenic functions of macrophage subsets. *Nat. Rev. Immunol.* **2011**, *11*, 723–737. [CrossRef]

166. Martinez, F.O.; Helming, L.; Milde, R.; Varin, A.; Melgert, B.N.; Draijer, C.; Thomas, B.; Fabbri, M.; Crawshaw, A.; Ho, L.P.; et al. Genetic programs expressed in resting and IL-4 alternatively activated mouse and human macrophages: Similarities and differences. *Blood* **2013**, *121*, e57–e69. [CrossRef]

167. Ginhoux, F.; Guilliams, M. Tissue-Resident Macrophage Ontogeny and Homeostasis. *Immunity* **2016**, *44*, 439–449. [CrossRef]

168. Epelman, S.; Lavine, K.J.; Randolph, G.J. Origin and functions of tissue macrophages. *Immunity* **2014**, *41*, 21–35. [CrossRef]

169. Ginhoux, F.; Jung, S. Monocytes and macrophages: Developmental pathways and tissue homeostasis. *Nat. Rev. Immunol.* **2014**, *14*, 392–404. [CrossRef]

170. Polasky, C.; Studt, T.; Steuer, A.K.; Loyal, K.; Ludtke-Buzug, K.; Bruchhage, K.L.; Pries, R. Impact of Superparamagnetic Iron Oxide Nanoparticles on THP-1 Monocytes and Monocyte-Derived Macrophages. *Front. Mol. Biosci.* **2022**, *9*, 811116. [CrossRef]

171. Wu, Q.; Miao, T.; Feng, T.; Yang, C.; Guo, Y.; Li, H. Dextrancoated superparamagnetic iron oxide nanoparticles activate the MAPK pathway in human primary monocyte cells. *Mol. Med. Rep.* **2018**, *18*, 564–570. [CrossRef] [PubMed]

172. Gonnissen, D.; Qu, Y.; Langer, K.; Ozturk, C.; Zhao, Y.; Chen, C.; Seebohm, G.; Dufer, M.; Fuchs, H.; Galla, H.J.; et al. Comparison of cellular effects of starch-coated SPIONs and poly(lactic-co-glycolic acid) matrix nanoparticles on human monocytes. *Int. J. Nanomed.* **2016**, *11*, 5221–5236. [CrossRef] [PubMed]

173. Sharkey, J.; Starkey Lewis, P.J.; Barrow, M.; Alwahsh, S.M.; Noble, J.; Livingstone, E.; Lennen, R.J.; Jansen, M.A.; Carrion, J.G.; Liptrott, N.; et al. Functionalized superparamagnetic iron oxide nanoparticles provide highly efficient iron-labeling in macrophages for magnetic resonance-based detection in vivo. *Cytotherapy* **2017**, *19*, 555–569. [CrossRef] [PubMed]

174. Mulens-Arias, V.; Rojas, J.M.; Perez-Yague, S.; Morales, M.P.; Barber, D.F. Polyethylenimine-coated SPIONs trigger macrophage activation through TLR-4 signaling and ROS production and modulate podosome dynamics. *Biomaterials* **2015**, *52*, 494–506. [CrossRef] [PubMed]

175. Jin, R.; Liu, L.; Zhu, W.; Li, D.; Yang, L.; Duan, J.; Cai, Z.; Nie, Y.; Zhang, Y.; Gong, Q.; et al. Iron oxide nanoparticles promote macrophage autophagy and inflammatory response through activation of toll-like Receptor-4 signaling. *Biomaterials* **2019**, *203*, 23–30. [CrossRef]

176. Rojas, J.M.; Sanz-Ortega, L.; Mulens-Arias, V.; Gutierrez, L.; Perez-Yague, S.; Barber, D.F. Superparamagnetic iron oxide nanoparticle uptake alters M2 macrophage phenotype, iron metabolism, migration and invasion. *Nanomedicine* **2016**, *12*, 1127–1138. [CrossRef]

177. Laskar, A.; Eilertsen, J.; Li, W.; Yuan, X.M. SPION primes THP1 derived M2 macrophages towards M1-like macrophages. *Biochem. Biophys. Res. Commun.* **2013**, *441*, 737–742. [CrossRef]

178. Yu, M.; Huang, S.; Yu, K.J.; Clyne, A.M. Dextran and polymer polyethylene glycol (PEG) coating reduce both 5 and 30 nm iron oxide nanoparticle cytotoxicity in 2D and 3D cell culture. *Int. J. Mol. Sci.* **2012**, *13*, 5554–5570. [CrossRef]

179. Wei, H.; Hu, Y.; Wang, J.; Gao, X.; Qian, X.; Tang, M. Superparamagnetic Iron Oxide Nanoparticles: Cytotoxicity, Metabolism, and Cellular Behavior in Biomedicine Applications. *Int. J. Nanomed.* **2021**, *16*, 6097–6113. [CrossRef]

180. Kodali, V.; Littke, M.H.; Tilton, S.C.; Teeguarden, J.G.; Shi, L.; Frevert, C.W.; Wang, W.; Pounds, J.G.; Thrall, B.D. Dysregulation of macrophage activation profiles by engineered nanoparticles. *ACS Nano* **2013**, *7*, 6997–7010. [CrossRef]

181. Robert, J. *Textbook of Cell Signalling in Cancer: An Educational Approach*; Springer: Berlin/Heidelberg, Germany, 2015. [CrossRef]

182. Kawai, T.; Akira, S. TLR signaling. *Cell Death Differ.* **2006**, *13*, 816–825. [CrossRef]

183. Heinsbroek, S.E.; Gordon, S. The role of macrophages in inflammatory bowel diseases. *Expert Rev. Mol. Med.* **2009**, *11*, e14. [CrossRef]

184. Lim, S.G.; Kim, J.K.; Suk, K.; Lee, W.H. Crosstalk between signals initiated from TLR4 and cell surface BAFF results in synergistic induction of proinflammatory mediators in THP-1 cells. *Sci. Rep.* **2017**, *7*, 45826. [CrossRef]

185. Broz, P.; Monack, D.M. Newly described pattern recognition receptors team up against intracellular pathogens. *Nat. Rev. Immunol.* **2013**, *13*, 551–565. [CrossRef]

186. Chen, H.; Li, P.; Yin, Y.; Cai, X.; Huang, Z.; Chen, J.; Dong, L.; Zhang, J. The promotion of type 1 T helper cell responses to cationic polymers in vivo via toll-like receptor-4 mediated IL-12 secretion. *Biomaterials* **2010**, *31*, 8172–8180. [CrossRef]

187. Zhao, Y.; Hu, X.; Liu, Y.; Dong, S.; Wen, Z.; He, W.; Zhang, S.; Huang, Q.; Shi, M. ROS signaling under metabolic stress: Cross-talk between AMPK and AKT pathway. *Mol. Cancer* **2017**, *16*, 79. [CrossRef]

188. Siegrist, S.; Kettiger, H.; Fasler-Kan, E.; Huwyler, J. Selective stimulation of the JAK/STAT signaling pathway by silica nanoparticles in human endothelial cells. *Toxicol. Vitr.* **2017**, *42*, 308–318. [CrossRef]

189. Nascimento, C.S.; Alves, E.A.R.; de Melo, C.P.; Correa-Oliveira, R.; Calzavara-Silva, C.E. Immunotherapy for cancer: Effects of iron oxide nanoparticles on polarization of tumor-associated macrophages. *Nanomedicine* **2021**, *16*, 2633–2650. [CrossRef]

190. Sukhbaatar, N.; Weichhart, T. Iron Regulation: Macrophages in Control. *Pharmaceuticals* **2018**, *11*, 137. [CrossRef]

191. Jennifer, B.; Berg, V.; Modak, M.; Puck, A.; Seyerl-Jiresch, M.; Kunig, S.; Zlabinger, G.J.; Steinberger, P.; Chou, J.; Geha, R.S.; et al. Transferrin receptor 1 is a cellular receptor for human heme-albumin. *Commun. Biol.* **2020**, *3*, 621. [CrossRef]

192. Nelson, N.R.; Port, J.D.; Pandey, M.K. Use of Superparamagnetic Iron Oxide Nanoparticles (SPIONs) via Multiple Imaging Modalities and Modifications to Reduce Cytotoxicity: An Educational Review. *JNT* **2020**, *1*, 8. [CrossRef]

193. Nairz, M.; Theurl, I.; Swirski, F.K.; Weiss, G. "Pumping iron"-how macrophages handle iron at the systemic, microenvironmental, and cellular levels. *Pflugers Arch.* **2017**, *469*, 397–418. [CrossRef] [PubMed]

194. Castaneda, O.A.; Lee, S.C.; Ho, C.T.; Huang, T.C. Macrophages in oxidative stress and models to evaluate the antioxidant function of dietary natural compounds. *J. Food Drug Anal.* **2017**, *25*, 111–118. [CrossRef] [PubMed]

195. Canton, M.; Sanchez-Rodriguez, R.; Spera, I.; Venegas, F.C.; Favia, M.; Viola, A.; Castegna, A. Reactive Oxygen Species in Macrophages: Sources and Targets. *Front. Immunol.* **2021**, *12*, 734229. [CrossRef]

196. Freeman, S.A.; Grinstein, S. Phagocytosis: Receptors, signal integration, and the cytoskeleton. *Immunol. Rev.* **2014**, *262*, 193–215. [CrossRef]

197. Tertrais, M.; Bigot, C.; Martin, E.; Poincloux, R.; Labrousse, A.; Maridonneau-Parini, I. Phagocytosis is coupled to the formation of phagosome-associated podosomes and a transient disruption of podosomes in human macrophages. *Eur. J. Cell Biol.* **2021**, *100*, 151161. [CrossRef]

198. Wheeler, K.C.; Jena, M.K.; Pradhan, B.S.; Nayak, N.; Das, S.; Hsu, C.D.; Wheeler, D.S.; Chen, K.; Nayak, N.R. VEGF may contribute to macrophage recruitment and M2 polarization in the decidua. *PLoS ONE* **2018**, *13*, e0191040. [CrossRef]

199. Richter, M.; Piwocka, O.; Musielak, M.; Piotrowski, I.; Suchorska, W.M.; Trzeciak, T. From Donor to the Lab: A Fascinating Journey of Primary Cell Lines. *Front. Cell Dev. Biol.* **2021**, *9*, 711381. [CrossRef]

 *pharmaceutics*

Review

An Overview on Immunogenic Cell Death in Cancer Biology and Therapy

Mosar Corrêa Rodrigues [1,2,†], José Athayde Vasconcelos Morais [1,2,†], Rayane Ganassin [1,2], Giulia Rosa Tavares Oliveira [1,2], Fabiana Chagas Costa [1,2], Amanda Alencar Cabral Morais [2], Ariane Pandolfo Silveira [2], Victor Carlos Mello Silva [2], João Paulo Figueiró Longo [2] and Luis Alexandre Muehlmann [1,2,*]

[1] Faculty of Ceilandia, University of Brasilia, Brasilia 72220-275, Brazil; mosarcr@gmail.com (M.C.R.); joseavmorais@gmail.com (J.A.V.M.); rayaneganassin@hotmail.com (R.G.); giulia_rosa12@hotmail.com (G.R.T.O.); fchagas16@gmail.com (F.C.C.)
[2] Laboratory of Nanobiotechnology, Department of Genetics and Morphology, Institute of Biological Sciences, University of Brasilia, Brasilia 70910-900, Brazil; amandaalencarcabral@gmail.com (A.A.C.M.); pandolfo.ariane@gmail.com (A.P.S.); victor@spctm.com (V.C.M.S.); jplongo82@gmail.com (J.P.F.L.)
* Correspondence: luisalex@unb.br
† These authors contributed equally to this work.

Abstract: Immunogenic cell death (ICD) is a modality of regulated cell death that is sufficient to promote an adaptive immune response against antigens of the dying cell in an immunocompetent host. An important characteristic of ICD is the release and exposure of damage-associated molecular patterns, which are potent endogenous immune adjuvants. As the induction of ICD can be achieved with conventional cytotoxic agents, it represents a potential approach for the immunotherapy of cancer. Here, different aspects of ICD in cancer biology and treatment are reviewed.

Keywords: DAMPs; immune system; chemotherapy; photodynamic therapy; radiotherapy; immunotherapy

Citation: Rodrigues, M.C.; Morais, J.A.V.; Ganassin, R.; Oliveira, G.R.T.; Costa, F.C.; Morais, A.A.C.; Silveira, A.P.; Silva, V.C.M.; Longo, J.P.F.; Muehlmann, L.A. An Overview on Immunogenic Cell Death in Cancer Biology and Therapy. *Pharmaceutics* **2022**, *14*, 1564. https://doi.org/10.3390/pharmaceutics14081564

Academic Editor: Jun Dai

Received: 27 June 2022
Accepted: 25 July 2022
Published: 27 July 2022

Publisher's Note: MDPI stays neutral with regard to jurisdictional claims in published maps and institutional affiliations.

1. Introduction

Both the innate and the adaptive branches of the immune system are involved in the elimination of malignant cells. Generally, for the successful elimination of tumors by immunity, the in situ presence of immunoadjuvants and antigens, as well as a non-immunosuppressor tumor microenvironment, are essential.

Tumor cells commonly express neoantigens, which are different from any other normal protein in the host, or tumor-associated antigens, which are expressed in an unusual tissue or in aberrantly high amounts [1,2]. As a result, one important feature of cancers is their ability to escape immunity, as the immune system is crucially involved in the defense against the development and progression of malignant cells [3].

In certain situations, the immunogenicity of tumors can be enhanced by increasing the local release and exposure of endogenous immunoadjuvants, such as damage-associated molecular patterns (DAMPs). The release of these molecules with a well-defined, immune-activating pattern is observed in a regulated cell death (RCD) modality known as immunogenic cell death (ICD). In this review we discuss how ICD can help to trigger or boost immune responses against cancer.

2. A Historical Overview on Cancer Immunotherapy

Activating or boosting adaptive immune responses against tumors has become a main pillar in the treatment of different cancers. Currently, immune-checkpoint inhibitors, chimeric antigen receptor (CAR) T cells, dendritic cell (DC)-based vaccines and immunostimulatory cytokines have been successfully used in clinical practice [4]. Different immune-checkpoint inhibitors approved by the FDA have significantly improved the treatment of advanced-stage melanoma, non-small-cell lung carcinoma, kidney carcinoma, urothelial carcinoma, hepatocellular carcinoma, and others [4,5].

The first evidence that the host immune system could be therapeutically targeted to treat cancer was brought to light in the 19th century, when erysipelas was reported to have an influence on tumor progression. This erythematous infection of the skin was relatively common in post-surgery patients given the poor sanitary conditions of surgery at that time. In 1867, the German physician Wilhelm Busch reported that a malignant tumor disappeared after the patient had contracted this infection [6].

In 1883, the German surgeon Friedrich Fehleisen demonstrated that the etiologic agent of erysipelas was *Streptococcus pyogenes* [7]. In his experiments, Fehleisen produced erysipelas by inoculating patients with pure cultures of *S. pyogenes*. On one occasion, out of six patients with inoperable malignancies who reacted to this inoculum, three showed only a slight change in their tumors. The other three patients, however, exhibited significant changes in the evolution of the disease. The fibrosarcoma nodules of one patient thus treated were reported to have disappeared, while two other patients with breast carcinoma had tumor remission. Similar clinical experiments were performed by other surgeons at that time [8].

In 1891, the American surgeon William Bradley Coley published an article citing the work of Fehleisen and his own observations on the curative effect of erysipelas upon malignancies [9]. Coley injected patients with *S. pyogenes*, inducing erysipelas on purpose. Some of these patients, after having recovered from local inflammation and fever, exhibited reduced tumor size or were even cured of the malignancy. According to Coley, some promptly responded to this therapy upon the first injection, while others needed several doses before some clinical improvement could be observed. As there were cases of death after the injection, Coley decided to inactivate the bacteria with heat before injecting them, in a preparation which was named Coley's toxin. The use of this bacterial preparation gave good results in bone and soft-tissue sarcomas [10]. This was one of the first anticancer immunotherapies described in the scientific literature.

Despite the reported success, many were quite skeptical about the results described by Coley. As recognized later, his work was probably the first formal study on immunotherapy and presented a rationale for inducing immune responses against tumors, but the lack of knowledge on immune mechanisms at the time his first patients were treated was a huge barrier to the acceptance of this approach [10,11]. However, as immunology advanced, many oncologists began to support the use of Coley's toxin. In 1935, the prestigious surgeon Ernest Amory Codman asserted that the results described by Coley were solid scientific evidence [10,12]. Codman was particularly impressed with six cases registered by Coley of five-year cures of patients with Ewing's sarcoma injected with the toxin [12]. Codman suggested that the increased production of lymphocytes induced by Coley's toxin might explain the "occasional miracle which follows this treatment".

In 1909, the renowned German scientist and physician Paul Ehrlich suggested that the immune system played an important role in eliminating cancer cells, speculating that it might be involved in suppressing the development and progression of carcinomas [13]. Later, as the role of the immune system in distinguishing self from non-self became clear, Frank Macfarlane Burnet developed the concept of immunosurveillance [14,15].

As Burnet wrote [15], cancer cells grow free from the normal control exercised by the organism as a whole. Cancer is then the consequence of the breakdown in one or more aspects in this control mechanism that holds the multitude of cells of an organism working together as a single functioning unit [15]. One key part of this normal control was the immune system, which is normally able to identify and to mount an effective reaction against tumor-specific or -associated antigens, eventually eliminating most of the potential cancer cells before they become a clinically apparent tumor [15].

In a favorable immune context, tumor antigens are processed and presented by antigen presenting cells (APCs), triggering the activation of $CD4^+$ and $CD8^+$ T cells, ultimately eliminating the tumor. However, malignant cells can suppress their own immunogenicity, thus avoiding being detected by the immune system. Cancer cells can, for instance, upregulate the expression of programmed cell death ligand 1 (PD-L1) molecules, present defects in the antigen presentation machinery, recruit immunosuppressive cells, such as myeloid-derived suppressor cells (MDSC) and T regulatory (Treg) cells, and lead to direct or indirect secretion of TGF-β and IL-10 [5,16].

The development of an immunoevasive phenotype is the result of a long, complex and dynamic interaction of transformed host cells and the immune system in what is known as cancer immunoediting [16,17]. This process can be divided into three stages: elimination, equilibrium and escape. If elimination is successful, the immune system eradicates aberrant cells and prevents the continuation of carcinogenesis. This event represents the fulfillment of immunosurveillance and is highly dependent on the immunogenicity of the abnormal cells. If the elimination is not complete, however, an equilibrium stage may occur, i.e., although these cells persist, they are not able to freely proliferate, and do not generate clinically apparent tumors. Occasionally, some of these aberrant cells may develop more efficient immune evasion abilities, then becoming poorly immunogenic or non-immunogenic. These cells eventually escape the immune system and generate a growing tumor mass [17,18].

Therefore, changing the balance towards the host immune system could be used therapeutically against cancers. In the 20th century, many advances in immunology contributed to the development of modern immunotherapy [19]. Some of these advances are particularly noteworthy: (i) the identification and demonstration of the role of T lymphocytes in animal models [20]; (ii) the demonstration of the presence of dendritic cells in peripheral lymphoid organs [21]; and (iii) the identification of natural killer cells (NK cells) [22]. Many approaches to target the immune system as an anticancer strategy followed, such as the use of cytokines, vaccines, adoptive cell therapies and antibodies against immune checkpoints [11].

The most recent milestone in modern immunotherapy, the discovery of immune checkpoint inhibitors, led to the development and approval by the FDA of anti-PD-1/PD-L1 and anti-CTLA-4 antibodies, which proved to be effective against melanoma and other different tumors [23].

Currently, many studies show that it is possible to increase the immunogenicity of tumors by triggering specific cell death modalities in cancer cells. In that regard, it has been observed that specific cytotoxic treatments, such as anthracycline-based chemotherapy [24] radiotherapy [25] and photodynamic therapy (PDT) [26,27], can induce immunogenic cell death (ICD), which can render cancers more efficient at triggering or boosting tumor antigen-specific immune responses. As this event can lead to the elimination of the primary tumor, and of occasional antigen-sharing metastatic foci as well, the induction of ICD has been suggested as a potential immunotherapeutic approach.

3. Immunogenic Cell Death

As defined by the Nomenclature Committee on Cell Death, ICD is a modality of RCD [28,29]. This implies that ICD activation depends on signaling transduction programs, and can thus be triggered or modulated by drugs and genetic components [28].

ICD differs from other RCD types, such as necroptosis, ferroptosis and pyroptosis, not only by the conditions under which it is triggered, but also by the fact that it activates an adaptive immune response in immunocompetent syngeneic hosts against antigens expressed by the dying cell [28]. The cells undergoing ICD exhibit morphological and molecular hallmarks of apoptosis with a well-defined pattern of release and exposure of DAMPs [29]. Thus, cells at ICD exhibit the two conditions necessary for eliciting an adaptive immune response: antigenicity and adjuvanticity [28,29].

DAMPs form a group of different molecules that normally perform structural and metabolic functions in living cells unrelated to their immune functions during ICD [30]. However, when emitted with a specific temporospatial profile during ICD, DAMPs are able to trigger or boost antigen-specific immune responses [31,32]. The activation of pattern recognition receptors (PRR) by DAMPs results in the maturation of dendritic cells (DCs) and the activation of CD4$^+$ and CD8$^+$ T cells [33], as represented in Figure 1. Different aspects of the activation of adaptive immune responses are fulfilled by the different DAMPs involved in ICD.

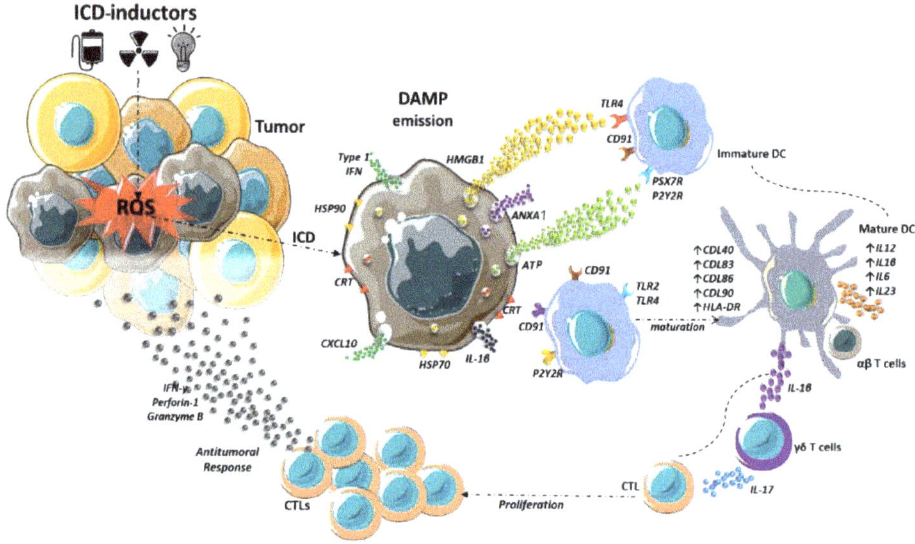

Figure 1. Engagement of adaptive immune response after immunogenic cell death in a tumor. Chemotherapy, radiotherapy, and photodynamic therapy can induce immunogenic cell death (ICD), which is a programmed cell death accompanied by the exposure of damage-associated molecular patterns (DAMPs). This can occur, for instance, as a consequence of oxidative stress in the endoplasmic reticulum. Some DAMPs, such as heat-shock protein (HSP)70, HSP90, and calreticulin are exposed on the plasma membrane, while others such as adenosine triphosphate (ATP), high mobility group box 1 protein (HMGB1), C-X-C motif chemokine ligand 10 (CXCL10) and annexin A1 (ANXA-1) are released to the extracellular medium. DAMPs then activate pattern recognition receptors of dendritic cells (DCs) and other antigen-presenting cells. This culminates in the maturation of the DCs and in the recruitment and activation of T cells. In this way, ICD can trigger or boost an adaptive antitumor immune response.

The time profile of the release and exposure, as well as the specific actions performed by the DAMPs, orchestrates the attraction, phagocytic activity and maturation of APCs in the tumor bed. The simple presence of a single DAMP or just a couple of them in the vicinity of tumor cells is generally not enough for the initiation or boosting of a cytotoxic, effective anticancer immune response. The absence of calreticulin or ANXA1, for example, is known to severely limit immune responses against tumor cells [34–36]. Moreover, the tumor microenvironment must permit immune cells to be activated and to perform their roles properly in order for DAMPs to exert their immunoadjuvant effects.

Specific conditions are known to induce ICD, such as chemotherapy [37–39], radiotherapy [40,41] and PDT [26,27,42–45]. Parameters such as the protocol of application and the drug used in these treatments are crucial to determine whether ICD is induced or not. In the case of chemotherapy, for instance, ICD can be induced by drugs such as mitoxantrone, oxaliplatin and cyclophosphamide, but not by cisplatin, etoposide and mitomycin C. In the case of PDT, the type and concentration of the photosensitizer, as well as the irradiation regimen, are key factors [26,27].

4. Endoplasmic Reticulum Stress

Different ICD-inducers act as stressors of the endoplasmic reticulum (ER) of target cells [31,33,46]. Indeed, the ER has been linked by many studies to programmed cell death [46–48]. Stressing conditions, such as oxidative stress and hyperthermia, can impair the folding of proteins. Unfolded proteins bind the luminal ER chaperone GRP78/BiP (Figure 2), triggering the activation of three ER transmembrane proteins: inositol-requiring enzyme 1α (IRE1α), protein kinase RNA-like endoplasmic reticulum kinase (PERK), and activating transcription factor 6 (ATF6). The downstream cellular events following the activation of these ER sensors can either restore normal ER metabolism or induce cell death [48].

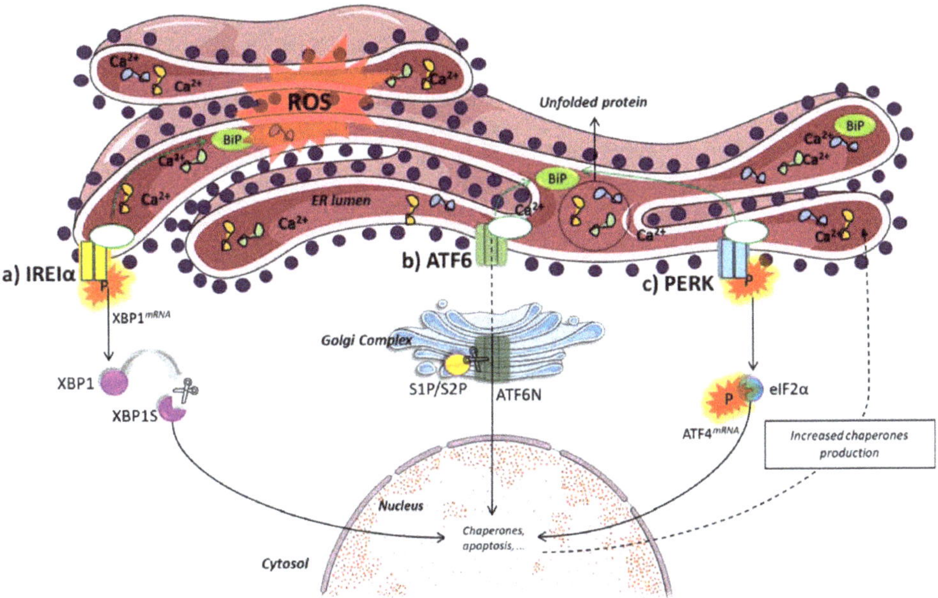

Figure 2. Activation of responses to protein malformation. These three pathways—IRE1α (**a**), ATF6 (**b**) and PERK (**c**)—are activated in unfolded protein response (UPR) to cope with disturbances in protein folding and to restore endoplasmic reticulum homeostasis after stress. The phosphorylation of eIF2α is a hallmark of immunogenic cell death. Legend: ROS: reactive oxygen species; BiP: Binding immunoglobulin protein; ER: endoplasmic reticulum; IRE1α: inositol-requiring enzyme 1α; XBP1: X box-binding protein 1 mRNA; XBP1S: spliced X box-binding protein 1 mRNA; ATF6: activating transcription factor 6; PERK: protein kinase RNA-like endoplasmic reticulum kinase; ATF4: activating transcription factor 4; S1P: site-1 protease; S2P: site-1 protease; eIF2α: eukaryotic initiation factor 2α.

The activated IRE1α dimerizes and is autophosphorylated, becoming able to catalyze the unconventional, cytosolic splicing of the mRNA for the transcription factor X box-binding protein 1 (uXBP1) to sXBP1, which is then translated. The XBP1 is a key protein in UPR, activating the transcription of different genes that can reinstate normal ER metabolism [49]. The PERK protein also dimerizes and auto-phosphorylates, activating its kinase domain to phosphorylate the eukaryotic initiation factor 2α (eIF2α) [50,51]. The phosphorylation of eIF2α is the key event in the integrated stress response (ISR), which is a part of the ER stress response, and is common to ICD induced by different cytotoxicants [34]. Thus, the presence of phosphorylated eIF2α is a hallmark of ICD and can be used as a biomarker of ISR in cell cultures and in biological samples [52].

Phosphorylated elF2α triggers the selective translation of the activating transcription factor (AFT4), which activates the expression of genes involved in protein folding, amino acid metabolism and regulation of oxidative stress [53]. Also, ATF6 is translocated to the Golgi apparatus, where it is then cleaved to ATF6N, releasing its transcription factor moiety that activates the transcription of genes for chaperones, XBP1, C/enhancer binding protein-homologous protein (CHOP), and others [48,54].

The UPR is important for cells under stress to survive and restore homeostasis. The activation of this response is often observed in cancer cells, as the tumor is often a stressing environment, with acidic pH and hypoxia, for instance. Thus, the phosphorylation of elF2α is frequently observed in cancer cells and plays an important role in tumor growth, invasion, and angiogenesis [55]. In this situation, tumors can cope with the constant stress and progress.

However, if the UPR is unable to restore the ER protein folding capacity, then cell death is triggered. In this context, the pro-apoptotic factor CHOP plays a key role [55–57]. The PERK/elF2α/ATF4 pathway activates the production of CHOP [53], which activates the expression of the BH3-only protein Bim. This leads to the activation of Bax/Bak and the release of cytochrome C to the cytosol, with the consequent induction of apoptosis [56,58]. It is worth mentioning that CHOP also acts on the Bcl2 protein (autophagy activation factor), inhibiting its action [59,60]. IRE1α and ATF6 will also activate CHOP via fragments of XBP1S or ATF6N, thereby triggering apoptosis [60,61].

An intense stress in the ER not only activates apoptosis but also promotes the exposure of ER chaperones, such as calreticulin and HSP70, and other DAMPs involved in ICD [62]. The main DAMPs are discussed in the next section.

5. Damage-Associated Molecular Patterns

Processes associated with ICD result in the emission of DAMPs, such as HSP70, HSP90, calreticulin, HMGB1, ATP, type I IFN, cancer cell-derived nucleic acids, ANXA1, and others [29,31,63]. The recognition of DAMPs by cells of the immune system triggers intracellular signaling pathways involved in responses to injury [33]. Frequently, many of these DAMPs perform specific functions in cells not related to immunity. Nevertheless, outside the cells, they can activate various immune cell receptors from the family of PRR. They are also capable of potentiating pro-inflammatory effects such as the maturation and activation of antigen-presenting cells, such as dendritic cells and macrophages, which ultimately activate T cells (CD8$^+$), creating an anti-tumor immunity [32].

The DAMPs are emitted by different mechanisms, such as the active exposure on plasma membrane (e.g., calreticulin, HSP70 and HSP90), released as end-stage degradation products (e.g., DNA and RNA), and secreted to the extracellular medium (e.g., ATP, HMGB1, uric acid, IL-1α and other pro-inflammatory cytokines) [33,64].

A well-studied DAMP is calreticulin, a soluble protein found mainly in the ER. Its main functions in this organelle are the buffering of Ca^{2+} [65], contributing to the homeostasis of this cation, and the lectin-like chaperone activity, essential to the folding of N-glycosylated proteins [66]. During ICD, it is exposed on the plasma membrane even before phosphatidylserine [67] and acts as an eat-me signal, inducing antigen-presenting cells to phagocytose the target cell [68]. Although calreticulin is essential to the immunogenicity of cells undergoing ICD, its presence alone on the cell surface is not enough for starting an immune response [36]. The presence of other DAMPs, therefore, is crucial for the immunogenicity of ICD. Other chaperones exposed on the plasma membrane in ICD are HSPs, which also act as eat-me signals for phagocytes [69,70].

In cancer cells undergoing ICD, type I interferons (IFN-Is) are released by mechanisms involving the detection of endogenous nucleic acids [28], i.e., dsRNA by TLR3 [71], or ds-DNA by cyclic GMP-AMP synthase (cGAS) [72]. The release of this DAMP has been shown to be crucial for the success of the ICD-inducing anthracycline-based chemotherapy [71]. IFN-Is are potent immunostimulatory proteins with significant anticancer effects, and they have been used to treat chronic myeloid leukemia, acute lymphoblastic leukemia, multiple myeloma, melanoma, Kaposi's sarcoma, and renal cell and bladder carcinoma [73]. It is notable that the nucleic acids released by cancer cells at ICD are also taken up by DCs and other immune cells, triggering potent IFN-I responses [28].

ANXA1 is also among the crucial DAMPs in ICD. This protein is expressed in granules of different cells, such as neutrophils, eosinophils and monocytes, and is found in different organs and tissues, such as the lungs the bone marrow and intestine [74,75]. Like other members of the annexins superfamily, ANXA1 binds phospholipids and can thus affect eicosanoid production [74]. It has an important role in the regulation and resolution of inflammation. Although ANXA1 has been described as a negative regulator of innate immunity, with neutrophils being its main target [74], it is known to be passively released by cells undergoing ICD, then activating the migration of APCs towards the dying cells, facilitating their engulfment and processing [34]. The importance of this DAMP in tumor immunology can be noted by the fact that the low expression of ANXA1 is often associated with poor DCs and T lymphocyte infiltration in the tumor bed and a higher ability of different human cancers to escape immunity [35].

Another DAMP released to the extracellular medium is ATP [43,76]. Extracellular ATP operates as a strong chemoattractant and promotes not only the recruitment of immune cells but also their maturation [77,78]. In addition to serving as one of the main sources of intracellular energy, ATP also acts in the extracellular signaling mechanisms [67]. ATP is secreted in response to the cytotoxic and cytostatic effects of some aggressive agents such as chemotherapy drugs [46]. ATP release occurs during the process of apoptosis, and different mechanisms can lead to the release of ATP from the intracellular to the extracellular environment, such as the caspase-dependent activation of Rho-associated, coiled-coil containing protein kinase 1 (ROCK1)-mediated, myosin II-dependent cellular blebbing, as well as the opening of pannexin 1 (PANX1) channels, which is also triggered by caspases and autophagy [79–81]. Of these three, opening pannexin 1 channels and autophagy appear to occur in ICD.

During the formation of apoptotic bodies, caspases 3 and/or 7 act in the cleavage in the C-terminal portion leading to the opening and activation of PANX 1, then releasing the ATP into the extracellular region [82]. In autophagy, the cytoplasmic constituents are degraded in double-membrane organelles, and the autophagosomes are subsequently fused with lysosomes, resulting in the degradation of the autophagocytic content by acid hydrolases and recycling towards energy metabolism or anabolic reactions, releasing the ATP for the extracellular portion [83]. ATP binds the purinergic receptors P2RY2 and P2RX7 of DCs, resulting in the influx of K^+ and Ca^{2+} ions, activating the inflammasome NLRP3, followed by activation of caspase-1, consequently stimulating the proteolytic maturation and secretion of interleukin 1 (IL-1) and interleukin 18 (IL-18), contributing to the immunogenic response [84].

HMGB1 is a non-histone nuclear protein that, once secreted, acts as an essential DAMP in the activation of DCs by ICD [76,85–87]. Its release occurs sometime after apoptosis [63,86]. It can also be released by cells of the innate immune system without the need for cell death in response to pathogens or secreted by cells in response to some damage in the late phase of apoptosis—necrosis [86,88,89]. HMGB1 activates toll-like receptors (TLR)4 and 2, and stimulates the production of pro-inflammatory factors by DCs [90], strongly contributing to the immunogenicity of ICD [33,91].

The exposure of calreticulin and the release of ANXA1, ATP and HMGB1 result in the attraction and maturation of DCs in the tumor microenvironment [91,92]. Once there, the DCs phagocytose the tumor cells and release pro-inflammatory mediators in the tumor, such as IL-1β, IL-18, IFN-γ, among others. Activated DCs also present the tumor antigens, culminating mainly in the activation of $CD8^+$ cytotoxic T lymphocytes, which are fundamental in the activation of the tumor retraction immune antitumor response [33,43]. The effects of the DAMPs discussed in this section are summarized in Table 1.

Table 1. Damage-associated molecular patterns involved in immunogenic cell death.

DAMP	Abbreviation	Effect on Immune Cells	References
Annexin A1	ANXA1	Expressed in different cells (neutrophils, eosinophils, and monocytes), ANXA1 has a role in the regulation and resolution of inflammation. It can act as a negative regulator of innate immunity, with neutrophils being its main target; it activates the migration of APCs towards the dying cells, facilitating their engulfment and processing.	[34,74]
Adenosine triphosphate	ATP	ATP acts as a strong chemoattractant and promotes not only the recruitment of immune cells but also their maturation.	[73,75]
Calreticulin	CRT	Calreticulin acts as phagocytosis inducer. Its exposure and the release of ANXA1, ATP and HMGB1 result in the attraction and maturation of DCs in the tumor microenvironment.	[36,39]
Deoxyribonucleic acid	DNA	With its accumulation in the cytoplasm, DNA can stimulate innate immune responses.	[93]
High mobility group box 1 protein	HMGB1	Acts as an essential DAMP in the DCs activation, stimulating the production of pro-inflammatory factors, strongly contributing to the immunogenicity of ICD.	[29,94]
Heat-shock protein	HSP 70 HSP 90	HSP act as eat-me signals for phagocytes. They can induce DC maturation and promote target engulfment by APC cells.	[29]
Type I interferon	IFN-I	IFN-Is acts as potent immunostimulatory proteins and have a crucial role in ICD. It can modulate the maturation, differentiation, and migration of DC cells, increase primary antibody responses, and activate B and T cells directly or indirectly.	[29,71]
Ribonucleic acid	RNA	It recruits leukocyte and M1-type macrophages.	[95]
Uric acid	UA	Crystalline UA can produce inflammatory mediators through macrophage activation and the enhancement of T cells.	[96]

6. ICD and DAMPs in Cancer Therapy

The literature on ICD in cancer biology has been significantly expanded over the last years, providing new possibilities for cancer treatment. Firstly, the discovery that ICD can be induced by certain classical anticancer agents has made it clear that some treatment regimens should be designed not only to directly eliminate tumor cells, but also to optimize their capacity to induce immune responses against tumor antigens. Secondly, ICD inducers can be used in combination with immunotherapeutics, such as immune checkpoint inhibitors.

Recently, Ganassin et al. (2022) demonstrated that curcumin, a polyphenol obtained from turmeric, causes ER stress and ICD in colorectal adenocarcinoma CT26 cells. The authors observed that, when treated with curcumin, these cells exhibited an initial increase of intracellular Ca^{2+}, which was followed by the activation of XBP1, a protein involved in the UPR response. The ER stress initiated by curcumin was accompanied by the induction of ICD.

Regarding chemotherapeutics, different drugs have been found to induce ICD in preclinical studies [36]. Moreover, chemotherapeutics such as doxorubicin, mitoxantrone, idarubicin, cyclophosphamide, bortezomib and oxaliplatin have already been clinically tested as ICD inducers, mainly in combination with immunotherapeutics [39]. For instance, the combination of doxorubicin, an ICD inducer, with the PD-1-targeting immune-checkpoint blocker, nivolumab, resulted in 35% objective response rates (ORRs) in the treatment of metastatic triple negative breast cancer patients [39,97]. When nivolumab was combined with cisplatin, a non-ICD inducer chemotherapeutic, ORRs were lower (23%) [97]. A number of other clinical studies are already under way to test this immunostimulatory combination using not only immune-checkpoint blockers, but also CAR-T cells, DC-based vaccines, immunostimulatory cytokines, and others [39].

Interesting results of the Phase II study published by Bota et al. (2018) also show that ICD-inducers can improve clinical outcomes in glioblastoma (WHO grade IV astrocytic glioma) immunotherapy [98]. The authors used an allogeneic/autologous therapeutic glioblastoma vaccine (ERC1671, Gliovac), which is a mixture of inactivated tumor cells and lysates of tumor cells derived from the treated patient and three other glioblastoma patients, combined with recombinant colony stimulating factor 2 (CSF2, best known as GM-CSF, used to activate immune responses). The patients received a short regimen of low-dose cyclophosphamide (50 mg/day for 4 days) prior to vaccination. Cyclophosphamide is an ICD-inducer, used in this case, according to the authors, for relaxing the immunosuppressive environment. Indeed, low-dose cyclophosphamide has been reported to suppress Treg cell response, helping to promote DC expansion and antitumor cytotoxic T cell-mediated response [99]. The protocol of treatment described by Bota et al. (2018) also included the treatment of these glioblastoma patients with bevacizumab, a monoclonal antibody specific to VEGF. The results showed increased survival for the patients thus treated (12 months vs. 7.5 months for patients receiving bevacizumab only). Although the authors did not discuss the occasional contribution of ICD to the outcomes, the benefit observed with this protocol makes it worth investigating the possible contribution of ICD in preclinical and clinical models, which could help to improve this combinatory therapy in the future.

Radiotherapy has been shown to induce ICD as well. In clinical practice, radiotherapy and chemoradiotherapy were shown to induce the exposure of the ICD-related DAMPs calreticulin, HSP70 and HMGB1 when applied either as a pre- or post-operative protocol of different cancers [100]. The results, however, are still contradictory. A possible cause for this can be the different experimental settings used and insufficient data regarding ICD hallmarks.

Lämmer et al. (2019) investigated the expression of cytosolic HSP70 in tumor tissue of 60 patients diagnosed with primary glioblastoma. As discussed before in this review, HSP70 is released during ICD, acting as a DAMP. In this study [101], the tumors were surgically resected and patients were then treated with radiotherapy and temozolomide chemotherapy. The progression-free survival and overall survival were significantly longer in patients exhibiting a higher expression of cytosolic HSP70. The authors hypothesized that this result may be due to the induction of ICD by the combination of radiotherapy and chemotherapy given to patients.

Rothammer et al. (2019) analyzed the concentration of HSP70 in the serum of 40 breast cancer patients, who received breast-conserving surgery and adjuvant radiotherapy [102]. The authors observed that patients with higher serum concentrations of HSP70 had an increased probability of developing contralateral recurrence or metastases within two years of receiving radiotherapy.

Protocols based on the combination of radiotherapy and immunotherapeutics have also been tested, and it remains unclear if patients benefit from radiotherapy-induced ICD, as the clinical data are contradictory. For instance, in the treatment of rectal cancer, the increase in HMGB1 was associated both with poorer [103] and with better [104] responses to chemoradiotherapy.

Hongo et al. (2015) studied 75 patients with lower rectal cancer who were treated with preoperative chemoradiotherapy, consisting of radiotherapy (1.8 Gy × 28 fractions) and chemotherapy with a 5-fluorouracil (FU) prodrug (300 mg/m^2/day) and leucovorin (75 mg/day). After chemoradiotherapy, tumors were surgically resected and analyzed for their expression of HMGB1. The patients with a higher expression of HMGB1 had a poorer response to chemoradiotherapy [103]. In the study by Huang et al. (2018), the presence of HMGB-1 in the cytosol, resulting from its translocation from the nucleus, was analyzed in the samples of locally advanced rectal cancer from 89 patients. The authors reported that patients whose cancer exhibited cytosolic HMGB-1 before neoadjuvant chemoradiotherapy had a better clinical outcome.

PDT has also been shown to induce ICD. This therapy is based on the pre-treatment of target cells with a photosensitizer, which is next photoactivated to generate reactive species and cause oxidative stress in situ [105]. As a consequence of the photoreactions thus induced, reactive species are produced, which can overwhelm the antioxidant defenses of the cell, generating oxidative stress and cell death [26]. This approach has been shown to induce ICD in specific conditions. The type and concentration of the photosensitizer [27,106], as well as the light dose [26], can significantly affect its ICD-inducing capacity. Moreover, oxygen supply in the target tissue can also affect PDT outcomes.

Doix et al. (2019) reported that the combination of a therapeutical vaccine based on DCs stimulated with PDT-killed cells and radiotherapy can delay the development of experimental squamous cell carcinoma in mice. Firstly, the authors showed that the non-porphyrinic photosensitizer OR141, at low doses, induced ICD in SCC7 squamous cell carcinoma cells in vitro. These PDT-killed SCC7 cells were then used to prime and stimulate the maturation of DCs in vitro, which were then employed as a therapeutic vaccine against xenografts of SCC7 cells developed in the flank of C3H mice. This DC vaccine was subcutaneously injected three times at one-week intervals near the tumor-draining lymph node. One week later, radiotherapy mediated by a Cesium-137 γ-ray irradiator was applied onto the tumors. Occasionally, a fourth injection of the DC vaccine was performed at the time of radiotherapy application (peri-radiotherapy). It was observed that the tumor growth was delayed only when the DC vaccine was applied during the peri-radiotherapy period.

Rodrigues et al. (2022) showed that PDT using the photosensitizer aluminum-phthalocyanine mainly induced necrosis at the highest photosensitizer concentration and light dose, while milder protocols of PDT were able to efficiently induce ICD in both colorectal CT26 and mammary 4T1 murine adenocarcinoma cells [26]. In vitro, the milder PDT protocol consisted of 12.2 nM aluminum-phthalocyanine for colorectal adenocarcinoma CT26 cells and 9.0 nM aluminum-phthalocyanine for breast adenocarcinoma 4T1 cells, both irradiated with 25 J/cm^2 red light dose. The exposure or release of calreticulin, HSP70, HSP90 and HMGB-1 close to that promoted by this PDT protocol was comparable to that observed with mitoxantrone treatment. Increasing the photosensitizer concentration and light dose resulted in a reduction on the release of these DAMPs. This shows that the PDT regimen has to be fine-tuned to induce ICD.

As commented previously, there are several ICD inducers described in the literature. Among them are classical approaches, such as chemotherapy, PDT and radiotherapy. These classical therapies are usually aimed at tumor destruction, mainly by inducing direct cytotoxicity to target tumor cells, regardless of the death mechanism underlying this effect. This cytotoxicity induced in a noncontrolled fashion can fail to trigger immune activation. Immune-targeted cytotoxic cancer treatments have to thus be designed to induce ICD, representing an adaptation in former protocols in order to increase the immunogenicity of cancer treatments.

Several ICD inducers are used in well-recognized therapies. Thus, we understand that the translation of this knowledge to clinical protocols can be easier if compared to the implementation of new anticancer drugs which are currently under development. As previously discussed, the literature shows evidence of abscopal effects in radiotherapy and

PDT protocols, for example, that could be due to ICD. However, due to the complexity of cancer cells and tumor tissues, it is probable that personalized approaches are necessary to the successful use of immune-activating strategies in the clinic. Another barrier to the development of clinical protocols for ICD induction is the lack of standardization in ICD studies, as evidenced in this section. Thus, the use of widely recognized, bona-fide parameters that demonstrate that an immunogenic response is being induced is essential for translating research into clinical protocols [107]. This can be supported by guidelines proposed in the literature [29].

7. Delivery of DAMPs and ICD-Inducers to Tumor Tissues

As commented previously, ICD is triggered by the combined release of a specific set-up of DAMPs that includes calreticulin, HMGB-1, and ATP, among others. These molecules are exposed to the external cell membrane, and/or released to the external media in dying cells. All of these events are part of a cyclic process which is finely regulated during this specific sub-type of apoptosis. In terms of cell biology, as previously mentioned, several therapeutic procedures can trigger this event and have been investigated in both pre-clinical and clinical applications [88].

However, due to the variety of these different types of stimuli, the exposure or release of DAMPs can vary among the therapeutic strategies. Moreover, it is possible that, for clinical applications, the control of DAMPs exposure and release can be different among patients and tumor stages due to internal sub-tumoral tissue organizations and the different phenotypes of cancer cells. This situation makes the clinical translation difficult, delaying this therapy for patients.

To reduce this potential drawback, some authors have discussed alternative strategies to burst the DAMPs present inside tumor tissues (Figure 3), thus ensuring the presence of these molecular triggers for ICD induction. In terms of efficacy, the simple delivery strategy of DAMPs to target tumor tissues may not be efficient because ICD induction depends both on the release of chemotactic agents and on the presence of recognition molecules, such as calreticulin, on the cell membrane surface. Moreover, as discussed earlier, calreticulin exposure is a key factor for specific cell recognition.

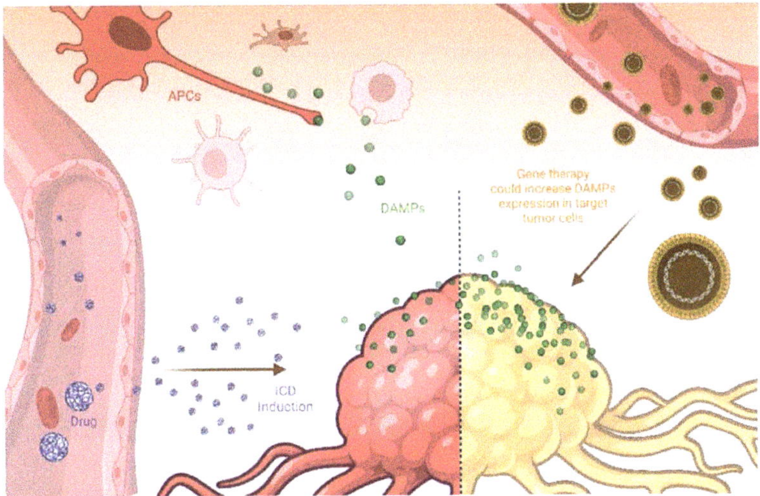

Figure 3. The induction of immunogenic cell death (ICD), the exposure of damage-associated molecular patterns (DAMPs), and the activation of antigen-presenting cells (APC) can be enhanced by delivery strategies based on nanotechnology.

Alternatively, some authors proposed the use of gene therapy to deliver DNA sequences that could increase DAMPs expression in target tumor cells. The rationale for this strategy is supported by the fact that some tumor types have a reduced expression of DAMPs, such as calreticulin. For instance, Garg et al. (2015) described a pre-clinical tumor model that reduces the constitutive calreticulin expression, thus reducing the effectiveness of immunogenic protocols [104]. The authors defined this tumor model as resistant to immunization against cancer cells. In terms of microevolution, it makes sense, as calreticulin is a molecule that activates tumor cell phagocytosis by APC. In this situation, selected tumor clones could reduce the expression of DAMPs, thus impairing ICD activation.

Due to the challenge of tumor targeting, nanoparticles or nanocarriers could be used to concentrate ICD inducers and DAMP molecules close to tumor tissues [26,27,108]. This is the classical argument for using nanotechnology for tumor therapy. Within this approach, nanocarriers are passively or actively delivered to the tumor regions and release the carried ICD activators to induce cell death and initiate the immune recognition and then the immune surveillance against malignant cells. This approach is somewhat different from the direct delivery of DAMPS to tumor tissues, but has been used successfully [108].

In the ICD approach widely proposed in the literature, conventional drugs, such as doxorubicin or mitoxantrone, for example, will lead tumor cells to succumb to ICD and release DAMPS. The preclinical results are promising; however, there are some concerns about the translation possibilities for this strategy, especially for the passive delivery of nanocarriers for tumor tissues. The main problems are the structural differences between preclinical induced tumor tissues and natural tumors that are developed in clinical conditions [109]. The argument is that passive targeting is not reproduced in clinical conditions. Despite all this discussion, there is some evidence that, at least in part, nanocarriers can increase the delivery of drugs and immunoadjuvant molecules to tumor tissues.

In the work by Zhou et al. (2022), murine melanoma (B16F10) cells were subjected to different treatments, such as hypoxia, cisplatin, radiotherapy, photodynamic therapy, and hypochlorous acid (HOCl). The cell-delivered secretions (CDS) of melanoma cells treated with HOCl activated dendritic cells and macrophages and produced the best antitumor immune response when compared to the other treatments. Aiming to increase the effectiveness of the treatment, the HOCl-CDS produced in vitro was then associated to nanofibers of a scaffold hydrogel containing melittin and RADA24 peptides. This nanosystem was then injected into the subcutaneous melanoma in vivo (C57BL/6 mice). The results indicated that the obtained hydrogel induced cell death, cytotoxicity in T lymphocytes, and increased the antitumor effect of the immune checkpoint inhibitors.

Another example of a successful use of nanocarriers for increasing the immunogenicity of tumors was published by Sethuraman et al. (2020) [105]. In this work, the authors describe a liposome nanocarrier with a DNA plasmid encoding calreticulin. They observed that this strategy increased the expression of calreticulin in target tumor cells, reducing tumor growth due to immune activation. Interestingly, when they combined this liposomal formulation with the application of focused ultrasound treatment, the results were better. This improvement is probably related to the temperature increase provided by the ultrasound. In higher temperatures, some amount of cell death could be triggered, which in combination with calreticulin superexpression could increase the immune activation.

The authors did not evaluate the modality of cell death induced by the treatment. However, it shows the potential of DAMPs delivery for immune system activation. As a potent phagocytosis inducer, calreticulin is a key factor for alternative types of immune activation. As noted previously, this protein was included as a molecular signature for the ICD, but its presence in external spaces may also promote tumor recognition by APC, thus contributing to immunological surveillance [106].

In this section, we presented potential uses of drug delivery strategies to optimize ICD induction. First, nanoparticles could deliver DAMPs directly to target tissues; second, nanocarriers could deliver ICD inducers; and third, nanoparticles could be used during gene therapy to increase the expression of DAMPS in the target tumor tissues. There is certainly no ideal strategy, but an effective therapy could be based on the combination of different approaches, and eventually on other strategies to optimize ICD induction in immunotherapies.

8. Conclusions

Although the induction of ICD has not been rationally used as a clinical treatment modality, it has potential for being exploited as an immunotherapy tool. The recent advances in immunotherapy have opened new fronts and possibilities in the fight against melanoma, for instance, with prolonged survival and durable responses in many patients. Studies on the immune evasion strategies deployed by tumors, as well as on the mechanisms underlying the activation of immune cells against abnormal host cells, can help to develop new weapons against cancer. The current knowledge allows us to suggest that the induction of ICD combined with other immunotherapeutic strategies, such as immune checkpoint blockade, can be explored to treat cancer. However, barriers such as the lack of standardization of ICD detection in clinical patients, as well as the need for more personalized protocols based on the detailed characterization of the tumor cells, have to be overcome to translate experimental protocols into the clinic.

Funding: This work was funded by the Brazilian agencies FAP-DF/Brazil (00193.00001053/2021-24), CNPq (403536/2021-9) and Coordenação de Aperfeiçoamento de Pessoal de Nível Superior—Brasil (CAPES)—Finance Code 001.

Institutional Review Board Statement: Not applicable.

Informed Consent Statement: Not applicable.

Data Availability Statement: Not applicable.

Conflicts of Interest: The authors declare no conflict of interest.

References

1. Jiang, T.; Shi, T.; Zhang, H.; Hu, J.; Song, Y.; Wei, J.; Ren, S.; Zhou, C. Tumor neoantigens: From basic research to clinical applications. *J. Hematol. Oncol.* **2019**, *12*, 93. [CrossRef]
2. Yamamoto, T.N.; Kishton, R.J.; Restifo, N.P. Developing neoantigen-targeted T cell–based treatments for solid tumors. *Nat. Med.* **2019**, *25*, 1488–1499. [CrossRef]
3. Weinberg, R.A. *The Biology of Cancer*, 2nd ed.; Garland Science, Taylor & Francis Group, LLC: New York, NY, USA, 2013.
4. Hegde, P.S.; Chen, D.S. Top 10 Challenges in Cancer Immunotherapy. *Immunity* **2020**, *52*, 17–35. [CrossRef]
5. O'Donnell, J.S.; Teng, M.W.L.; Smyth, M.J. Cancer immunoediting and resistance to T cell-based immunotherapy. *Nat. Rev. Clin. Oncol.* **2019**, *16*, 151–167. [CrossRef]
6. Busch, W. Aus der Sitzung der medicinischen Section vom 13 November 1867. *Berl. Klin. Wochenschr.* **1868**, *5*, 137.
7. Fehleisen, F. *Die Aetiologie des Erysipels*; Theodor Fischer: Berlin, Germany, 1883.
8. Bruns, P. Die Heilwirkung des Erysipels auf Geschwülste. Bruns' Beiträge zur klinische Chirurgie. 1887. Available online: https://books.google.com.br/books?id=0Gw1AQAAMAAJ&printsec=frontcover&source=gbs_ge_summary_r&cad=0#v=onepage&q&f=false (accessed on 5 May 2022).
9. Coley, W.B., II. Contribution to the Knowledge of Sarcoma. *Ann. Surg.* **1891**, *14*, 199–220. [CrossRef] [PubMed]
10. McCarthy, E.F. The toxins of William B. Coley and the treatment of bone and soft-tissue sarcomas. *Iowa Orthop. J.* **2006**, *26*, 154–158. Available online: https://www.ncbi.nlm.nih.gov/pmc/articles/PMC1888599/ (accessed on 6 May 2022). [PubMed]
11. Oiseth, S.J.; Aziz, M.S. Cancer immunotherapy: A brief review of the history, possibilities, and challenges ahead. *J. Cancer Metastasis Treat.* **2017**, *3*, 250. [CrossRef]
12. Codman, E. Symposium on the Treatment of Primary Malignant Bone Tumors: The Memorial Hospital Conference on the Treatment of Bone Sarcoma. *Am. J. Surg.* **1935**, *27*, 3–6. [CrossRef]
13. Ehrlich, P. "Über den Jetzigen Stand der Karzinomforschung" Beiträge zur Experimentellen. Pathologie und Chemotherapie. 1909. Available online: https://curiosity.lib.harvard.edu/contagion/catalog/36-990061083080203941 (accessed on 5 May 2022).
14. Dunn, G.P.; Old, L.J.; Schreiber, R.D. The Immunobiology of Cancer Immunosurveillance and Immunoediting. *Immunity* **2004**, *21*, 137–148. [CrossRef]
15. Burnet, M. Cancer—A biological approach. I. The processes of control. *Br. Med. J.* **1957**, *1*, 779–786. [CrossRef]
16. Schreiber, R.D.; Old, L.J.; Smyth, M.J. Cancer Immunoediting: Integrating Immunity's Roles in Cancer Suppression and Promotion. *Science* **2011**, *331*, 1565–1570. [CrossRef]

17. Pernot, S.; Terme, M.; Voron, T.; Colussi, O.; Marcheteau, E.; Tartour, E.; Taieb, J. Colorectal cancer and immunity: What we know and perspectives. *World J. Gastroenterol.* **2014**, *20*, 3738–3750. [CrossRef]
18. Sanchez-Castañón, M.; Er, T.; Bujanda, L.; Herreros-Villanueva, M. Immunotherapy in colorectal cancer: What have we learned so far? *Clin. Chim. Acta* **2016**, *460*, 78–87. [CrossRef]
19. Dobosz, P.; Dzieciątkowski, T. The Intriguing History of Cancer Immunotherapy. *Front. Immunol.* **2019**, *10*, 2965. [CrossRef]
20. Miller, J.F.A.P.; Mitchell, G.F.; Weiss, N.S. Cellular Basis of the Immunological Defects in Thymectomized Mice. *Nature* **1967**, *214*, 992–997. [CrossRef]
21. Steinman, R.M.; Cohn, Z.A. Identification of a novel cell type in peripheral lymphoid organs of mice: I. morphology, quantitation, tissue distribution. *J. Exp. Med.* **1973**, *137*, 1142–1162. [CrossRef]
22. Kiessling, R.; Klein, E.; Wigzell, H. "Natural" killer cells in the mouse. I. Cytotoxic cells with specificity for mouse Moloney leukemia cells. Specificity and distribution according to genotype. *Eur. J. Immunol.* **1975**, *5*, 112–117. [CrossRef]
23. Yousefi, H.; Yuan, J.; Keshavarz-Fathi, M.; Murphy, J.F.; Rezaei, N. Immunotherapy of cancers comes of age. *Expert Rev. Clin. Immunol.* **2017**, *13*, 1001–1015. [CrossRef]
24. Casares, N.; Pequignot, M.O.; Tesniere, A.; Ghiringhelli, F.; Roux, S.; Chaput, N.; Schmitt, E.; Hamai, A.; Hervas-Stubbs, S.; Obeid, M.; et al. Caspase-dependent immunogenicity of doxorubicin-induced tumor cell death. *J. Exp. Med.* **2005**, *202*, 1691–1701. [CrossRef]
25. Krombach, J.; Hennel, R.; Brix, N.; Orth, M.; Schoetz, U.; Ernst, A.; Schuster, J.; Zuchtriegel, G.; Reichel, C.A.; Bierschenk, S.; et al. Priming anti-tumor immunity by radiotherapy: Dying tumor cell-derived DAMPs trigger endothelial cell activation and recruitment of myeloid cells. *OncoImmunology* **2018**, *8*, e1523097. [CrossRef]
26. Rodrigues, M.C.; Júnior, W.T.D.S.; Mundim, T.; Vale, C.L.C.; de Oliveira, J.V.; Ganassin, R.; Pacheco, T.J.A.; Morais, J.A.V.; Longo, J.P.F.; Azevedo, R.B.; et al. Induction of Immunogenic Cell Death by Photodynamic Therapy Mediated by Aluminum-Phthalocyanine in Nanoemulsion. *Pharmaceutics* **2022**, *14*, 196. [CrossRef]
27. Morais, J.A.V.; Almeida, L.R.; Rodrigues, M.C.; Azevedo, R.B.; Muehlmann, L.A. The induction of immunogenic cell death by photodynamic therapy in B16F10 cells in vitro is effected by the concentration of the photosensitizer. *Photodiagn. Photodyn. Ther.* **2021**, *35*, 102392. [CrossRef]
28. Galluzzi, L.; Vitale, I.; Aaronson, S.A.; Abrams, J.M.; Adam, D.; Agostinis, P.; Alnemri, E.S.; Altucci, L.; Amelio, I.; Andrews, D.W.; et al. Molecular mechanisms of cell death: Recommendations of the Nomenclature Committee on Cell Death 2018. *Cell Death Differ.* **2018**, *25*, 486–541. [CrossRef]
29. Galluzzi, L.; Vitale, I.; Warren, S.; Adjemian, S.; Agostinis, P.; Martinez, A.B.; Chan, T.A.; Coukos, G.; Demaria, S.; Deutsch, E.; et al. Consensus guidelines for the definition, detection and interpretation of immunogenic cell death. *J. Immunother. Cancer* **2020**, *8*, e000337. [CrossRef]
30. Minton, K. DAMP-driven metabolic adaptation. *Nat. Rev. Immunol.* **2019**, *20*, 1. [CrossRef]
31. Collett, G.P.; Redman, C.W.; Sargent, I.L.; Vatish, M. Endoplasmic reticulum stress stimulates the release of extracellular vesicles carrying danger-associated molecular pattern (DAMP) molecules. *Oncotarget* **2018**, *9*, 6707–6717. [CrossRef]
32. Gong, T.; Liu, L.; Jiang, W.; Zhou, R. DAMP-sensing receptors in sterile inflammation and inflammatory diseases. *Nat. Rev. Immunol.* **2019**, *20*, 95–112. [CrossRef]
33. Serrano-del Valle, A.; Anel, A.; Naval, J.; Marzo, I. Immunogenic cell death and immunotherapy of multiple myeloma. *Front. Cell Dev. Biol.* **2019**, *7*, 50. [CrossRef]
34. Kroemer, G.; Galassi, C.; Zitvogel, L.; Galluzzi, L. Immunogenic cell stress and death. *Nat. Immunol.* **2022**, *23*, 487–500. [CrossRef]
35. Baracco, E.E.; Stoll, G.; Van Endert, P.; Zitvogel, L.; Vacchelli, E.; Kroemer, G. Contribution of annexin A1 to anticancer immunosurveillance. *OncoImmunology* **2019**, *8*, e1647760. [CrossRef]
36. Obeid, M.; Tesniere, A.; Ghiringhelli, F.; Fimia, G.M.; Apetoh, L.; Perfettini, J.-L.; Castedo, M.; Mignot, G.; Panaretakis, T.; Casares, N.; et al. Calreticulin exposure dictates the immunogenicity of cancer cell death. *Nat. Med.* **2006**, *13*, 54–61. [CrossRef]
37. Chattopadhyay, S.; Liu, Y.-H.; Fang, Z.-S.; Lin, C.-L.; Lin, J.-C.; Yao, B.-Y.; Hu, C.-M.J. Synthetic Immunogenic Cell Death Mediated by Intracellular Delivery of STING Agonist Nanoshells Enhances Anticancer Chemo-immunotherapy. *Nano Lett.* **2020**, *20*, 2246–2256. [CrossRef]
38. Jafari, S.; Lavasanifar, A.; Hejazi, M.S.; Maleki-Dizaji, N.; Mesgari, M.; Molavi, O. STAT3 inhibitory stattic enhances immunogenic cell death induced by chemotherapy in cancer cells. *DARU J. Pharm. Sci.* **2020**, *28*, 159–169. [CrossRef]
39. Vanmeerbeek, I.; Sprooten, J.; De Ruysscher, D.; Tejpar, S.; Vandenberghe, P.; Fucikova, J.; Spisek, R.; Zitvogel, L.; Kroemer, G.; Galluzzi, L.; et al. Trial watch: Chemotherapy-induced immunogenic cell death in immuno-oncology. *OncoImmunology* **2020**, *9*, 1703449. [CrossRef]
40. Doix, B.; Trempolec, N.; Riant, O.; Feron, O. Low Photosensitizer Dose and Early Radiotherapy Enhance Antitumor Immune Response of Photodynamic Therapy-Based Dendritic Cell Vaccination. *Front. Oncol.* **2019**, *9*, 811. [CrossRef]
41. Rapoport, B.L.; Anderson, R. Realizing the Clinical Potential of Immunogenic Cell Death in Cancer Chemotherapy and Radiotherapy. *Int. J. Mol. Sci.* **2019**, *20*, 959. [CrossRef]
42. Li, W.; Yang, J.; Luo, L.; Jiang, M.; Qin, B.; Yin, H.; Zhu, C.; Yuan, X.; Zhang, J.; Luo, Z.; et al. Targeting photodynamic and photothermal therapy to the endoplasmic reticulum enhances immunogenic cancer cell death. *Nat. Commun.* **2019**, *10*, 3349. [CrossRef]

43. Turubanova, V.D.; Balalaeva, I.V.; Mishchenko, T.A.; Catanzaro, E.; Alzeibak, R.; Peskova, N.N.; Efimova, I.; Bachert, C.; Mitroshina, E.V.; Krysko, O.; et al. Immunogenic cell death induced by a new photodynamic therapy based on photosens and photodithazine. *J. Immunother. Cancer* **2019**, *7*, 350. [CrossRef]

44. He, H.; Liu, L.; Liang, R.; Zhou, H.; Pan, H.; Zhang, S.; Cai, L. Tumor-targeted nanoplatform for in situ oxygenation-boosted immunogenic phototherapy of colorectal cancer. *Acta Biomater.* **2020**, *104*, 188–197. [CrossRef]

45. Deng, H.; Zhou, Z.; Yang, W.; Lin, L.-S.; Wang, S.; Niu, G.; Song, J.; Chen, X. Endoplasmic Reticulum Targeting to Amplify Immunogenic Cell Death for Cancer Immunotherapy. *Nano Lett.* **2020**, *20*, 1928–1933. [CrossRef]

46. Ganassin, R.; Oliveira, G.R.T.; da Rocha, M.C.O.; Morais, J.A.V.; Rodrigues, M.C.; Motta, F.N.; Azevedo, R.B.; Muehlmann, L.A. Curcumin induces immunogenic cell death in murine colorectal carcinoma CT26 cells. *Pharmacol. Res.-Mod. Chin. Med.* **2022**, *2*, 100046. [CrossRef]

47. Dulloo, I.; Atakpa-Adaji, P.; Yeh, Y.-C.; Levet, C.; Muliyil, S.; Lu, F.; Taylor, C.W.; Freeman, M. iRhom pseudoproteases regulate ER stress-induced cell death through IP3 receptors and BCL-2. *Nat. Commun.* **2022**, *13*, 1257. [CrossRef]

48. Iurlaro, R.; Muñoz-Pinedo, C. Cell death induced by endoplasmic reticulum stress. *FEBS J.* **2015**, *283*, 2640–2652. [CrossRef]

49. Acosta-Alvear, D.; Zhou, Y.; Blais, A.; Tsikitis, M.; Lents, N.H.; Arias, C.; Lennon, C.J.; Kluger, Y.; Dynlacht, B.D. XBP1 Controls Diverse Cell Type- and Condition-Specific Transcriptional Regulatory Networks. *Mol. Cell* **2007**, *27*, 53–66. [CrossRef]

50. Wang, Q.; Sun, A.-Z.; Chen, S.-T.; Chen, L.-S.; Guo, F.-Q. SPL6 represses signalling outputs of ER stress in control of panicle cell death in rice. *Nat. Plants* **2018**, *4*, 280–288. [CrossRef]

51. Shibusawa, R.; Yamada, E.; Okada, S.; Nakajima, Y.; Bastie, C.C.; Maeshima, A.; Kaira, K.; Yamada, M. Dapagliflozin rescues endoplasmic reticulum stress-mediated cell death. *Sci. Rep.* **2019**, *9*, 9887. [CrossRef]

52. Bezu, L.; Sauvat, A.; Humeau, J.; Leduc, M.; Kepp, O.; Kroemer, G. eIF2α phosphorylation: A hallmark of immunogenic cell death. *OncoImmunology* **2018**, *7*, e1431089. [CrossRef]

53. Anspach, L.; Tsaryk, R.; Seidmann, L.; Unger, R.E.; Jayasinghe, C.; Simiantonaki, N.; Kirkpatrick, C.J.; Pröls, F. Function and mutual interaction of BiP-, PERK-, and IRE1α-dependent signalling pathways in vascular tumours. *J. Pathol.* **2020**, *251*, 123–134. [CrossRef]

54. Yoshida, H.; Okada, T.; Haze, K.; Yanagi, H.; Yura, T.; Negishi, M.; Mori, K. ATF6 Activated by Proteolysis Binds in the Presence of NF-Y (CBF) Directly to the *cis* -Acting Element Responsible for the Mammalian Unfolded Protein Response. *Mol. Cell. Biol.* **2000**, *20*, 6755–6767. [CrossRef]

55. Rozpedek, W.; Pytel, D.; Mucha, B.; Leszczynska, H.; Diehl, J.A.; Majsterek, I. The Role of the PERK/eIF2α/ATF4/CHOP Signaling Pathway in Tumor Progression During Endoplasmic Reticulum Stress. *Curr. Mol. Med.* **2016**, *16*, 533–544. [CrossRef]

56. Klymenko, O.; Huehn, M.; Wilhelm, J.; Wasnick, R.; Shalashova, I.; Ruppert, C.; Henneke, I.; Hezel, S.; Guenther, K.; Mahavadi, P.; et al. Regulation and role of the ER stress transcription factor CHOP in alveolar epithelial type-II cells. *Klin. Wochenschr.* **2019**, *97*, 973–990. [CrossRef]

57. Yang, H.; Niemeijer, M.; van de Water, B.; Beltman, J.B. ATF6 Is a Critical Determinant of CHOP Dynamics during the Unfolded Protein Response. *iScience* **2020**, *23*, 100860. [CrossRef]

58. Sheng, X.; Nenseth, H.Z.; Qu, S.; Kuzu, O.F.; Frahnow, T.; Simon, L.; Greene, S.; Zeng, Q.; Fazli, L.; Rennie, P.S.; et al. IRE1α-XBP1s pathway promotes prostate cancer by activating c-MYC signaling. *Nat. Commun.* **2019**, *10*, 323. [CrossRef] [PubMed]

59. Zhou, W.; Fang, H.; Wu, Q.; Wang, X.; Liu, R.; Li, F.; Xiao, J.; Yuan, L.; Zhou, Z.; Ma, J.; et al. Ilamycin E, a natural product of marine actinomycete, inhibits triple-negative breast cancer partially through ER stress-CHOP-Bcl-2. *Int. J. Biol. Sci.* **2019**, *15*, 1723–1732. [CrossRef] [PubMed]

60. Singh, R.; Letai, A.; Sarosiek, K. Regulation of apoptosis in health and disease: The balancing act of BCL-2 family proteins. *Nat. Rev. Mol. Cell Biol.* **2019**, *20*, 175–193. [CrossRef]

61. Cubillos-Ruiz, J.R.; Bettigole, S.E.; Glimcher, L.H. Tumorigenic and Immunosuppressive Effects of Endoplasmic Reticulum Stress in Cancer. *Cell* **2017**, *168*, 692–706. [CrossRef]

62. Wang, Q.; Ju, X.; Wang, J.; Fan, Y.; Ren, M.; Zhang, H. Immunogenic cell death in anticancer chemotherapy and its impact on clinical studies. *Cancer Lett.* **2018**, *438*, 17–23. [CrossRef] [PubMed]

63. Coleman, L.G.; Maile, R.; Jones, S.W.; Cairns, B.A.; Crews, F.T. HMGB1/IL-1β complexes in plasma microvesicles modulate immune responses to burn injury. *PLoS ONE* **2018**, *13*, e0195335. [CrossRef]

64. Mishchenko, T.; Mitroshina, E.; Balalaeva, I.; Krysko, O.; Vedunova, M.; Krysko, D.V. An emerging role for nanomaterials in increasing immunogenicity of cancer cell death. *Biochim. Biophys. Acta* **2019**, *1871*, 99–108. [CrossRef]

65. Michalak, M.; Groenendyk, J.; Szabo, E.; Gold, L.I.; Opas, M. Calreticulin, a multi-process calcium-buffering chaperone of the endoplasmic reticulum. *Biochem. J.* **2009**, *417*, 651–666. [CrossRef]

66. Spiro, R.G.; Zhu, Q.; Bhoyroo, V.; Söling, H.-D. Definition of the Lectin-like Properties of the Molecular Chaperone, Calreticulin, and Demonstration of Its Copurification with Endomannosidase from Rat Liver Golgi. *J. Biol. Chem.* **1996**, *271*, 11588–11594. [CrossRef] [PubMed]

67. Ahmed, A.; Tait, S.W. Targeting immunogenic cell death in cancer. *Mol. Oncol.* **2020**, *14*, 2994–3006. [CrossRef] [PubMed]

68. Schcolnik-Cabrera, A.; Oldak, B.; Juárez, M.; Cruz-Rivera, M.; Flisser, A.; Mendlovic, F. Calreticulin in phagocytosis and cancer: Opposite roles in immune response outcomes. *Apoptosis* **2019**, *24*, 245–255. [CrossRef]

69. Calvet, C.Y.; Famin, D.; André, F.M.; Mir, L.M. Electrochemotherapy with bleomycin induces hallmarks of immunogenic cell death in murine colon cancer cells. *OncoImmunology* **2014**, *3*, e28131. [CrossRef]

70. Pitt, J.M.; Kroemer, G.; Zitvogel, L. Immunogenic and Non-immunogenic Cell Death in the Tumor Microenvironment. *Adv. Exp. Med. Biol.* **2017**, *1036*, 65–79. [CrossRef]

71. Sistigu, A.; Yamazaki, T.; Vacchelli, E.; Chaba, K.; Enot, D.P.; Adam, J.; Vitale, I.; Goubar, A.; Baracco, E.E.; Remédios, C.; et al. Cancer cell–autonomous contribution of type I interferon signaling to the efficacy of chemotherapy. *Nat. Med.* **2014**, *20*, 1301–1309. [CrossRef] [PubMed]

72. MacKenzie, K.J.; Carroll, P.; Martin, C.-A.; Murina, O.; Fluteau, A.; Simpson, D.J.; Olova, N.; Sutcliffe, H.; Rainger, J.K.; Leitch, A.; et al. cGAS surveillance of micronuclei links genome instability to innate immunity. *Nature* **2017**, *548*, 461–465. [CrossRef]

73. Kumar, A.; Khani, A.T.; Swaminathan, S. Type I interferons: One stone to concurrently kill two birds, viral infections and cancers. *Curr. Res. Virol. Sci.* **2021**, *2*, 100014. [CrossRef]

74. Bruschi, M.; Petretto, A.; Vaglio, A.; Santucci, L.; Candiano, G.; Ghiggeri, G.M. Annexin A1 and Autoimmunity: From Basic Science to Clinical Applications. *Int. J. Mol. Sci.* **2018**, *19*, 1348. [CrossRef] [PubMed]

75. Gavins, F.N.E.; Hickey, M.J. Annexin A1 and the regulation of innate and adaptive immunity. *Front. Immunol.* **2012**, *3*, 354. [CrossRef]

76. Hernández, I.B.B.; Angelier, M.L.; D'Ondes, T.D.B.D.B.; Di Di Maggio, A.; Yu, Y.; Oliveira, S. The Potential of Nanobody-Targeted Photodynamic Therapy to Trigger Immune Responses. *Cancers* **2020**, *12*, 978. [CrossRef]

77. Tripathi, D.; Tanaka, K. A crosstalk between extracellular ATP and jasmonate signaling pathways for plant defense. *Plant Signal. Behav.* **2018**, *13*, e1432229. [CrossRef]

78. Rodríguez-Nuevo, A.; Zorzano, A. The sensing of mitochondrial DAMPs by non-immune cells. *Cell Stress* **2019**, *3*, 195–207. [CrossRef] [PubMed]

79. Martins, I.; Wang, Y.; Michaud, M.; Ma, Y.; Sukkurwala, A.Q.; Shen, S.; Kepp, O.; Metivier, D.; Galluzzi, L.; Perfettini, J.-L.; et al. Molecular mechanisms of ATP secretion during immunogenic cell death. *Cell Death Differ.* **2013**, *21*, 79–91. [CrossRef] [PubMed]

80. Di Virgilio, F.; Sarti, A.C.; Falzoni, S.; De Marchi, E.; Adinolfi, E. Extracellular ATP and P2 purinergic signalling in the tumour microenvironment. *Nat. Rev. Cancer* **2018**, *18*, 601–618. [CrossRef]

81. Deng, Z.; He, Z.; Maksaev, G.; Bitter, R.M.; Rau, M.; Fitzpatrick, J.A.J.; Yuan, P. Cryo-EM structures of the ATP release channel pannexin 1. *Nat. Struct. Mol. Biol.* **2020**, *27*, 373–381. [CrossRef]

82. Redza-Dutordoir, M.; Averill-Bates, D.A. Activation of apoptosis signalling pathways by reactive oxygen species. *Biochim. Biophys. Acta-Mol. Cell Res.* **2016**, *1863*, 2977–2992. [CrossRef]

83. Borges Da Silva, H.; Beura, L.K.; Wang, H.; Hanse, E.A.; Gore, R.; Scott, M.C.; Walsh, D.A.; Block, K.E.; Fonseca, R.; Yan, Y.; et al. The purinergic receptor P2RX7 directs metabolic fitness of long-lived memory CD8$^+$ T cells. *Nature* **2018**, *559*, 264–268. [CrossRef]

84. Amores-Iniesta, J.; Barberà-Cremades, M.; Martínez, C.M.; Pons, J.A.; Revilla-Nuin, B.; Martínez-Alarcón, L.; Di Virgilio, F.; Parrilla, P.; Baroja-Mazo, A.; Pelegrín, P. Extracellular ATP Activates the NLRP3 Inflammasome and Is an Early Danger Signal of Skin Allograft Rejection. *Cell Rep.* **2017**, *21*, 3414–3426. [CrossRef]

85. Andersson, U.; Yang, H.; Harris, H. High-mobility group box 1 protein (HMGB1) operates as an alarmin outside as well as inside cells. In *Seminars in Immunology*; Academic Press: New York, NY, USA, 2018.

86. Tanaka, M.; Kataoka, H.; Yano, S.; Sawada, T.; Akashi, H.; Inoue, M.; Suzuki, S.; Inagaki, Y.; Hayashi, N.; Nishie, H.; et al. Immunogenic cell death due to a new photodynamic therapy (PDT) with glycoconjugated chlorin (G-chlorin). *Oncotarget* **2016**, *7*, 47242–47251. [CrossRef]

87. Wang, Y.-J.; Fletcher, R.; Yu, J.; Zhang, L. Immunogenic effects of chemotherapy-induced tumor cell death. *Gene Funct. Dis.* **2018**, *5*, 194–203. [CrossRef]

88. Liu, P.; Zhao, L.; Loos, F.; Iribarren, K.; Lachkar, S.; Zhou, H.; da Silva, L.C.G.; Chen, G.; Bezu, L.; Boncompain, G.; et al. Identification of pharmacological agents that induce HMGB1 release. *Sci. Rep.* **2017**, *7*, 14915. [CrossRef]

89. Rufo, N.; Garg, A.; Agostinis, P. The Unfolded Protein Response in Immunogenic Cell Death and Cancer Immunotherapy. *Trends Cancer* **2017**, *3*, 643–658. [CrossRef] [PubMed]

90. Jin, Y.; Guan, Z.; Wang, X.; Wang, Z.; Zeng, R.; Xu, L.; Cao, P. ALA-PDT promotes HPV-positive cervical cancer cells apoptosis and DCs maturation via miR-34a regulated HMGB1 exosomes secretion. *Photodiagn. Photodyn. Ther.* **2018**, *24*, 27–35. [CrossRef] [PubMed]

91. Antón, M.; Alén, F.; Gómez de Heras, R.; Serrano, A.; Pavón, F.J.; Leza, J.C.; García-Bueno, B.; Rodríguez de Fonseca, F.; Orio, L. Oleoylethanolamide prevents neuroimmune HMGB1/TLR4/NF-kB danger signaling in rat frontal cortex and depressive-like behavior induced by ethanol binge administration. *Addict. Biol.* **2017**, *22*, 724–741. [CrossRef]

92. Radogna, F.; Dicato, M.; Diederich, M. Natural modulators of the hallmarks of immunogenic cell death. *Biochem. Pharmacol.* **2019**, *162*, 55–70. [CrossRef] [PubMed]

93. Nastasi, C.; Mannarino, L.; D'Incalci, M. DNA Damage Response and Immune Defense. *Int. J. Mol. Sci.* **2020**, *21*, 7504. [CrossRef]

94. Venereau, E.; Casalgrandi, M.; Schiraldi, M.; Antoine, D.J.; Cattaneo, A.; De Marchis, F.; Liu, J.; Antonelli, A.; Preti, A.; Raeli, L.; et al. Mutually exclusive redox forms of HMGB1 promote cell recruitment or proinflammatory cytokine release. *J. Gen. Physiol.* **2012**, *140*, i6. [CrossRef]

95. Preissner, K.T.; Fischer, S.; Deindl, E. Extracellular RNA as a Versatile DAMP and Alarm Signal That Influences Leukocyte Recruitment in Inflammation and Infection. *Front. Cell Dev. Biol.* **2020**, *8*, 619221. [CrossRef] [PubMed]

96. Wang, Y.; Ma, X.; Su, C.; Peng, B.; Du, J.; Jia, H.; Luo, M.; Fang, C.; Wei, Y. Uric acid enhances the antitumor immunity of dendritic cell-based vaccine. *Sci. Rep.* **2015**, *5*, 16427. [CrossRef]
97. Voorwerk, L.; Slagter, M.; Horlings, H.M.; Sikorska, K.; Van De Vijver, K.K.; De Maaker, M.; Nederlof, I.; Kluin, R.J.C.; Warren, S.; Ong, S.; et al. Immune induction strategies in metastatic triple-negative breast cancer to enhance the sensitivity to PD-1 blockade: The TONIC trial. *Nat. Med.* **2019**, *25*, 920–928. [CrossRef]
98. Bota, D.A.; Chung, J.; Dandekar, M.; Carrillo, J.A.; Kong, X.-T.; Fu, B.D.; Hsu, F.P.; Schönthal, A.H.; Hofman, F.M.; Chen, T.C.; et al. Phase II study of ERC1671 plus bevacizumab versus bevacizumab plus placebo in recurrent glioblastoma: Interim results and correlations with $CD4^+$ T-lymphocyte counts. *CNS Oncol.* **2018**, *7*, CNS22. [CrossRef]
99. Le, D.T.; Jaffee, E.M. Regulatory T-cell Modulation Using Cyclophosphamide in Vaccine Approaches: A Current Perspective. *Cancer Res.* **2012**, *72*, 3439–3444. [CrossRef]
100. Vaes, R.; Hendriks, L.; Vooijs, M.; De Ruysscher, D. Biomarkers of Radiotherapy-Induced Immunogenic Cell Death. *Cells* **2021**, *10*, 930. [CrossRef]
101. Lämmer, F.; Delbridge, C.; Würstle, S.; Neff, F.; Meyer, B.; Schlegel, J.; Kessel, K.A.; Schmid, T.E.; Schilling, D.; Combs, S.E. Cytosolic Hsp70 as a biomarker to predict clinical outcome in patients with glioblastoma. *PLoS ONE* **2019**, *14*, e0221502. [CrossRef]
102. Rothammer, A.; Sage, E.K.; Werner, C.; Combs, S.E.; Multhoff, G. Increased heat shock protein 70 (Hsp70) serum levels and low NK cell counts after radiotherapy—Potential markers for predicting breast cancer recurrence? *Radiat. Oncol.* **2019**, *14*, 78. [CrossRef]
103. Hongo, K.; Kazama, S.; Tsuno, N.H.; Ishihara, S.; Sunami, E.; Kitayama, J.; Watanabe, T. Immunohistochemical detection of high-mobility group box 1 correlates with resistance of preoperative chemoradiotherapy for lower rectal cancer: A retrospective study. *World J. Surg. Oncol.* **2015**, *13*, 7. [CrossRef]
104. Huang, C.-Y.; Chiang, S.-F.; Ke, T.-W.; Chen, T.-W.; Lan, Y.-C.; You, Y.-S.; Shiau, A.-C.; Chen, W.T.-L.; Chao, K.S.C. Cytosolic high-mobility group box protein 1 (HMGB1) and/or PD-1+ TILs in the tumor microenvironment may be contributing prognostic biomarkers for patients with locally advanced rectal cancer who have undergone neoadjuvant chemoradiotherapy. *Cancer Immunol. Immunother.* **2018**, *67*, 551–562. [CrossRef]
105. Muehlmann, L.; Ma, B.; Longo, J.P.; Santos, M.; Azevedo, R. Aluminum–phthalocyanine chloride associated to poly(methyl vinyl ether-co-maleic anhydride) nanoparticles as a new third-generation photosensitizer for anticancer photodynamic therapy. *Int. J. Nanomed.* **2014**, *9*, 1199–1213. [CrossRef]
106. Garg, A.D.; Agostinis, P. ER stress, autophagy and immunogenic cell death in photodynamic therapy-induced anti-cancer immune responses. *Photochem. Photobiol. Sci.* **2014**, *13*, 474–487. [CrossRef]
107. Fucikova, J.; Kepp, O.; Kasikova, L.; Petroni, G.; Yamazaki, T.; Liu, P.; Zhao, L.; Spisek, R.; Kroemer, G.; Galluzzi, L. Detection of immunogenic cell death and its relevance for cancer therapy. *Cell Death Dis.* **2020**, *11*, 1013. [CrossRef] [PubMed]
108. Sun, Y.; Feng, X.; Wan, C.; Lovell, J.F.; Jin, H.; Ding, J. Role of nanoparticle-mediated immunogenic cell death in cancer immunotherapy. *Asian J. Pharm. Sci.* **2020**, *16*, 129–132. [CrossRef]
109. Longo, J.P.F.; Muehlmann, L.A. Nanomedicine beyond tumor passive targeting: What next? *Nanomedicine* **2020**, *15*, 1819–1822. [CrossRef]

pharmaceutics

Review

Activation of Cellular Players in Adaptive Immunity via Exogenous Delivery of Tumor Cell Lysates

Jihyun Seong and Kyobum Kim *

Department of Chemical and Biochemical Engineering, Dongguk University, 30 Pildong-ro 1-gil, Jung-gu, Seoul 22012, Korea; jh.seong520@gmail.com
* Correspondence: kyobum.kim@dongguk.edu; Tel.: +82-2-2260-8597

Abstract: Tumor cell lysates (TCLs) are a good immunogenic source of tumor-associated antigens. Since whole necrotic TCLs can enhance the maturation and antigen-presenting ability of dendritic cells (DCs), multiple strategies for the exogenous delivery of TCLs have been investigated as novel cancer immunotherapeutic solutions. The TCL-mediated induction of DC maturation and the subsequent immunological response could be improved by utilizing various material-based carriers. Enhanced antitumor immunity and cancer vaccination efficacy could be eventually achieved through the in vivo administration of TCLs. Therefore, (1) important engineering methodologies to prepare antigen-containing TCLs, (2) current therapeutic approaches using TCL-mediated DC activation, and (3) the significant sequential mechanism of DC-based signaling and stimulation in adaptive immunity are summarized in this review. More importantly, the recently reported developments in biomaterial-based exogenous TCL delivery platforms and co-delivery strategies with adjuvants for effective cancer vaccination and antitumor effects are emphasized.

Keywords: tumor cell lysate; adjuvant; dendritic cell; exogenous delivery system; cancer immunotherapy

Citation: Seong, J.; Kim, K. Activation of Cellular Players in Adaptive Immunity via Exogenous Delivery of Tumor Cell Lysates. *Pharmaceutics* **2022**, *14*, 1358. https://doi.org/10.3390/pharmaceutics14071358

Academic Editors: Lúis Alexandre Muehlmann and João Paulo Figueiró Longo

Received: 29 April 2022
Accepted: 22 June 2022
Published: 27 June 2022

Publisher's Note: MDPI stays neutral with regard to jurisdictional claims in published maps and institutional affiliations.

1. Introduction

Cancer immunotherapy is an emerging antitumor treatment technique, which works via specific antigen-mediated modulation in the patient's immune system [1]. Conventional anticancer therapies, including chemotherapeutic drugs and targeted treatments, have clinical limitations and adverse effects, such as non-specificity, drug resistance, and low efficacy in cancer mutation and metastasis [2–4]. Hence, the precise control and modulation of adaptive immune responses for pre-existing intratumoral therapy is the most important engineering parameter for developing effective cancer immunotherapeutic approaches [5–7].

Based on the interplay between T cell populations and other immune cellular components, engineering modulation in adaptive immunity could effectively eliminate cancer cells and inhibit tumor growth. As depicted in Figure 1, in particular, CD4+ T cells differentiate into various T helper cell subsets, including T helper (Th)1, Th2, Th9, Th17, and T follicular helper cells, in which Th1 cells react with antigen-presenting cells (APCs) and indirectly assist in the differentiation of CD8+ T cells into cytotoxic T lymphocytes (CTLs) by secreting a cytokine, such as interferon (INF)-γ. Additionally, interleukin (IL)-2 secreted from Th1 induces the proliferation of CD8+ T cells [8,9]. Thus, both activated CD4+ and CD8+ T cells augment long-lasting and strong antitumor immune responses by generating memory T cells to persist in anamnestic immune responses. APCs such as macrophages and dendritic cells (DCs) are crucial mediators for inducing T cell activation by antigen presentation on their surfaces. After taking up the tumor antigen molecules, these antigens are processed in the proteasome or phagosomes in the cytosol and presented in the form of peptides via major histocompatibility complex (MHC) class I or II molecules on the cellular surface of APCs. APCs with the MHC–peptide complex travel to secondary lymphoid organs to stimulate T cells. Upon contacting T cells, APCs initiate the priming of naïve T cells by interacting with the MHC–peptide complex and T cell receptor (TCR), and secrete cytokines to activate T cells.

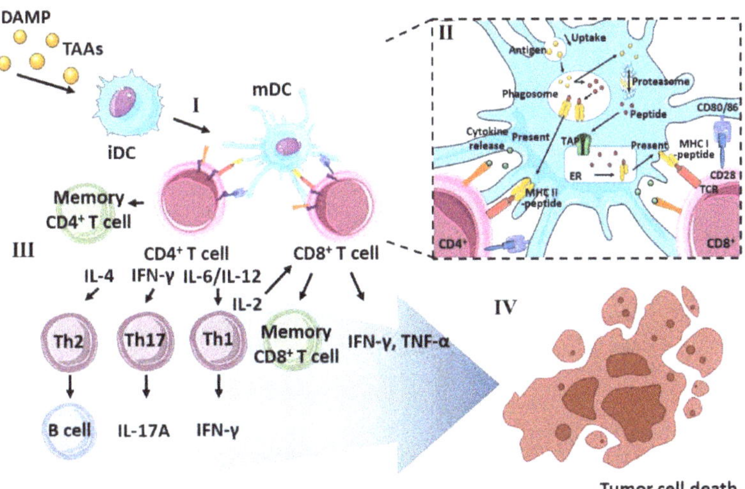

Figure 1. Overview of DC and T cell interplay for anticancer immunotherapy. (**I**) Differentiation of stimulated immature DCs (iDCs) by DAMP molecules and TAAs to mature DCs (mDCs), (**II**) process of antigen presentation via MHC molecules in DCs, (**III**) priming of T cells to effector T cells, (**IV**) induction of the tumor cell death by various types of T cells. Reproduced with permission from [10]; images used are from Servier Medical Art.

Previous studies have experimentally demonstrated that APC–T cell interaction by three distinct signals effectively induced antigen-specific T cell activation [11,12]: (1) the interaction between antigenic peptides presented by the MHC and the TCR; (2) co-stimulatory signals induced by the interaction between B7 molecules (e.g., CD80 and CD86) in APCs and CD28 in T cells, which trigger stronger immune responses; and (3) polarizing signals mediated by the production and secretion of multiple cytokines (e.g., IL-12 and tumor necrosis factor (TNF)-α by APCs [13]. Among these signaling interactions, MHC-mediated antigen cross-presentation is the most critical for the initiation of antigen-specific immune responses. Presented MHC I-immunogenic peptide complexes can be recognized by CTLs [14], whereas T helper cells can be activated by MHC II-mediated extracellular (or exogenous), immunogenic, antigenic peptide complex presentation [15]. Finally, activated T cell populations migrate toward the tumor microenvironment to kill specific tumor cells [16]. Cancer cells overexpress universal tumor-associated antigens (TAAs) and individual mutant neo-antigens [17,18]. To incorporate these TAAs for facilitating APC-dependent antigen presentation and subsequent T cell activation, lysed tumor cell bodies containing soluble tumor antigen molecules (e.g., tumor cell lysates (TCLs)) have been investigated for cancer therapy.

Therefore, cell-based engineering techniques to control and stimulate adaptive immune responses and further tumor suppression have recently been developed for designing efficient cancer therapeutics and vaccinations. TCLs containing various epitope sources are utilized for the induction of both CD8+ and CD4+ T cells [19,20] and potential personalized therapy. Endogenous damage-associated molecular patterns (DAMPs) are released by dying or damaged cells (i.e., host biomolecules that can initiate non-inflammatory responses to infection), and these specific TAAs interact with the pattern recognition receptors (PRRs) of DCs. Sequentially, activated antigen presentation on DCs induces proper T cell priming toward Th1 cells and differentiation of T lymphocytes into CTLs. Moreover, secreted cytokines, such as IL-12, IL-15, and IFN, from DCs are also able to stimulate T cell activation [20]. These T cells primed by APCs with TCL-derived antigens are key effectors of anticancer immunity. Antigen-specific memory T cells, which exert immediate effector functions without the need for further differentiation, sufficiently suppress tumor recurrence [21].

A growing body of research has explored the potential capability of TCLs for cancer vaccination. Because of the high antitumor immunity effects of TCLs, the majority of recent cancer immunotherapies utilize the TCL-mediated activation of APCs and T cells, along with the MHC pathway. However, due to the technical drawbacks of the naked form of TCLs, including a short half-life and the limited availability of various antigens, a lower therapeutic effect in immunity than the treatment of a specific antigen has been frequently observed. To strengthen the clinical efficacy of TCLs, particularly in the case of in vivo administration, precisely designed delivery systems should be utilized for increasing the stability of cargo TCLs and facilitating the co-administration of adjuvants. Therefore, the TAA-mediated activation of TCLs as an immune activator could be applied to induce in vitro necrosis of cancer cells, and stimulate downstream antitumor responses and immunological memory generation.

To this end, this review focuses on the current progress in engineered cancer immunotherapies by exogenous TCL delivery, emphasizing (1) practical applications using TCL-mediated DC activation and sequential stimulation in adaptive immunity in various cancer types, (2) the significance of adaptive immunological functions, and (3) the utilization of a series of delivery platforms for the co-administration of multiple adjuvants for effective cancer vaccination and antitumor treatment.

2. Preparation of Tumor Cell Lysates

2.1. Physical Disruption and Stimulation of Tumor Cells to Obtain Whole Tumor Cells

TCLs prime antitumor immunity and exhibit immune tolerance against self-antigens. Live tumor cells as a source of antigens could be less immunogenic since these cells contain or secrete factors such as vascular endothelial growth factor, soluble FAS ligand, and MHC class I chain-related proteins A and B, which suppress the function of DCs and T cells [7]. Figure 2 demonstrates the summary of cell lysis by external factors, and their conditions are indicated in Table 1.

Simply, TCLs are generated by repeated freeze-and-thaw cycles (Figure 2A), and protein fragments from the whole tumor cell population are obtained. The development of ice crystals during freezing, and the subsequent concentration upon thawing, results in the physical rupture of cellular bodies [37]. This repeated process facilitates the large-scale release of inflammatory proteins [38]. In general, these protein fragments are DAMPs, including heat shock protein (HSP) and high-mobility group box-1 (HMGB-1), which are classified as class I DAMPs [39]. HSPs and HMGB-1 directly bind and trigger Toll-like receptor (TLR) 2 and TLR4, which are the PRRs located in immune cell membranes. Activated TLR2 and 4 initiate NF-kB and interferon regulatory factors via the myeloid differentiation primary response 88-dependent pathway and toll/interleukin-1 receptor domain-containing adapter-inducing interferon-ß-dependent pathways. Through this, pro-inflammatory cytokines (e.g., IL-1ß, IL-6, and IFN) are released from DCs and stimulate T cell immunity with presented tumor-specific antigens using MHC molecules. [40].

The subsequent physical treatment of tumor cells, such as sonication, is optionally introduced to facilitate the homogeneity of the prepared TCLs. Nano-scale TCLs could be obtained using only sonication (Figure 2B) [29]. Additionally, ultraviolet (UV) irradiation is also commonly used to prepare TCLs by inducing immunogenic cell death (ICD) (Figure 2C) [30]. UV irradiation (1500 μW/cm^2 for 10 min) of TC-1 tumor cells results in both apoptosis and necrosis, and the TCLs from these UV-pulsed DCs exhibit significant surface expression of CD86, CD80, and MHC II molecules. The UV irradiation of tumor cells also generates effective TCL modulators for inducing an antitumor immune response by further enhancing CD8+ cell populations.

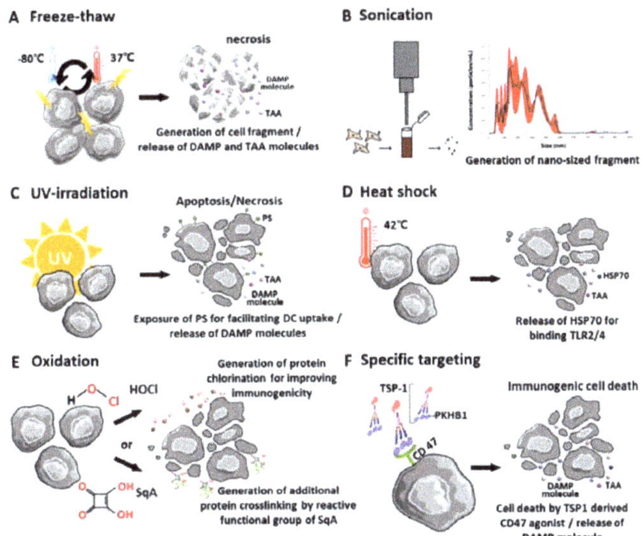

Figure 2. Schematic illustration of preparation of TCLs via various conditions. (**A**) Repeated freeze–thaw cycle could induce necrosis of tumor cells and release immunogenetic molecules (e.g., DAMPs, TAAs), (**B**) sonication generates nano-sized fragments, average particle size as measured by Nano Sight, (**C**) UV irradiation for facilitated cellular uptake, (**D**) heat shock evokes release of HSP70 in tumor cells, (**E**) oxidation rapidly induces tumor cell death and generates advantageous immunogenic molecules, (**F**) PKHB1 peptide-derived TSP-1 interacts with CD47 and activates the atypical caspase-independent and calcium-dependent signaling in cell death. (**B**,**F**) are reproduced with permission from Refs. [29,33,36].

Table 1. Condition for preparation of TCLs.

Classification of Process		Condition	Ref.
Physical disruption	Freeze–thaw cycle	Freeze at −80 °C and thaw at 37 °C (repeat)	[22–28]
	Sonication	Sonicate 3 times for 10 s	[29]
	UV irradiation	Irradiate with 1500 $\mu W/cm^2$ UVB	[30]
Pretreatment of source tumor cells	Heat shock	1. Treat at 42 °C for 1 h and 37 °C for 2 h 2. Additional physical disruption	[31,32]
	CD47 agonist	Treat 150 or 300 μM of PKHB1 for 2 h	[33]
	Phyllanthus amarus	1. Treat 1000 $\mu g/mL$ *Phyllanthus amarus* 2. Additional physical disruption	[34]
Cell membrane isolation	Sucrose-dependent	1. Mix 0.0759 M sucrose and 0.225 M D-mannitol-containing buffer 2. Centrifuge at $10,000\times g$ for 25 min 3. Centrifuge the supernatant at $150,000\times g$ for 35 min	[10]
	Sucrose-independent	1. Centrifuge at $10,000\times g$ for 25 min 2. Centrifuge the supernatant at $150,000\times g$ for 40 min	[35]

2.2. Pretreatment of Source Tumor Cells

2.2.1. Heat Shock

The induction of early necrosis using heat shock could be an alternative approach to obtaining TCLs (Figure 2D). A temperature of 42–43 °C could induce optimal cell death in antitumor immune outcomes, and maximum HSP production in the extracellular spaces of necrotic tumor cells, which could activate an adaptive antitumor immune response [41]. Mild hyperthermia (around 40 °C) induces thermotolerance [42], whereas high hyperthermia (over 45 °C) induces protein denaturation. In particular, HSP70 directly binds to CD40 receptors of DCs, and promotes the release of co-stimulatory signals [43]. Heat treatment of tumor cells also increases the expression of other DAMPs (such as HMGB-1 and ATP), and these molecules are recognized as danger signals by DCs.

For example, the heat shock treatment of three human melanoma cell lines at 42 °C for 1 h resulted in an allogeneic TCL mixture (TRIMEL) containing antigen components. The administration of TRIMEL significantly upregulated the release of the pro-inflammatory cytokine IFN-γ in DCs compared to the application of TCLs without heat shock treatment. Consequently, a previous study reported that TRIMEL showed clinical vaccination effects by developing a delayed type of hypersensitivity response in 64% of patients [31].

2.2.2. Oxidation

The oxidation of source tumor cells prior to the preparation of TCLs could facilitate necrosis and augment the immunogenicity of the antigenic components in TCLs by increasing oxidative stress (Figure 2E). Through this modulation, DCs could boost the uptake of antigenic danger signals and antigen processing mechanisms [44]. Therefore, hypochlorous acid (HOCl)-mediated oxidation is used for the generation of effective TCL contents, since protein chlorination enhances proteolytic vulnerability and improves the immunogenicity of the antigenic components [45]. HOCl-mediated oxidation also produces aldehyde-modified antigens with higher immunogenicity than that of unmodified antigens [46]. Chiang et al. [47] compared the in vitro efficacy of DCs pulsed with various TCLs obtained by HOCl-mediated oxidation, UVB irradiation, and six freeze–thaw cycles. Here, both HOCl-mediated oxidation and UVB irradiation efficiently induced the necrosis of tumor cells expressing ovalbumin (OVA) antigens, and the MHC-1-dependent presentation of the peptide SIINFEKL was achieved in DCs treated with the TCLs. In vivo tumor suppression in ID8 ovarian tumor models also demonstrated the enhanced immunogenic capability of the antigen contents in TCLs obtained from HOCl-mediated oxidation.

As a more stable molecule than HOCl, squaric acid (SqA) has been clinically approved for the treatment of skin papillomas [32]. SqA was also shown to induce the complete necrosis of source tumor cells and induce subsequent chemical changes in tumor antigens by combining with them via mechanisms including redox alteration, additional crosslinking, and aggregation through the reactive functional group. The resulting DAMPs from SqA-treated TCLs stimulated DCs, and these activated DCs elicited significant cytokine (IL-12 and IFN-γ) secretion and antigen presentation ability, indicating a more potent Th1 response.

2.2.3. Specific Targeting

Furthermore, the incorporation of biological substances into source tumor cells could also augment TCL-mediated immune activation. One of these stimulatory substances is known to act as an agonist peptide to activate CD47 in cancer cells. Previous reports demonstrated that CD47 activation using soluble peptides derived from thrombospondin-1(TSP-1) effectively induced cell death in several types of cancer cells (Figure 2F) [48,49]. Particularly, ICD induced by a TSP1-derived CD47 agonist (PKHB1 peptide: KRFYVVMWKK), and DC activation using TCLs obtained from PKHB1-treated L5178YR tumor cells (PKHB1-TCL), has been reported [33]. The sequential mechanisms of (1) CD47 activation by PKHB1, (2) exposure to several DAMPs by atypical caspase-independent and calcium-dependent signaling in cell death, (3) the enhanced maturation of bone marrow-derived DCs with proper antigen presentation, and (4) the stimulation of antitumor T cell responses in an in vivo L5178Y-R tumor model using syngeneic BALB/c mice, were obtained using PKHB1 TCLs.

2.2.4. Treatment with Natural Compounds

A natural compound was also used to modulate source tumor cells to facilitate the apoptosis of cancer cells. Pretreatment of both HCT 116 and MCF-7 cancer cell lines with an ethanol extract of *Phyllanthus amarus* induced the reactive oxygen species (ROS)-mediated apoptosis of tumor cells [34]. The TCLs from these apoptotic cancer cells effectively activated monocyte-derived DCs, showing significantly facilitated gene expression levels of IL-12 and IL-6 cytokines compared to TCLs from lipopolysaccharide (LPS)-treated cancer cells. The subsequent maturation of DCs was also determined by the enhanced immune functions of antigen presentation, chemotaxis capacity, phagocytic activity, T cell proliferation, and cytokine release.

2.3. Preparation of Tumor Cell Membranes

As a source of TCLs, several studies primarily focused on the production of tumor cell membrane components. Since cell membrane contents participate in protein–protein interactions in immune system processes, inflammatory responses, and chemokine signaling pathways, tumor cell membrane proteins (e.g., CD44, MUC, CD98, and integrin) could be used as tumor-specific antigens and receptors to effectively trigger immune responses in cancer therapy [50]. Centrifugation has been used to isolate cell membrane proteins from tumor cells. For purification, (1) the physical disruption of collected cells by homogenization or freeze–thaw cycles with lysis buffer, (2) centrifugation at low speed (1000–2000 RCF) to separate cellular debris and nuclei contained in the pellet, and (3) ultracentrifugation at high speed (100,000–200,000 RCF) to separate all membrane fractions from soluble proteins in the supernatant, have generally been performed (Figure 3A). Additional sucrose treatment provides a density gradient for obtaining membrane fractions (Figure 3B). The resuspension of membrane fractions within sucrose results in further separation of cell surface membranes, mitochondrial membranes, and other types of membrane components [51]. Isolated cell membrane components obtained through this centrifugation process could be further incorporated into various template biomaterials. For instance, a biomimetic antitumor nanovaccine was fabricated via the coating of membrane components onto calcium pyrophosphate inorganic NP templates (Figure 3C) [10]. This inorganic carrier platform to deliver TCL membrane-derived antigens consisted of (1) cell membrane fragments, isolated from sucrose-dependent separation, that promoted specific immune reactions as antigens, and (2) biocompatible calcium phosphate templates as immune adjuvants that stimulated innate immunity by activating the NLRP-3 inflammasome and the production of cytokines (e.g., IL-1ß) [52] for T cell-based responses. Therefore, the dual functionality of calcium pyrophosphate nanoparticles coated with antigen-rich TCL membranes could improve the antigen presentation of DCs, as well as provide adjuvant effects, dramatically increasing the expression of DC surface markers and the subsequent proliferation of CD8$^+$ T cells (Figure 3D).

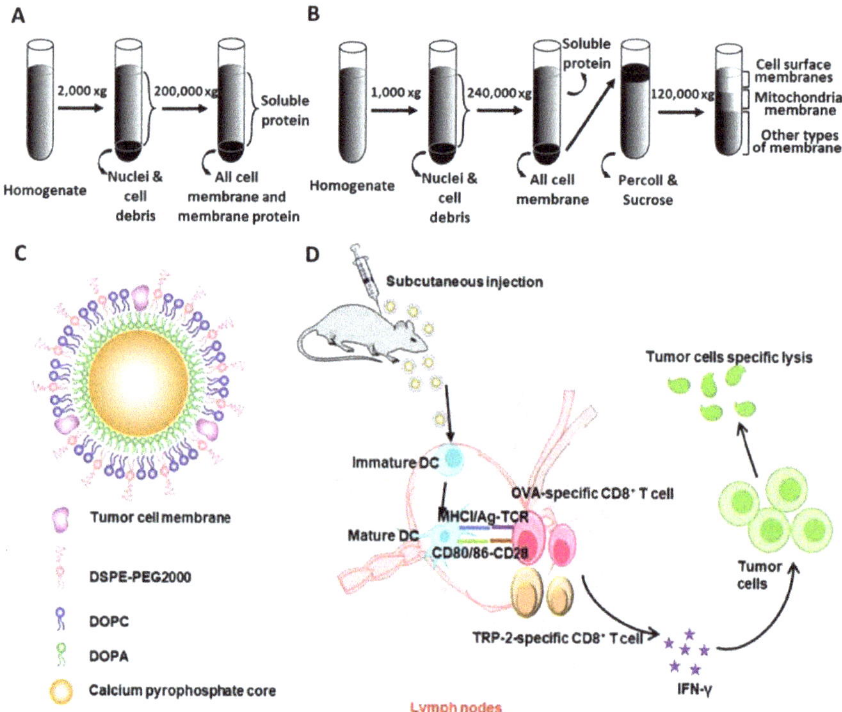

Figure 3. Schematic illustration of tumor cell membrane isolation and application in anticancer immunotherapy. (**A**) Cell membrane isolation using ultracentrifugation to obtain the total cell membrane component, (**B**) sucrose-dependent isolation for separating the cell surface membrane components, (**C**) design of tumor cell membrane-coated CaPyro nanoparticle, and (**D**) anticancer immunity mechanism using these NPs. All subfigures were reproduced with permission from Refs. [10,51].

3. Role of DCs in Cancer Immunotherapy

3.1. Phenotype of Dendritic Cells

DCs located in the spleen and various lymphoid tissues generally exhibit unique immune functions in activating T cells through antigen presentation [53,54]. However, the interaction between DCs and T cells occurs only in mature stages of DCs, which depends upon successful antigen uptake. DCs mostly exist in an immature state, but sufficient antigen uptake initiates a change to the mature state. During the functional maturation process, changes in the morphological and phenotypic characteristics of DCs influence immune system activity [55]. Mature DCs (mDCs) with a rough surface and multiple pseudopodia, and immature DCs (iDCs) with a spherical and smooth structure, exhibit different phagocytotic and migration abilities [56]. Therefore, when phagocytosis and endocytosis preferentially occur in the immature state, the morphological conversion (i.e., more dendritic structure) and optimization of antigen presentation by DCs occur sequentially. Then, these mDCs with a higher level of MHC molecules quickly migrate to the lymph nodes for 2–3 days, while maintaining their presentation ability, and are ready to stimulate other immune cells [57]. Consequently, mDCs can initiate and maintain adaptive immunity (including antigen specificity, humoral immunity mediated by antibodies, antigen-specific cellular immunity and memory) through a pathophysiological network with other immune cells, such as T cells, B cells, and NK cells [58].

3.2. Antigen Presentation by MHC Molecules

The recognition of MHC I and II molecules is crucial for the communication that leads to DC-induced immune responses. The major MHC-dependent antigen process in DCs can be identified as followed: MHC II aids in the presentation of exogenous antigens internalized into DCs, whereas MHC I helps in the presentation of peptides generated from reprocessed proteins and peptides through proteasome-mediated degradation in the cytosol [59–61].

DCs provide pathogenic information that "alerts" the immune system to an infection by increasing MHC II production, or regulating MHC II degradation, by the following mechanisms [62]: (1) after synthesis in the endoplasmic reticulum (ER) of APCs, MHC II molecules are delivered to the plasma through the Golgi network, or by direct transport to late endosomal compartments, (2) plasma-loaded MHC II molecules internalize exogenous protein antigens by clathrin-mediated endocytosis [63], (3) the internalized antigenic proteins are processed to peptides via endosomal and lysosomal proteolysis, (4) the processed peptide molecules are then combined with MHC II on the late endosomal surface, and these immunodominant MHC II–peptide complexes migrate to the cellular surface membranes of APCs for identification by CD4[+] T lymphocytes, and the initiation of T helper immune responses [64,65], and (5) MHC II–peptide complexes are recycled through ubiquitination in proteasomes, and further degradation processes in lysosomes, until DC maturation is complete [66,67].

MHC I-mediated cross-presentation in the immune system occurs via immune proteasomes [68]. For the cross-presentation of exogenous TAAs using MHC I molecules: (1) exogenous antigenic proteins (such as viral proteins produced during infection) internalized by phagocytosis are transferred to proteasomes via the ubiquitin–proteasome pathway, and degraded by proteolytic enzymes; (2) the resulting peptides are transported into the ER by the transporter associated with antigen processing (TAP) [69] and an ATP-dependent transporter; (3) MHC I molecules are fabricated in the ER and connected with the TAP, and subsequent binding of MHC I to the transported peptides occurs; (4) MHC I–peptide complexes are then delivered to cell surface membranes for cross-presentation to activate antigen-specific CTLs; and (5) completely equipped CTLs kill prospective target cells, such as virus-infected cells or tumor cells [70]. Although it does not contribute as much as the proteasome pathway, the vacuolar pathway, which does not rely on proteasomes and TAP, also participates in cross-presentation via MHC I [71]: (1) internalized exogenous antigens are degraded by protein catabolism using cathepsin S as a protease within the endocytic compartment, (2) MHC I molecules are generated from the ER and transferred to the endosome, and (3) MHC I-containing endosomes are loaded with the peptides, and then, peptide–MHC I complexes are presented on the cellular plasma membrane [72].

3.3. Downstream T Cell Commitment by mDCs

After successful antigen presentation by DCs, the interaction between mDCs and T cells in lymph nodes occurs to initiate cell-mediated adaptive immune responses. Further T cell commitments, such as proliferation and differentiation, are regulated by the level of TCRs triggered by antigen-presenting mDCs and the effectiveness of the signal amplification the T cells receive [57]. Several types of important mDC-mediated signals in lymph nodes are required for the activation and differentiation of naïve T cells. (1) The peptide–MHC complex initiates antigen-dependent signal transduction, (2) costimulatory molecules (i.e., B7 molecules, CD40, or ICAM-1) amplify the signaling process, and even a low level of available antigens effectively induces TCR-dependent T cell commitment [73,74], and (3) soluble cytokines facilitate further T cell activity

One of the crucial cytokine signals, IL-2 produced by activated Th1 cells, upregulates T cell proliferation. The direct activation of CD8[+] T cells, and the subsequent expansion of T cell populations upon TCR activation, is mediated by autocrine and paracrine IL-2 signaling [75,76]. IL-2 also promotes the differentiation of effector T cells [77]. Moreover, the duration of sustained TCR stimulation is controlled by the secretion of IL-12 by mDCs, which promote the progression of T cell differentiation and the subsequent formation of terminally differentiated effector cells. Specifically, in the presence of IL-12, T cells can develop into Th1 cells or Th2 cells, and these T helper cell populations gain the ability to move to inflamed organs to perform their own roles as effectors [78]. The stability of the mDC–T cell synapse maintains the duration of the stimulation during the signaling and transduction processes [79]. For instance, CD4$^+$ T cells need to be in contact with mDCs for 24 h to induce efficient cell division [80]. Even when naïve CD8$^+$ T lymphocytes interacted with mDCs for only 8 h, they exhibited a stronger proclivity for differentiating into effector and memory T cells [81,82].

3.4. Limitations of Ex Vivo Manipulation and the In Vivo Administration of DCs

Previous immunotherapeutic strategies have used the direct administration of ex vivo pulsed autologous DCs to activate T cell populations. One of the representative APC-based administrations was first approved by the Food and Drug Administration (FDA) as a cancer vaccine (Sipuleucel-T; Dendreon, CA, USA) in 2010 for late-stage castration refractory prostate cancer. This method includes APC isolation from patient blood, the co-incubation of APCs with prostatic acid phosphatase antigen and a granulocyte-macrophage colony-stimulating factor (GM-CSF), and reperfusion into the patient [17]. To develop an engineering manipulation of DCs using whole TCLs, the ex vivo differentiation of monocyte-derived autologous DCs was achieved by the incorporation of GM-CSF, IL-4, and additional stimuli components (e.g., LPS or TNF-α) to increase the potency of DC activation. Pulsing DCs by incubating them with TCLs also facilitates the production of mDCs [83,84]. Therefore, the vaccination platform involving ex vivo DC pulsing has also been applied to several cancer types, and the potential immunological response against specific cancers, with suitable safety for clinical trials, has been demonstrated. The ex vivo manipulation of DCs using melanoma-derived antigenic TCLs effectively induced signals for melanoma-associated antigen-1 (MAGE-1)-specific CTL responses, and two out of sixteen patients showed long-lasting immune responses over 6 months by successfully modulating antitumor immunity [85]. When using a HOCl-treated TCL mixture (derived from three ovarian tumor lines), DCs also exhibited Th1-dependent antitumor effects and tumor growth delays in stage II/IV ovarian cancer patients [47].

However, the therapeutic efficacy of the ex vivo manipulation and in vivo administration of DCs depends upon the administration route [86], sufficient numbers of delivered DCs [87], and the DC subset [88,89]. It has been reported that more than 90% of ex vivo engineered DCs died or were lost to non-targeted sites, and therefore, only a small fraction of the delivered DCs could home in on a lymph node, resulting in an insufficient T cell response [90,91]. Additionally, the optimization of ex vivo culture conditions, the expansion process, and the loading efficiency of tumor antigens for proper antigen presentation, are all required [92,93]. Due to these technical limitations in obtaining sufficient in vivo immune responses, direct injections of TCLs targeting in vivo resident DC populations without ex vivo DC control have been extensively studied to facilitate antigen-specific immune responses against cancer. Therefore, recent progress in biomaterial-mediated in vivo TCL administration and successful T cell pathway activation by antigen-presenting DCs are emphasized in the following sections.

4. Therapeutic Outcomes of Exogenous TCL Delivery Using Various Biomaterials

Exogenous TCL delivery using various multifunctional biomaterials through in vivo administration has been utilized in cancer immunotherapy. In particular, exogenous TCL delivery induced APC-dependent enhancement and the effective orchestration of adaptive immune responses by (1) augmenting in vivo DC maturation and activation, (2) increasing antigen presentation in DCs, and (3) further inducing T cells by interacting with multiple DCs. However, weak immunogenicity can be caused by a variety of factors, including (1) a lack of appropriate immunological DAMP signals [94], (2) inefficient delivery of relevant TAAs to resident in vivo DCs, and (3) the undesired degradation of antigen molecules during migration in the bloodstream and lymphatic system. Therefore, various delivery platforms with protective efficacy for cargo TCLs, and additional functionality to improve the immunogenicity of TCLs, have been investigated. As well as the intrinsic stimulation by biomaterial via direct immune cell regulation (i.e., DC activation and T cell proliferation) through the recognition of exogenous substances and their following interactions with immune cells [95], cargo TCL protection (i.e., preservation of its bioactivity upon in vivo administration) and subsequent augmentation under sustained DC activation, are technical advantages of biomaterial-based delivery platforms. The successful development of an efficient TCL delivery platform could represent a novel immune modulatory strategy for anticancer treatments, through the sequential accurate targeting processes of longer circulation with improved colloidal stability, sufficient delivery of TAAs in TCLs to lymph nodes, the sustained release of cargo TCLs, preservation of in vivo TCL bioactivity, and the enhanced cellular uptake of TAAs to DCs [96,97]. Hence, recent therapeutic approaches have focused on material-assisted TCL delivery platforms. This section reviews the current progress in TCL-mediated immune activation, anticancer treatment, and prospective applications of cancer vaccines.

4.1. Nanoparticles

4.1.1. Design Parameters for TCL Carriers

The most representative TCL delivery platform comprises nanoparticle (NP)-based carriers, which use several types of materials decorated with functional moieties to boost their delivery efficacy. The encapsulation of TCLs in the NP core can protect cargos from degradation during in vivo circulation, and regulate their release. These particle-based carriers can also be easily modified with functional ligands or molecules on their surface [98,99]. The efficiency of antigen-containing TCLs in draining lymph nodes can be influenced by the characteristics of the template particles, such as size, morphology, and charge.

For example, the efficiency of nano-sized polystyrene particles for activating APC subsets was higher than that of micro-sized particles in terms of cellular uptake [100]. NPs can easily infiltrate cells, and their resident particle populations in lymph nodes are three fold larger than those of micro-sized particles. Thus, a potential T cell immune response can be effectively induced by DC maturation. Moreover, this size-dependent immunogenicity was also observed in vaccination efficacy against tumors [101]. The delivery of human papillomavirus peptides using 40–50 nm Ag NPs resulted in higher uptake into DCs in draining lymph nodes, in vivo localization in C57BL/6 mice models, and immunological responses. Protection in tumor challenge models and the clearance of established tumors was also found.

In terms of the morphology and geometry of particulate carriers, the spherical shape of NPs exhibits (1) reduced adhesion to vessel walls, and longer circulation time [102], and (2) facile ligand conjugation onto larger surface areas. Spherical NPs with surface-conjugated ligands can be fully enveloped by target cellular membranes via strong ligand–receptor interactions, and consequently facilitate receptor-mediated endocytosis [103]. Spherical NPs could overcome a minimal membrane binding energy barrier, resulting in low free energy change for internalization into target cells [104]. Shape-dependent immune adjuvant efficacy was also observed in the delivery of AuNPs coated with virus envelope proteins (VEPs) [105]. The in vivo inoculation of these NP-VEPs into mice resulted in shape-dependent cytokine production in DCs. Rod-shaped NPs induced pro-inflammatory IL-1β and IL-18 production by activating the inflammasome-dependent process as adjuvants for eliciting immunity. In contrast, the same antigen delivery system using spherical or cube-shaped NPs induced the secretion of other types of inflammatory cytokines, including TNF-R, IL-6, IL-12, and GM-CSF.

An optimal surface charge and charge density of NPs is also required in order to increase the duration of blood circulation and prevent their loss to untargeted regions [106]. In general, positively charged NPs interact more efficiently with negatively charged cell membranes, and higher cellular uptake occurs [107]. A previous study reported charge-dependent NP uptake by 3T3 fibroblasts [108], indicating that the cellular internalization of trimethylammonium-coated AuNPs with positive surface charges was faster than that of negatively charged phosphonate-coated particles. The interaction with various in vivo protein components and delivered NPs resulted in the formation of a protein corona, which might reduce NP uptake regardless of the charge of the NPs. Moreover, a higher concentration of positively charged NPs (>5 nM) caused oxidative stress and cell death. Thus, the charge property of NP carriers should be also optimized to improve colloidal stability and interaction with target cells, and, therefore, effective exogenous delivery of antigen molecules.

4.1.2. Polymer-Based Materials

Among the various template materials used to fabricate NP cores, polymer-based NPs have shown a series of technical advantages for carrier development. Such improvements in functional polymeric NPs for TCL delivery include the controllability of the sustained release of various TCL cargos, cargo-protective efficacy through encapsulation, increased half-life and bioavailability of antigens, and a compatibility with vaccine adjuvant delivery, which is beneficial for inducing long-lasting immunity [109].

Poly(lactic-co-glycolic acid) (PLGA) NPs have become a popular candidate for drug delivery systems due to their biodegradability via hydrolysis and easy surface functionalization [110,111], and they can also be used for the delivery of antigens or adjuvant to improve DC-mediated immune responses [112]. Table 2 summarizes the TCL delivery platforms using polymer-based materials.

Table 2. Polymer-based material delivery platforms for exogenous TCL delivery.

Material	TCL Type	Specificity	Material Platform	Target Cancer	Outcome	Ref.
PLGA	Whole TCLs	Human	TCL-loaded PLGA NPs	Gastric cancer	Increased IL-12 and IFN-γ in DCs Th1 immune system pathway activation	[113]
	CM	Mouse	Cell membrane coated-CpG-PLGA NPs	Melanoma	Stability and longer circulation High recognition of specific tumor antigens 86% survival in vaccination group	[114]
	CM	Mouse	Cell membrane coated-R848-PLGA NP–mannose moiety conjugate	Melanoma	Specific binding by mannose Homotypic targeting on cancer cell surface antigens	[115]
PEG	CM	Mouse	Co-delivery of PEGylated cell membrane and CpG	Melanoma	Enhanced serum stability Efficient trafficking to LNs 63% tumor regression	[26]
PEGylated LM	CM	Mouse	Cell membrane coated-PEG-LM NPs	Breast	Immune adjuvant effect and photothermal conversion efficacy with irradiation Metal-induced NF-kB immune pathway activation	[116]
CTS	Whole TCLs	Mouse	Mannose-coated TCLs-CTS NPs	Melanoma	Mitochondrial stress, ROS generation, and cGAS-STING pathway activation Improvement in NP uptake efficacy	[22]
PDA	Whole TCLs	Mouse	TCL-loaded PDA NPs	Colorectal cancer	Reacted with dopamine receptor Increased the subpopulation of T cells	[24]

PLGA, poly(lactic-co-glycolic acid); TCL, tumor cell lysate; IL, interleukin; IFN, interferon; Th, T helper cell; CM, cell membrane; PEG, polyethylene glycol; R848, resiquimod; LN, lymph node; LM, liquid metal; CTS, chitosan; ROS, reactive oxygen species; PDA, polydopamine.

The potential application of PLGA NPs loaded with gastric TCLs for antigastric tumor immunotherapy has been demonstrated [113]. In this instance, TCLs were prepared from primary gastric tumor cells obtained from gastric cancer patients, and encapsulated into PLGA NPs as DC antigen delivery vehicles. Upon delivery to mDCs, a higher expression of HLA-DR and co-stimulatory molecules (e.g., CD80 and CD86), and increased levels of IL-12 and IFN-γ were achieved than from bolus TCL treatment, leading to Th1 immune system pathway activation and augmented T lymphocyte proliferation. In addition to the conventional advantages of polymeric NPs for TCL delivery, PEGylation can also be applied to increase the retention time of therapeutic antigens, thus avoiding in vivo degradation by various proteases, and providing steric stabilization through the formation of a hydration layer on the particle surfaces [117]. Flexible PEG linker-mediated functionalization of NP surfaces also facilitates the adjustment of the chain length to improve cell recognition and uptake [118]. PEGlyated cancer cell membrane vesicles (PEG-CCVs) have also been developed to enhance serum stability and efficient trafficking to lymph nodes (Figure 4A) [26]. This PEGlyation was carried out using 5 kDa DSPE-PEG, and the resulting PEG-CCVs maintained in vitro stability (i.e., size and PDI) in 10% fetal bovine serum (FBS) conditions for 3 days at 37 °C, as well as in vivo draining efficiency to local lymph nodes upon subcutaneous administration.

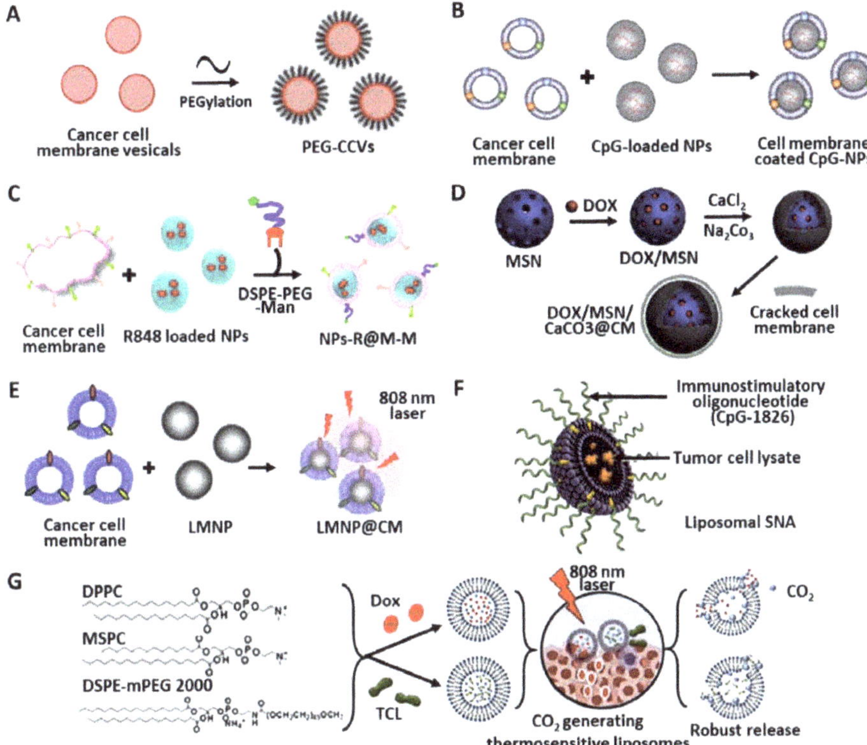

Figure 4. Schematic illustration of various biomaterial-based TCL delivery platforms. (**A**) PEGylated cancer cell membrane vesicles (CCVs) for steric stabilization, (**B,C**) PLGA nanoparticle-mediated delivery, (**D**) cancer cell membrane-coated inorganic material-based designs, (**E**) cell membrane-coated liquid metal nanoparticle with NIR irradiation, (**F,G**) cargo encapsulated within liposomal nanoparticles with lipid-mediated surface modification (All figures were reproduced with permission from Refs. [26,114–116,119–121]).

4.1.3. Camouflage Using Cancer Cell Membranes

As previously discussed, endogenous plasma membranes from whole TCLs are a good source as antigens, mimicking the surface architecture of cancer cells and inducing interplay with immune cells by the presence of membrane-bound tumor antigens. Hence, the artificial coating of cancer cell membrane components onto NP surfaces has also been developed. Such PLGA NPs covered with cancer cell membranes (CCNPs) exhibited colloidal stability and longer circulating properties, and effectively trained the immune system to recognize and fight tumors [122]. In addition, the incorporation of cancer cell membranes onto CpG-containing NPs showed synergistic anticancer vaccination efficacy (Figure 4B) [114]. The concurrent presentation of both immunostimulatory tumor antigens and adjuvant could enhance the effective antigen presentation and the activation of downstream immune processes. Based on the facilitated expression level of co-stimulatory receptors on DCs, cancer cell membrane-associated specific antigen presentation, and higher CD8$^+$ T cell proliferation to recognize specific melanoma antigens (i.e., gp100 and TRP2), in vivo vaccination resulted in survival rates of 86% in a B16-F10 tumor model with mice.

Similarly, the combination of a mannose (Man) moiety and a TLR 7 agonist (R837) with CCNPs (Man-R837-CCNPs) showed enhanced cellular uptake and antitumor immune responses (Figure 4C) [115]. Through (1) specific binding between Man and its receptors on DCs, (2) activation of innate immunity by R837 adjuvant, and (3) stimulation by melanoma cell membranes, BMDCs treated with Man-R837-CCNPs achieved higher maturation, with the enhanced expression of CD80 and CD86 and the significantly increased secretion of cytokines (IL-12p40 and TNF-α). Although template CCNPs, R837-loaded PLGA NPs without membrane coating, and R837-CCNPs without a Man moiety slightly inhibited tumor progression compared to untreated controls in B16-OVA tumor models, Man-R837-CCNPs exhibited the strongest antitumor efficacy and vaccination through homotypic targeting mediated by cancer cell surface antigens, and increased numbers of CD8$^+$ T cells.

4.1.4. Inorganic Templates for TCL Delivery

In addition to polymeric NPs, inorganic porous particles, such as calcium carbonate (CaCO$_3$) and mesoporous silica NPs (MSNs), have also been used as templates for the encapsulation of proteins and peptide antigens. Lybaert et al. [23] utilized CaCO$_3$ particles covered with a polymeric TLR7/8 agonist (CL264) to encapsulate TCLs. CaCO$_3$ particles with highly porous inner architecture showed a high loading capacity for macromolecules via surface adsorption and encapsulation into the inner core. Surface coating with polycations of the copolymer of N -(hydroxypropyl) methacrylamide (HPMA) and N-(3-aminopropyl) methacrylamide (APMA) modulated the surface charges to adsorb the TLR 7/8 agonist by the combination of electrostatic interaction and physisorption. Additionally, TCLs were prepared from the Lewis lung cancer cell line expressing OVA antigens, coprecipitated with CaCl$_2$ and Na$_2$CO$_3$ during the fabrication of CaCO$_3$ particles, and were encapsulated into the core.

The delivered OVA-containing TCLs using TCL-TLR-CaCO$_3$ particles resulted in the cross-presentation of OVA by DCs after the migration of the particles into phagosomes and fusion with acidic lysosomes [123]. The results of the co-delivery of tumor-associated antigens using TCLs and the TLR7/8 agonist indicate the higher efficiency of cross-presentation and in vivo antitumor responses via enhanced immunogenicity, compared to any single treatment.

Since TLR is one of the PRRs in DCs [124], this co-delivery strategy using TCLs-TLR-CaCO$_3$ particles could (1) activate PRRs by pathogen-associated molecular patterns (PAMPs) and DAMPs derived from necrotic cells (i.e., TCLs), and (2) upregulate antigen presentation by the additional efficacy of the TLR agonist as a potent activator.

A similar surface coating was also used to fabricate cancer cell membrane-coated MSNs [119]. Likewise, the chemotherapeutic drug doxorubicin (DOX) was entrapped in the inner porous structure of the MSNs (i.e., DOX-MSNs), and membrane fragments from

LNCaP-AI prostate cancer cell lines (CMs) were then adsorbed onto the DOX-MSNs (i.e., DOX-MSN-CM) (Figure 4D). Along with the induced apoptosis of prostate cancer cells, the co-administration of DOX and CMs using MSNs significantly suppressed tumor growth in LNCaP-AI tumor models.

Recently, liquid metal (LM) has also been utilized as a template core for the development of a nanovaccine for tumor prevention [116]. In this study, CMs derived from 4T1 murine breast tumor cells were coated onto mPEG$_{5000}$-SH-modified eutectic gallium–indium LM NPs (Figure 4E). As well as the antigenic efficacy of CMs and the immune adjuvant effect of LM, the additional photothermal conversion efficacy of LM NPs irradiated by an 808-nm laser facilitated local inflammation, and the subsequent recruitment of APCs, by the increased secretion of pro-inflammatory factors (i.e., IL-6 and TNF-α) and metal-induced NF-kB immune activation pathways [125]. In addition to the effective in vivo delivery of antigens to lymph nodes, three vaccinations within 15 days before the inoculation of 4T1 tumor cells in a mouse model also indicated the significant tumor prophylactic efficacy of CM-coated LM NPs with laser irradiation.

4.1.5. Adjuvant Activities of NPs

Some materials have shown potent adjuvant efficacy to stimulate cellular immunity and modulate immune responses. For instance, aluminum phosphate (AP) was discovered in 1926 as an adjuvant, and was later approved by the United States FDA [126,127]. Therefore, aluminum-containing adjuvants could also be used as cancer vaccines by antigen adsorption via electrostatic attraction and ligand exchange. In particular, CpG-loaded AP NPs coated with B16F10 tumor cell membranes have been developed for cancer vaccination in melanoma models [35]. Again, the surface-incorporated cancer cell membranes enhanced the colloidal dispersion of AP NPs and functioned as native tumor antigens. The dual functions of the AP-mediated adjuvant effects and immunogenicity of antigens effectively mDCs activation, improved lymph node targeting, and facilitated strong tumor-specific cellular immune responses after subcutaneous injection in mice.

Chitosan, a cationic polysaccharide, is also widely used as a vaccine delivery vehicle due to its adjuvant efficacy to promote IFN secretion in mature bone marrow-derived cells (BMDCs), and thus, enhances antigen-specific Th1 responses [128]. Chitosan adjuvants delivered to DCs could induce mitochondrial stress and generate ROS. Subsequent activation of the cGAS-STING pathway triggers the production of type I interferons, and further DC maturation occurs. In addition to the adjuvant effect of chitosan, the Man-based surface functionalization of core chitosan NPs (Man-CTS NPs) facilitates the targeting efficacy of TCL delivery to APCs by binding to Man receptors located on DC membranes [22]. This Man coating also enhances the in vitro bone marrow DC uptake of antigens in TCLs through receptor targeting [129]. Therefore, treatment using B16 melanoma TCL-loaded Man-CTS NPs augmented DC maturation and the related antigen presentation, indicated by the enhanced expression levels of surface markers (i.e., MHC I, MHC II, CCR7, CD80, CD86, and CD40) in vitro and in vivo. An elicited adjuvant effect and T cell priming were further observed with the increased proliferation of both CD8$^+$ and CD4$^+$ T cells, and the upregulated expression levels of serum IFN-γ and IL-4, confirming in vivo T cell activation in melanoma mice models. Vaccination efficacy and therapeutic effects of TCL-loaded Man-CTS NPs were proven by tumor growth inhibition and reductions in tumor weight.

Additionally, the neurotransmitter dopamine (DA) has also been used for the immune system activation of effector T cells and the suppression of regulatory T (Treg) cells by reacting with DA receptors. DA activates NF-κB to upregulate pro-inflammatory cytokines and chemokines (e.g., IL-6, IL-1β, IL-18, CCL2, and CXCL8) [130]. Wang et al. synthesized polydopamine (PDA)-based NPs covalently conjugated with colorectal cancer TCLs (TCL@PDA NPs) by the interaction between catechols in DA and the amine/thiol groups of antigens in TCLs [24]. PDA-based NPs showed potential as an antigen carrier, exhibiting (1) PDA-mediated pro-inflammation, with increased secretion of IFN-γ and TNF-α, and (2) DC maturation, with the enhanced expression of MHC II and secretion of Th1-related cytokines. In a C57BL/6 mouse model, three (day 4, 10, and 18 after cancer inoculation) subcutaneous vaccinations with TCL@PDA NPs significantly increased the subpopulations of CD4$^+$ and CD8$^+$ T cells in the spleen and LNs, as well as the memory T cell response. Therefore, both in vivo antitumor efficacy and tumor prevention effects were sufficiently achieved by the combination of PDA and TCLs.

4.2. Liposome

Liposomes are another type of exogenous TCL delivery platform. Due to the characteristic structure and composition of liposomes, the entrapment of hydrophilic cargo into the inner core of the liposomes, and additional lipid-mediated surface modification with functional moieties, are possible [131]. Based on these liposomal design strategies, Callmann et al. developed TCL-loaded liposomal spherical nucleic acids (Lys-SNAs) (Figure 4F) [120]. For their fabrication, TCLs from triple-negative breast cancer cells were encapsulated in the core of liposomes, while cholesteryl-modified immunostimulatory oligonucleotide adjuvants (CpG-1826) were immobilized on the surface. As described in the previous section, the oxidation of tumor cells prior to lysate generation using HOCl (OxLys) increases immunogenic aldehyde-modified antigens. After peritumoral administration into an EMT6 mouse mammary carcinoma model, OxLys-SNAs significantly increased the population of cytotoxic CD8$^+$ T cells, and simultaneously decreased that of myeloid derived-suppressor cells within the tumor microenvironment compared to Lys-SNAs and simple mixtures of OxLys. The enhanced therapeutic efficacy of the OxLys-SNA formulation was also indicated by antitumor activity, prolonged survival, and the inhibition of tumor regeneration. Therefore, the proper packaging and presentation of adjuvant and human-specific TCL-derived antigens into the liposomal structure is also an important design parameter for exogenous TCL delivery.

In addition to tumor-specific antigen delivery, leading to the maturation and activation of DCs, additional functions of liposomal carriers could facilitate immune modulatory responses. Won et al. [121] developed CO$_2$-generating thermosensitive liposomes (BG-TSLs) that encapsulate melanoma-derived whole TCLs (Figure 4G). The lipid layers (a combination of DPPC/MSPC/DSPE-mPEG 2000) of these liposomal TCL carriers were fabricated using a thin lipid film hydration method [132]. Triggering TCL payload release by external near-infrared (NIR) irradiation increased anticancer responses through effective antigen presentation and maturation of DCs, T cell activation, and the proliferation of cytotoxic CD8$^+$ T cell populations. Moreover, CO$_2$ bubbles generated by the decomposition of the NH$_4$HCO$_3$ co-payload enhanced the expression of pro-inflammatory cytokines, and suppressed tumor growth in tumor-bearing C57/BL6 mice models. Therefore, the combination of multiple cargo molecules with TCLs and the stimuli-responsive modulation of the liposomal architecture could be employed not only for in vivo DC activation, but also for therapeutic anticancer treatment with CpG-1826, which showed complete tumor remission after 100 days in 45% of the animals tested.

4.3. 3D Polymeric Gel

The hydrogel-mediated co-delivery of multiple immune modulators has also been investigated. As an injectable vaccination platform, Song et al. developed poly(L-valine) (PEV)-based 3D peptide hydrogels for the co-delivery of melanoma-derived TCLs and a TLR3 agonist (Figure 5A) [133]. The sustained release of both tumor antigens and immune potentiators promoted DC maturation. The injected peptide hydrogel was able to maintain the localization of encapsulated TCLs at the in vivo vaccination site, and the expression of CD86 and MHC II antigens on DCs, and the CD8$^+$ T cell response, was significantly elevated compared to the administration of free TCLs or gels without the agonist molecule. Further tumor suppression also suggests that the formulation of peptide hydrogels encapsulated with TCL-derived tumor antigens and a TLR agonist could be utilized as a cancer vaccine platform.

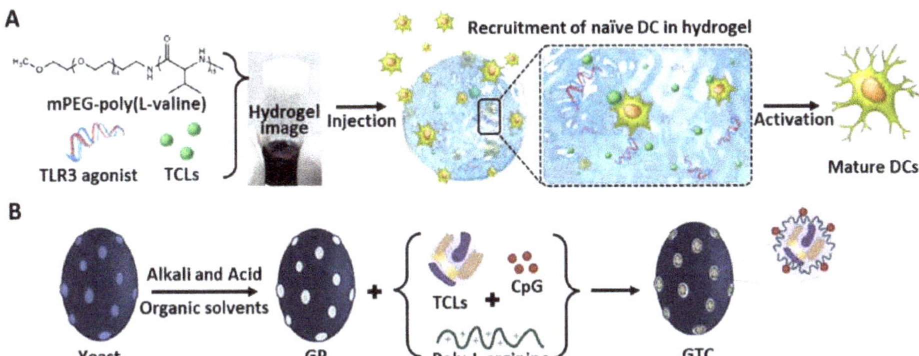

Figure 5. Schematic illustration of TCLs and adjuvant co-delivery using exogenous delivery platforms. (**A**) After injection of TLR3 agonist and TCL-loaded mPEG-poly(L-valine) hydrogels, naïve DCs aggregate around hydrogel. (**B**) Natural component, β-glucan particle (GP)-based TCL and CpG delivery. All figures were reproduced with permission from Refs. [28,133].

A similar peptide hydrogel formulation has also been applied for the delivery and in vivo localization of multiple immune stimulants. mPEG-poly (L-alanine) (PEA)-based injectable peptide hydrogel could effectively encapsulate (1) melanoma-derived TCLs, (2) GM-CSF, and (3) dual immune checkpoint inhibitors (anti-CTLA-4/PD-1 antibody) during the spontaneous self-assembly of the polypeptide and subsequent gel formation via hydrophobic interactions [25]. Hence, persistent and synergistic DC activation by released TCL antigens and GM-CSF was achieved in C57BL/6J mice models with enhanced T cell responses. Especially, the augmented expansion of effector CD8$^+$ T cells within the spleens and tumors of immunized mice by immune checkpoint blockade was observed. This hydrogel-based combination therapy showed superior immune modulation and anticancer efficacy compared to any single cargo delivery, demonstrating prolonged in vivo antigen-specific T cell immune responses.

Furthermore, cryogels (i.e., supermacroporous polymeric network obtained from the ice crystal formations through the steps of phase separation, crosslinking, and polymerization [134]) were also developed as a similar co-delivery platform for the in vivo administration of GM-CSF (DC enhancement factor) and CpG ODN (DC-activating factor) [135]. This cryogel-mediated vaccination platform effectively enhanced DC activation and leukocyte recruitment, and showed higher survival rates in melanoma-challenged C57BL/6 mice models than bolus treatment with immunoactive factors.

4.4. Natural Components

Some natural compounds possess sufficient adjuvant efficacy to trigger DC activation. Previous studies have used LPS, a membrane component of Gram-negative bacterial cell walls, because of its adjuvant effect on the activation of TLR4 signaling pathways and the CD4$^+$ T cell response [136]. Hence, a series of studies have investigated exogenous signaling via TLR4 on immune cells, and have tried to design TLR4 agonists as vaccine adjuvants [137–139]. LPS was reported to interact with TLR-4 in DCs, inducing multiple intracellular signaling cascades to express extracellular signal-regulated kinase, c-Jun N-terminal kinase, p38 mitogen-activated protein kinases, and NF-κB, and affected the production of IL-12 [140]. However, the single-use of LPS for immune activation might evoke vaccine reactogenicity, and induce improper signaling direction for DC activation and further vaccination [141,142].

Despite LPS-mediated immune activation, a high level of immunosuppressive cytokine secretion (such as IL-10) is usually observed. Therefore, other cellular components in bacterial cells could be used for the upregulated expression of immunoactivators, with reductions in immunosuppressive cytokines to deliver the TCLs [27]. For example, the empty envelope of Gram-negative bacteria (i.e., bacterial ghosts (BGs)) with intact cell surface structures exhibited strong adjuvant properties for the induction of DC maturation, and carried TCLs as immune adjuvants in the empty inner core. Facilitated by co-administration with IFN-γ, these TCL-loaded BGs showed superior DC maturation (i.e., upregulated expression of DC maturation markers, including CD86, CD80, and MHC II) compared to treatment with LPS. The secretion of Th1-polarizing cytokine IL-12p70 in DCs was also increased by TCL-loaded BGs with IFN-γ, whereas the level of pro-tolerogenic cytokine IL-10 was decreased. Moreover, the expression of immunoglobulin-like transcript 3, an inhibitory receptor used to establish suppressor T cells by inducing tolerance [143], was also decreased in DCs treated with TCL-loaded BGs. These results demonstrate that the TCL-loaded BGs could potentially overcome immunosuppressive and pro-tolerogenic effects on various cancer types as an effective inducer of Th1-polarized CD4$^+$ and associated CD8$^+$ T cell-mediated antitumor immunity.

The β-glucan particles (GPs) derived from yeast (e.g., *Saccharomyces cerevisiae*) are another example of natural compound-based fabrication of a TCL carrier (Figure 5B) [144]. Since the 1,3-β-glucan outer shell can provide receptor-mediated phagocytic uptake by cells expressing β-glucan receptors, GPs can be used for the APC-targeted delivery of soluble payloads [145]. Various potential functions of GPs, such as the stimulation of pathogens invading the body, sustained antigen release, facile internalization into APCs, and PAMP-like signaling, could induce robust immune activation. Through a similar encapsulation of antigens into the inner hollow cavity of GPs, the induction of safe immunogenicity by an engineered pathogen-mimicking system, and long-term interaction via the sustained release of cargo antigens, could be achieved. Therefore, Hou et al. [28], developed GPs encapsulating murine colon adenocarcinoma cell (MC38) lysates with additional stimulation provided by a CpG TLR9 agonist. In addition, the co-incorporation of poly-L-arginine improved the protection against challenge from live tumor cells in animal models when co-injected with tumor antigens, and also promoted the in vivo charging of MHC II$^+$ APCs [146,147]. This GP platform was internalized in up to 70% of the DCs by energy-dependent and dectin-1 receptor-mediated endocytosis, and the sustained release of the cargos resulted in the significantly higher expression of CD86 than that of the LPS controls. Moreover, NLR pyrin domain-containing protein 3 inflammasome-mediated DC activation was also confirmed by increased cleaved caspase-1 p10 (10 kDa) levels in GP-treated BMDCs, and the correlated IL-1ß secretion [148]. A summary of whole-TCL delivery platforms using liposomes, 3D polymeric gel, and natural components are indicated in Table 3.

Table 3. Biomaterial-mediated whole-TCL delivery platform.

Platform	Material	Specificity	Material Platform	Target Cancer	Outcome	Ref.
Liposome	Liposomal spherical nucleic acids	Mouse	CpG-1826-coated and TCL-loaded liposome	Triple-negative breast cancer cell	Increased population of CTLs Decreased population of myeloid derived suppressor cells	[120]
	CO_2-generating thermosensitive liposomes	Mouse	Co-delivery of DOX-loaded liposome and TCL-loaded liposome	Melanoma	High expression of pro-inflammatory cytokines and suppressed tumor growth by external NIR irradiation and generated CO_2 bubbles	[121]
3D polymeric gel	PEV-based hydrogel	Mouse	TCL- and TLR3-loaded PEV hydrogel	Melanoma	Localization of injectable hydrogel and induction of sustained release Highest percentage of CTLs in LN	[133]
	PEA-based hydrogel	Mouse	TCL, GM-CSF, and anti-CTLA4/PD-1 Ab-loaded PEA hydrogel	Melanoma	Persistent and synergistic DCs activation Augmented expansion of effector CD8$^+$ T cells	[25]
	Cryogel	Mouse	CpG ODN, GM-CSF, and RGD-loaded cryogel-containing TCLs	Melanoma	Enhanced DC activation Leukocyte recruitment Greater survival rates	[135]
Natural component	Empty envelope of bacterial ghost	Human	Combination of TCL-loaded bacterial ghost and IFN-γ	Melanoma, renal cell carcinoma, glioblastoma	Decreased expression of ILT3 and inhibitory receptor	[27]
	Yeast derived ß-glucan particle	Mouse	TCL, CpG, and poly-L-arginine-loaded ß-glucan	Colorectal cancer	High internalization in DC NLRP3 inflammasome-mediated DC activation	[28]

3D, three dimensional; CTL, cytotoxic T lymphocytes; DOX, doxorubicin; TCL, tumor cell lysate; NIR, near-infrared radiation; TLR, Toll-like receptor; PEV, poly(L-valine); PEA, poly(L-alanine); LN, lymph node; GM-CSF, granulocyte-macrophage colony-stimulating factor; Ab, antibody; DCs, dendritic cells; RGD, Arg-Gly-Asp; IFN, interferon; ILT3, immunoglobulin-like transcript 3; NLRP3, nucleotide-binding oligomerization domain 3.

4.5. Future Progress of Cancer Immunotherapy Using TCLs

The study of the relationship between cancer and immune responses has increased rapidly over the last few decades, among which TCLs have demonstrated their utility to elicit sustained CTL responses and vaccine effectiveness in cancer therapy. Moreover, since TCLs do not induce a strong enough CTL response against cancer, additional immune agonists or adjuvants have been utilized in combination, as previously described [149]. A series of delivery platforms described in this review possess the necessary functionalities, including an effective cargo protective carrier, immune agonistic property, and/or adjuvant efficacy. However, it should be also considered that there might be a possible risk of overreaction, such as cytokine storm activation, during periods of high immune activity [150]. Therefore, in order to develop more effective strategies in TCL delivery, the optimization for clinical safety, and the combination with an additional immune agonist or adjuvant, is necessary for inducing selective activation of T cells to respond to specific tumor antigens, rather than broad activation of various immune cells, which could cause deleterious side effects [151]. It should be also emphasized that there is still work to be done in developing combination therapy and optimizing vaccine platforms before TCL-based treatment becomes a viable immune modulatory and therapeutic strategy [152].

5. Conclusions

TCL-mediated cancer immunotherapy has been shown to involve the activation of tumor-specific CD8$^+$ and CD4$^+$ T cells via a vast array of immunogenic epitopes. However, an in-depth understanding of the physiological functions of DCs and in vivo interactions with other immune cell populations are needed to improve therapeutic effectiveness and establish optimal modulation in adaptive immunity. To emphasize the efficacy of TCL-mediated anticancer therapy, we reviewed (1) various experimental methods for preparing TCLs as a major immunomodulatory source, (2) TCL-mediated augmentation in DC-T cell interaction, and the subsequently induced activation of T cells, and (3) the recent progress in the biomaterial-based in vivo administration of TCLs. With the aid of co-stimulatory adjuvants, biomaterial-mediated exogenous TCL delivery could be an efficient therapeutic strategy to enhance the stability and sustained release of cargo TCLs, improve the specificity of DC targeting, and activate DCs synergistically. As a result of sufficient DC activation (i.e., increased antigen presentation and cytokine release), antigen-specific T cell-mediated tumor suppression and vaccination can be upregulated through the dynamic interplay of immune responses. Therefore, exogenous TCL delivery techniques could be a promising treatment for enhancing the DC-mediated activation of adaptive immune responses, vaccination, and tumor-specific suppression.

Funding: This research was supported by the National Research Foundation of Korea (NRF) grant funded by the Korea government (MSIT) (2021R1A4A3024237, 2019R1A2C1084828, and 2017M3A7B8061942).

Institutional Review Board Statement: Not applicable.

Informed Consent Statement: Not applicable.

Data Availability Statement: The data presented in this study are contained within the article.

Acknowledgments: Not applicable.

Conflicts of Interest: The authors disclose no conflict of interest in this work.

Abbreviation

Th	T helper
APC	Antigen-presenting cell
CTL	Cytotoxic T lymphocyte
IFN	Interferon
IL	Interleukin
DC	Dendritic cell
MHC	Major histocompatibility complex
TCR	T cell receptor
iDC	Immature DC
mDC	Mature DC
TNF	Tumor necrosis factor
TAA	Tumor-associated antigens
TCL	Tumor cell lysate
DAMP	Damage-associated molecular patterns
PRR	Pattern recognition receptor
HSP	Heat shock protein
HMGB-1	High-mobility group box-1
TLR	Toll-like receptor
UV	Ultraviolet
ICD	Immunogenic cell death
HOCl	Hypochlorous acid
OVA	Ovalbumin
SqA	Squaric acid
TSP-1	Thrombospondin-1
ROS	Reactive oxygen species
LPS	Lipopolysaccharide
ER	Endoplasmic reticulum
TAP	Transporter associated with antigen processing

FDA	Food and Drug Administration
GM-CSF	Granulocyte-macrophage colony-stimulating factor
MAGE-1	Melanoma-associated antigen-1
NP	Nanoparticle
VEP	Virus envelope protein
PLGA	Poly(lactic-co-glycolic acid)
PEG-CCV	PEGlyated cancer cell membrane vesicle
FBS	Fatal bovine serum
CCNP	Cancer cell membrane nanoparticle
Man	Mannose
$CaCO_3$	Calcium carbonate
MSN	Mesoporous silica NP
HPMA	N -(hydroxypropyl) methacrylamide
APMA	N-(3-aminopropyl) methacrylamide
PAMP	Pathogen associated molecular pattern
DOX	Doxorubicin
LM	Liquid metal
AP	Aluminum phosphate
BMDC	Bone marrow dendritic cell
DA	Dopamine
PDA	Polydopamine
Lys-SNA	TCL-loaded liposomal spherical nucleic acid
CpG-1826	Cholesteryl-modified immunostimulatory oligonucleotide adjuvants
BG-TSLs	CO_2-generating thermosensitive liposomes
NIR	Near-infrared
PEV	Poly(L-valine)
PEA	Poly(L-alanine)
BGs	Bacterial ghosts
GPs	ß-glucan particles
MC38	Murine colon adenocarcinoma cell

References

1. Coulie, P.G.; Van den Eynde, B.J.; van der Bruggen, P.; Boon, T. Tumour antigens recognized by T lymphocytes: At the core of cancer immunotherapy. *Nat. Rev. Cancer* **2014**, *14*, 135–146. [CrossRef] [PubMed]
2. Skipper, H.E.; Heidelberger, C.; Welch, A.D. Some Biochemical Problems of Cancer Chemotherapy. *Nature* **1957**, *179*, 1159–1162. [CrossRef] [PubMed]
3. Love, R.R.; Leventhal, H.; Ma, D.V.E.; Nerenz, D.R. Side effects and emotional distress during cancer chemotherapy. *Cancer* **1989**, *63*, 604–612. [CrossRef]
4. Huang, A.; Garraway, L.A.; Ashworth, A.; Weber, B. Synthetic lethality as an engine for cancer drug target discovery. *Nat. Rev. Drug Discov.* **2019**, *19*, 23–38. [CrossRef] [PubMed]
5. O'Donnell, J.S.; Teng, M.W.L.; Smyth, M.J. Cancer immunoediting and resistance to T cell-based immunotherapy. *Nat. Rev. Clin. Oncol.* **2019**, *16*, 151–167. [CrossRef]
6. Kraehenbuehl, L.; Weng, C.-H.; Eghbali, S.; Wolchok, J.D.; Merghoub, T. Enhancing immunotherapy in cancer by targeting emerging immunomodulatory pathways. *Nat. Rev. Clin. Oncol.* **2021**, *19*, 37–50. [CrossRef]
7. Harari, A.; Graciotti, M.; Bassani-Sternberg, M.; Kandalaft, L.E. Antitumour dendritic cell vaccination in a priming and boosting approach. *Nat. Rev. Drug Discov.* **2020**, *19*, 635–652. [CrossRef]
8. Huang, H.; Hao, S.; Li, F.; Ye, Z.; Yang, J.; Xiang, J. CD4$^+$Th1 cells promote CD8$^+$Tc1 cell survival, memory response, tumor localization and therapy by targeted delivery of interleukin 2 via acquired pMHC I complexes. *Immunology* **2007**, *120*, 148–159. [CrossRef]
9. Li, H.; Edin, M.L.; Gruzdev, A.; Cheng, J.; Bradbury, J.A.; Graves, J.P.; DeGraff, L.M.; Zeldin, D.C. Regulation of T helper cell subsets by cyclooxygenases and their metabolites. *Prostaglandins Other Lipid Mediat.* **2012**, *104–105*, 74–83. [CrossRef]
10. Li, M.; Qin, M.; Song, G.; Deng, H.; Wang, D.; Wang, X.; Dai, W.; He, B.; Zhang, H.; Zhang, Q. A biomimetic antitumor nanovaccine based on biocompatible calcium pyrophosphate and tumor cell membrane antigens. *Asian J. Pharm. Sci.* **2020**, *16*, 97–109. [CrossRef]
11. Waldman, A.D.; Fritz, J.M.; Lenardo, M.J. A guide to cancer immunotherapy: From T cell basic science to clinical practice. *Nat. Rev. Immunol.* **2020**, *20*, 651–668. [CrossRef] [PubMed]
12. Lamberti, M.J.; Nigro, A.; Mentucci, F.M.; Vittar, N.B.R.; Casolaro, V.; Col, J.D. Dendritic Cells and Immunogenic Cancer Cell Death: A Combination for Improving Antitumor Immunity. *Pharmaceutics* **2020**, *12*, 256. [CrossRef] [PubMed]

13. Lee, H.-G.; Cho, M.-Z.; Choi, J.-M. Bystander CD4+ T cells: Crossroads between innate and adaptive immunity. *Exp. Mol. Med.* **2020**, *52*, 1255–1263. [CrossRef]

14. Dersh, D.; Hollý, J.; Yewdell, J.W. A few good peptides: MHC class I-based cancer immunosurveillance and immunoevasion. *Nat. Rev. Immunol.* **2020**, *21*, 116–128. [CrossRef] [PubMed]

15. Wieczorek, M.; Abualrous, E.T.; Sticht, J.; Álvaro-Benito, M.; Stolzenberg, S.; Noé, F.; Freund, C. Major Histocompatibility Complex (MHC) Class I and MHC Class II Proteins: Conformational Plasticity in Antigen Presentation. *Front. Immunol.* **2017**, *8*, 292. [CrossRef] [PubMed]

16. Saxena, M.; van der Burg, S.H.; Melief, C.J.M.; Bhardwaj, N. Therapeutic cancer vaccines. *Nat. Rev. Cancer* **2021**, *21*, 360–378. [CrossRef]

17. Hu, Z.; Ott, P.A.; Wu, C.J. Towards personalized, tumour-specific, therapeutic vaccines for cancer. *Nat. Rev. Immunol.* **2017**, *18*, 168–182. [CrossRef]

18. Vormehr, M.; Türeci, Ö.; Sahin, U. Harnessing Tumor Mutations for Truly Individualized Cancer Vaccines. *Annu. Rev. Med.* **2019**, *70*, 395–407. [CrossRef]

19. Khodaei, T.; Sadri, B.; Nouraein, S.; Vahedi, N.; Mohammadi, J. Cancer vaccination: Various platforms and recent advances. *J. Immun. Biol.* **2020**, *5*, 151.

20. Jorgovanovic, D.; Song, M.; Wang, L.; Zhang, Y. Roles of IFN-gamma in tumor progression and regression: A review. *Biomark Res.* **2020**, *8*, 49. [CrossRef]

21. Woodland, D.L.; Kohlmeier, J.E. Migration, maintenance and recall of memory T cells in peripheral tissues. *Nat. Rev. Immunol.* **2009**, *9*, 153–161. [CrossRef] [PubMed]

22. Shi, G.-N.; Zhang, C.-N.; Xu, R.; Niu, J.-F.; Song, H.-J.; Zhang, X.-Y.; Wang, W.-W.; Wang, Y.-M.; Li, C.; Wei, X.-Q.; et al. Enhanced antitumor immunity by targeting dendritic cells with tumor cell lysate-loaded chitosan nanoparticles vaccine. *Biomaterials* **2017**, *113*, 191–202. [CrossRef]

23. Lybaert, L.; Ryu, K.A.; Nuhn, L.; De Rycke, R.; De Wever, O.; Chon, A.C.; Esser-Kahn, A.P.; De Geest, B.G. Cancer Cell Lysate Entrapment in CaCO$_3$ Engineered with Polymeric TLR-Agonists: Immune-Modulating Microparticles in View of Personalized Antitumor Vaccination. *Chem. Mater.* **2017**, *29*, 4209–4217. [CrossRef]

24. Wang, X.; Wang, N.; Yang, Y.; Wang, X.; Liang, J.; Tian, X.; Zhang, H.; Leng, X. Polydopamine nanoparticles carrying tumor cell lysate as a potential vaccine for colorectal cancer immunotherapy. *Biomater. Sci.* **2019**, *7*, 3062–3075. [CrossRef] [PubMed]

25. Song, H.; Yang, P.; Huang, P.; Zhang, C.; Kong, D.; Wang, W. Injectable polypeptide hydrogel-based co-delivery of vaccine and immune checkpoint inhibitors improves tumor immunotherapy. *Theranostics* **2019**, *9*, 2299–2314. [CrossRef]

26. Ochyl, L.J.; Bazzill, J.D.; Park, C.; Xu, Y.; Kuai, R.; Moon, J.J. PEGylated tumor cell membrane vesicles as a new vaccine platform for cancer immunotherapy. *Biomaterials* **2018**, *182*, 157–166. [CrossRef]

27. Dobrovolskienė, N.; Pašukonienė, V.; Darinskas, A.; Kraśko, J.; Žilionytė, K.; Mlynska, A.; Gudlevičienė, Ž.; Mišeikytė-Kaubrienė, E.; Schijns, V.; Lubitz, W.; et al. Tumor lysate-loaded Bacterial Ghosts as a tool for optimized production of therapeutic dendritic cell-based cancer vaccines. *Vaccine* **2018**, *36*, 4171–4180. [CrossRef]

28. Hou, Y.; Liu, R.; Hong, X.; Zhang, Y.; Bai, S.; Luo, X.; Zhang, Y.; Gong, T.; Zhang, Z.; Sun, X. Engineering a sustained release vaccine with a pathogen-mimicking manner for robust and durable immune responses. *J. Control. Release* **2021**, *333*, 162–175. [CrossRef]

29. Dombroski, J.A.; Jyotsana, N.; Crews, D.W.; Zhang, Z.; King, M.R. Fabrication and Characterization of Tumor Nano-Lysate as a Preventative Vaccine for Breast Cancer. *Langmuir* **2020**, *36*, 6531–6539. [CrossRef]

30. Benencia, F.; Courrèges, M.C.; Coukos, G. Whole tumor antigen vaccination using dendritic cells: Comparison of RNA electroporation and pulsing with UV-irradiated tumor cells. *J. Transl. Med.* **2008**, *6*, 21. [CrossRef]

31. Aguilera, R.; Saffie, C.; Tittarelli, A.; González, F.E.; Ramírez, M.; Reyes, D.; Pereda, C.; Hevia, D.; García, T.; Salazar, L.; et al. Heat-Shock Induction of Tumor-Derived Danger Signals Mediates Rapid Monocyte Differentiation into Clinically Effective Dendritic Cells. *Clin. Cancer Res.* **2011**, *17*, 2474–2483. [CrossRef] [PubMed]

32. Mookerjee, A.; Graciotti, M.; Kandalaft, L.E. A cancer vaccine with dendritic cells differentiated with GM-CSF and IFN alpha and pulsed with a squaric acid treated cell lysate improves T cell priming and tumor growth control in a mouse model. *Bioimpacts* **2018**, *8*, 211. [CrossRef] [PubMed]

33. Martínez-Torres, A.C.; Calvillo-Rodríguez, K.M.; Uscanga-Palomeque, A.C.; Gómez-Morales, L.; Mendoza-Reveles, R.; Caballero-Hernandez, D.; Karoyan, P.; Rodríguez-Padilla, C. PKHB1 Tumor Cell Lysate Induces Antitumor Immune System Stimulation and Tumor Regression in Syngeneic Mice with Tumoral T Lymphoblasts. *J. Oncol.* **2019**, *2019*, 1–11. [CrossRef] [PubMed]

34. Mohamed, S.I.A.; Jantan, I.; Nafiah, M.A.; Seyed, M.A.; Chan, K.M. Dendritic cells pulsed with generated tumor cell lysate from Phyllanthus amarus Schum. & Thonn. induces anti-tumor immune response. *BMC Complement. Altern. Med.* **2018**, *18*, 232. [CrossRef]

35. Gan, J.; Du, G.; He, C.; Jiang, M.; Mou, X.; Xue, J.; Sun, X. Tumor cell membrane enveloped aluminum phosphate nanoparticles for enhanced cancer vaccination. *J. Control. Release* **2020**, *326*, 297–309. [CrossRef]

36. Chiang, C.L.; Coukos, G.; Kandalaft, L.E. Whole Tumor Antigen Vaccines: Where Are We? *Vaccines (Basel)* **2015**, *3*, 344–372. [CrossRef]

37. Johnson, B.H.; Hecht, M.H. Recombinant Proteins Can Be Isolated from E. coli Cells by Repeated Cycles of Freezing and Thawing. *Nat. Biotechnol.* **1994**, *12*, 1357–1360. [CrossRef]

38. Melcher, A.; Gough, M.; Todryk, S.; Vile, R. Apoptosis or necrosis for tumor immunotherapy: What's in a name? *J. Mol. Med. (Berl.)* **1999**, *77*, 824–833. [CrossRef]

39. Land, W.G. The Role of Damage-Associated Molecular Patterns in Human Diseases: Part I—Promoting inflammation and immunity. *Sultan Qaboos Univ. Med. J.* **2015**, *15*, e9–e21.

40. Pouwels, S.D.; Heijink, I.H.; Hacken, N.H.T.; Vandenabeele, P.; Krysko, D.V.; Nawijn, M.C.; van Oosterhout, A.J. DAMPs activating innate and adaptive immune responses in COPD. *Mucosal Immunol.* **2013**, *7*, 215–226. [CrossRef]

41. Lin, F.-C.; Hsu, C.-H.; Lin, Y.-Y. Nano-therapeutic cancer immunotherapy using hyperthermia-induced heat shock proteins: Insights from mathematical modeling. *Int. J. Nanomed.* **2018**, *13*, 3529–3539. [CrossRef] [PubMed]

42. Bettaieb, A.; Averill-Bates, D.A. Thermotolerance induced at a mild temperature of 40 degrees C alleviates heat shock-induced ER stress and apoptosis in HeLa cells. *Biochim. Biophys Acta* **2015**, *1853*, 52–62. [CrossRef] [PubMed]

43. Salimu, J.C.; Spary, L.; Al-Taei, S.; Clayton, A.; Mason, M.D.; Staffurth, J.; Tabi, Z. Cross-presentation of the oncofoetal tumor antigen 5T4 from irradiated prostate cancer cells—A key role for Hsp70. *J. Immunother. Cancer* **2014**, *2*, P161. [CrossRef]

44. Grant, M.L.; Shields, N.; Neumann, S.; Kramer, K.; Bonato, A.; Jackson, C.; A Baird, M.; Young, S.L. Combining dendritic cells and B cells for presentation of oxidised tumour antigens to CD8+ T cells. *Clin. Transl. Immunol.* **2017**, *6*, e149. [CrossRef]

45. Marcinkiewicz, J.; Chain, B.; Olszowska, E.; Olszowski, S.; Zgliczyński, J. Enhancement of immunogenic properties of ovalbumin as a result of its chlorination. *Int. J. Biochem.* **1991**, *23*, 1393–1395. [CrossRef]

46. Allison, M.; Fearon, D.T. Enhanced immunogenicity of aldehyde-bearing antigens: A possible link between innate and adaptive immunity. *Eur. J. Immunol.* **2000**, *30*, 2881–2887. [CrossRef]

47. Chiang, C.L.-L.; Kandalaft, L.E.; Tanyi, J.; Hagemann, A.R.; Motz, G.T.; Svoronos, N.; Montone, K.; Mantia-Smaldone, G.M.; Smith, L.; Nisenbaum, H.L.; et al. A Dendritic Cell Vaccine Pulsed with Autologous Hypochlorous Acid-Oxidized Ovarian Cancer Lysate Primes Effective Broad Antitumor Immunity: From Bench to Bedside. *Clin. Cancer Res.* **2013**, *19*, 4801–4815. [CrossRef]

48. Martinez-Torres, A.C.; Quiney, C.; Attout, T.; Boullet, H.; Herbi, L.; Vela, L.; Barbier, S.; Chateau, D.; Chapiro, E.; Nguyen-Khac, F.; et al. CD47 agonist peptides induce programmed cell death in refractory chronic lymphocytic leukemia B cells via PLCgamma1 activation: Evidence from mice and humans. *PLoS Med.* **2015**, *12*, e1001796. [CrossRef]

49. Denèfle, T.; Boullet, H.; Herbi, L.; Newton, C.; Martinez-Torres, A.C.; Guez, A.; Pramil, E.; Quiney, C.; Pourcelot, M.; Levasseur, M.; et al. Thrombospondin-1 Mimetic Agonist Peptides Induce Selective Death in Tumor Cells: Design, Synthesis, and Structure–Activity Relationship Studies. *J. Med. Chem.* **2016**, *59*, 8412–8421. [CrossRef]

50. Leth-Larsen, R.; Lund, R.; Hansen, H.V.; Laenkholm, A.-V.; Tarin, D.; Jensen, O.N.; Ditzel, H.J. Metastasis-related Plasma Membrane Proteins of Human Breast Cancer Cells Identified by Comparative Quantitative Mass Spectrometry. *Mol. Cell. Proteom.* **2009**, *8*, 1436–1449. [CrossRef]

51. Lund, R.; Leth-Larsen, R.; Jensen, O.N.; Ditzel, H.J. Efficient Isolation and Quantitative Proteomic Analysis of Cancer Cell Plasma Membrane Proteins for Identification of Metastasis-Associated Cell Surface Markers. *J. Proteome Res.* **2009**, *8*, 3078–3090. [CrossRef] [PubMed]

52. Hayashi, M.; Aoshi, T.; Kogai, Y.; Nomi, D.; Haseda, Y.; Kuroda, E.; Kobiyama, K.; Ishii, K.J. Optimization of physiological properties of hydroxyapatite as a vaccine adjuvant. *Vaccine* **2016**, *34*, 306–312. [CrossRef] [PubMed]

53. Steinman, R.M. Dendritic cells: Versatile controllers of the immune system. *Nat. Med.* **2007**, *13*, 1155–1159. [CrossRef] [PubMed]

54. Collin, M.; Bigley, V. Human dendritic cell subsets: An update. *Immunology* **2018**, *154*, 3–20. [CrossRef]

55. Dalod, M.; Chelbi, R.; Malissen, B.; Lawrence, T. Dendritic cell maturation: Functional specialization through signaling specificity and transcriptional programming. *EMBO J.* **2014**, *33*, 1104–1116. [CrossRef]

56. Kim, M.K.; Kim, J. Properties of immature and mature dendritic cells: Phenotype, morphology, phagocytosis, and migration. *RSC Adv.* **2019**, *9*, 11230–11238. [CrossRef]

57. Lanzavecchia, A.; Sallusto, F. Regulation of T Cell Immunity by Dendritic Cells. *Cell* **2001**, *106*, 263–266. [CrossRef]

58. Fang, P.; Li, X.; Dai, J.; Cole, L.; Camacho, J.A.; Zhang, Y.; Ji, Y.; Wang, J.; Yang, X.-F.; Wang, H. Immune cell subset differentiation and tissue inflammation. *J. Hematol. Oncol.* **2018**, *11*, 1–22. [CrossRef]

59. Roche, P.A.; Cresswell, P. Antigen Processing and Presentation Mechanisms in Myeloid Cells. *Microbiol. Spectr.* **2016**, *4*. [CrossRef]

60. Albert, M.L.; Sauter, B.; Bhardwaj, N. Dendritic cells acquire antigen from apoptotic cells and induce class I-restricted CTLs. *Nature* **1998**, *392*, 86–89. [CrossRef]

61. Embgenbroich, M.; Burgdorf, S. Current Concepts of Antigen Cross-Presentation. *Front. Immunol.* **2018**, *9*, 1643. [CrossRef] [PubMed]

62. Neefjes, J.; Jongsma, M.L.M.; Paul, P.; Bakke, O. Towards a systems understanding of MHC class I and MHC class II antigen presentation. *Nat. Rev. Immunol.* **2011**, *11*, 823–836. [CrossRef] [PubMed]

63. Hiltbold, E.M.; A Roche, P. Trafficking of MHC class II molecules in the late secretory pathway. *Curr. Opin. Immunol.* **2002**, *14*, 30–35. [CrossRef]

64. Turley, S.J.; Inaba, K.; Garrett, W.S.; Ebersold, M.; Unternaehrer, J.; Steinman, R.M.; Mellman, I. Transport of Peptide-MHC Class II Complexes in Developing Dendritic Cells. *Science* **2000**, *288*, 522–527. [CrossRef] [PubMed]

65. Decker, W.K.; Xing, D.; Shpall, E.J. Dendritic Cell Immunotherapy for the Treatment of Neoplastic Disease. *Biol. Blood Marrow Transplant.* **2006**, *12*, 113–125. [CrossRef] [PubMed]

66. Cho, K.-J.; Ishido, S.; Eisenlohr, L.C.; Roche, P.A. Activation of Dendritic Cells Alters the Mechanism of MHC Class II Antigen Presentation to CD4 T Cells. *J. Immunol.* **2020**, *204*, 1621–1629. [CrossRef]

67. Landsverk, O.J.B.; Bakke, O.; Gregers, T.F. MHC II and the Endocytic Pathway: Regulation by Invariant Chain. *Scand. J. Immunol.* **2009**, *70*, 184–193. [CrossRef]
68. Kloetzel, P.-M.; Ossendorp, F. Proteasome and peptidase function in MHC-class-I-mediated antigen presentation. *Curr. Opin. Immunol.* **2003**, *16*, 76–81. [CrossRef]
69. Shen, L.; Sigal, L.J.; Boes, M.; Rock, K.L. Important Role of Cathepsin S in Generating Peptides for TAP-Independent MHC Class I Crosspresentation In Vivo. *Immunity* **2004**, *21*, 155–165. [CrossRef]
70. Guermonprez, P.; Saveanu, L.; Kleijmeer, M.J.; Davoust, J.; van Endert, P.; Amigorena, S. ER–phagosome fusion defines an MHC class I cross-presentation compartment in dendritic cells. *Nature* **2003**, *425*, 397–402. [CrossRef]
71. Tang-Huau, T.-L.; Gueguen, P.; Goudot, C.; Durand, M.; Bohec, M.; Baulande, S.; Pasquier, B.; Amigorena, S.; Segura, E. Human in vivo-generated monocyte-derived dendritic cells and macrophages cross-present antigens through a vacuolar pathway. *Nat. Commun.* **2018**, *9*, 1–12. [CrossRef] [PubMed]
72. Cruz, F.M.; Colbert, J.D.; Merino, E.; Kriegsman, B.A.; Rock, K.L. The Biology and Underlying Mechanisms of Cross-Presentation of Exogenous Antigens on MHC-I Molecules. *Annu. Rev. Immunol.* **2017**, *35*, 149–176. [CrossRef] [PubMed]
73. Tai, Y.; Wang, Q.; Korner, H.; Zhang, L.; Wei, W. Molecular Mechanisms of T Cells Activation by Dendritic Cells in Autoimmune Diseases. *Front. Pharmacol.* **2018**, *9*, 642. [CrossRef] [PubMed]
74. Sallusto, F.; Lanzavecchia, A. The instructive role of dendritic cells on T-cell responses. *Arthritis Res. Ther.* **2002**, *4*, S127–S132. [CrossRef]
75. Basu, A.; Ramamoorthi, G.; Albert, G.; Gallen, C.; Beyer, A.; Snyder, C.; Koski, G.; Disis, M.L.; Czerniecki, B.J.; Kodumudi, K. Differentiation and Regulation of TH Cells: A Balancing Act for Cancer Immunotherapy. *Front. Immunol.* **2021**, *12*. [CrossRef]
76. Owen, D.; Mahmud, S.; Vang, K.B.; Kelly, R.M.; Blazar, B.R.; Smith, K.A.; Farrar, M.A. Identification of Cellular Sources of IL-2 Needed for Regulatory T Cell Development and Homeostasis. *J. Immunol.* **2018**, *200*, 3926–3933. [CrossRef]
77. Cheng, L.E.; Greenberg, P.D. Selective Delivery of Augmented IL-2 Receptor Signals to Responding CD8$^+$T Cells Increases the Size of the Acute Antiviral Response and of the Resulting Memory T Cell Pool. *J. Immunol.* **2002**, *169*, 4990–4997. [CrossRef]
78. Lanzavecchia, A.; Sallusto, F. Dynamics of T Lymphocyte Responses: Intermediates, Effectors, and Memory Cells. *Science* **2000**, *290*, 92–97. [CrossRef]
79. Lanzavecchia, A.; Sallusto, F. Antigen decoding by T lymphocytes: From synapses to fate determination. *Nat. Immunol.* **2001**, *2*, 487–492. [CrossRef]
80. Miller, M.J.; Safrina, O.; Parker, I.; Cahalan, M.D. Imaging the Single Cell Dynamics of CD4+ T Cell Activation by Dendritic Cells in Lymph Nodes. *J. Exp. Med.* **2004**, *200*, 847–856. [CrossRef]
81. Mempel, T.R.; Henrickson, S.E.; Von Adrian, U.H. T-cell priming by dendritic cells in lymph nodes occurs in three distinct phases. *Nature* **2004**, *427*, 154–159. [CrossRef] [PubMed]
82. Kaech, S.M.; Ahmed, R. Memory CD8+ T cell differentiation: Initial antigen encounter triggers a developmental program in naïve cells. *Nat. Immunol.* **2001**, *2*, 415–422. [CrossRef] [PubMed]
83. Rainone, V.; Martelli, C.; Ottobrini, L.; Biasin, M.; Borelli, M.; Lucignani, G.; Trabattoni, D.; Clerici, M. Immunological characterization of whole tumour lysate-loaded dendritic cells for cancer immunotherapy. *PLoS ONE* **2016**, *11*, e0146622. [CrossRef] [PubMed]
84. Weigel, B.J.; Nath, N.; Taylor, P.A.; Panoskaltsis-Mortari, A.; Chen, W.; Krieg, A.M.; Brasel, K.; Blazar, B.R. Comparative analysis of murine marrow–derived dendritic cells generated by Flt3L or GM-CSF/IL-4 and matured with immune stimulatory agents on the in vivo induction of antileukemia responses. *Blood* **2002**, *100*, 4169–4176. [CrossRef]
85. Nestle, F.O.; Alijagic, S.; Gilliet, M.; Sun, Y.; Grabbe, S.; Dummer, R.; Burg, G.; Schadendorf, D. Vaccination of melanoma patients with peptide- or tumorlysate-pulsed dendritic cells. *Nat. Med.* **1998**, *4*, 328–332. [CrossRef]
86. Aarntzen, E.H.; Srinivas, M.; Bonetto, F.; Cruz, L.J.; Verdijk, P.; Schreibelt, G.; van de Rakt, M.; Lesterhuis, W.J.; van Riel, M.; Punt, C.J.; et al. Targeting of 111In-Labeled Dendritic Cell Human Vaccines Improved by Reducing Number of Cells. *Clin. Cancer Res.* **2013**, *19*, 1525–1533. [CrossRef]
87. Celli, S.; Day, M.; Müller, A.J.; Molina-Paris, C.; Lythe, G.; Bousso, P. How many dendritic cells are required to initiate a T-cell response? Blood. *J. Am. Soc. Hematol.* **2012**, *120*, 3945–3948.
88. Wculek, S.K.; Cueto, F.J.; Mujal, A.M.; Melero, I.; Krummel, M.F.; Sancho, D. Dendritic cells in cancer immunology and immunotherapy. *Nat. Rev. Immunol.* **2020**, *20*, 7–24. [CrossRef]
89. Anguille, S.; Smits, E.L.; Bryant, C.; Van Acker, H.H.; Goossens, H.; Lion, E.; Fromm, P.D.; Hart, D.N.; Van Tendeloo, V.F.; Berneman, Z.N. Dendritic Cells as Pharmacological Tools for Cancer Immunotherapy. *Pharmacol. Rev.* **2015**, *67*, 731–753. [CrossRef]
90. Ali, O.A.; Huebsch, N.; Cao, L.; Dranoff, G.; Mooney, D.J. Infection-mimicking materials to program dendritic cells in situ. *Nat. Mater.* **2009**, *8*, 151–158. [CrossRef]
91. Steinman, R.M.; Banchereau, J. Taking dendritic cells into medicine. *Nature* **2007**, *449*, 419–426. [CrossRef] [PubMed]
92. Palucka, K.; Banchereau, J. Cancer immunotherapy via dendritic cells. *Nat. Cancer* **2012**, *12*, 265–277. [CrossRef]
93. Gu, Y.-Z.; Zhao, X.; Song, X.-R. Ex vivo pulsed dendritic cell vaccination against cancer. *Acta Pharmacol. Sin.* **2020**, *41*, 959–969. [CrossRef] [PubMed]
94. Wei, Q.; Su, Y.; Xin, H.; Zhang, L.; Ding, J.; Chen, X. Immunologically Effective Biomaterials. *ACS Appl. Mater. Interfaces* **2021**, *13*, 56719–56724. [CrossRef] [PubMed]

95. Vandenberk, L.; Belmans, J.; Van Woensel, M.; Riva, M.; Van Gool, S.W. Exploiting the Immunogenic Potential of Cancer Cells for Improved Dendritic Cell Vaccines. *Front. Immunol.* **2016**, *6*. [CrossRef] [PubMed]

96. Senapati, S.; Mahanta, A.K.; Kumar, S.; Maiti, P. Controlled drug delivery vehicles for cancer treatment and their performance. *Signal Transduct. Target. Ther.* **2018**, *3*, 7. [CrossRef]

97. Fang, R.H.; Kroll, A.V.; Gao, W.W.; Zhang, L.F. Cell Membrane Coating Nanotechnology. *Adv. Mater.* **2018**, *30*, e1706759. [CrossRef]

98. Bachmann, M.F.; Jennings, G.T. Vaccine delivery: A matter of size, geometry, kinetics and molecular patterns. *Nat. Rev. Immunol.* **2010**, *10*, 787–796. [CrossRef]

99. Kwon, Y.J.; James, E.; Shastri, N.; Fréchet, J.M. In vivo targeting of dendritic cells for activation of cellular immunity using vaccine carriers based on pH-responsive microparticles. *Proc. Natl. Acad. Sci. USA* **2005**, *102*, 18264–18268. [CrossRef]

100. Hardy, C.L.; LeMasurier, J.S.; Mohamud, R.; Yao, J.; Xiang, S.D.; Rolland, J.M.; O'Hehir, R.E.; Plebanski, M. Differential Uptake of Nanoparticles and Microparticles by Pulmonary APC Subsets Induces Discrete Immunological Imprints. *J. Immunol.* **2013**, *191*, 5278–5290. [CrossRef]

101. Fifis, T.; Gamvrellis, A.; Crimeen-Irwin, B.; Pietersz, G.A.; Li, J.; Mottram, P.L.; McKenzie, I.F.C.; Plebanski, M. Size-Dependent Immunogenicity: Therapeutic and Protective Properties of Nano-Vaccines against Tumors. *J. Immunol.* **2004**, *173*, 3148–3154. [CrossRef] [PubMed]

102. Cooley, M.; Sarode, A.; Hoore, M.; Fedosov, D.A.; Mitragotri, S.; Gupta, A.S. Influence of particle size and shape on their margination and wall-adhesion: Implications in drug delivery vehicle design across nano-to-micro scale. *Nanoscale* **2018**, *10*, 15350–15364. [CrossRef] [PubMed]

103. Huang, C.; Zhang, Y.; Yuan, H.; Gao, H.; Zhang, S. Role of Nanoparticle Geometry in Endocytosis: Laying Down to Stand Up. *Nano Lett.* **2013**, *13*, 4546–4550. [CrossRef] [PubMed]

104. Li, Y.; Kröger, M.; Liu, W.K. Shape effect in cellular uptake of PEGylated nanoparticles: Comparison between sphere, rod, cube and disk. *Nanoscale* **2015**, *7*, 16631–16646. [CrossRef]

105. Niikura, K.; Matsunaga, T.; Suzuki, T.; Kobayashi, S.; Yamaguchi, H.; Orba, Y.; Kawaguchi, A.; Hasegawa, H.; Kajino, K.; Ninomiya, T.; et al. Gold Nanoparticles as a Vaccine Platform: Influence of Size and Shape on Immunological Responses *in Vitro* and in Vivo. *ACS Nano* **2013**, *7*, 3926–3938. [CrossRef]

106. Neek, M.; Kim, T.I.; Wang, S.-W. Protein-based nanoparticles in cancer vaccine development. *Nanomed. Nanotechnol. Biol. Med.* **2018**, *15*, 164–174. [CrossRef]

107. Verma, A.; Stellacci, F. Effect of surface properties on nanoparticle–cell interactions. *Small* **2010**, *6*, 12–21. [CrossRef]

108. Hühn, D.; Kantner, K.; Geidel, C.; Brandholt, S.; De Cock, I.; Soenen, S.J.H.; Rivera_Gil, P.; Montenegro, J.-M.; Braeckmans, K.; Müllen, K.; et al. Polymer-Coated Nanoparticles Interacting with Proteins and Cells: Focusing on the Sign of the Net Charge. *ACS Nano* **2013**, *7*, 3253–3263. [CrossRef]

109. Riley, R.S.; June, C.H.; Langer, R.; Mitchell, M.J. Delivery technologies for cancer immunotherapy. *Nat. Rev. Drug Discov.* **2019**, *18*, 175–196. [CrossRef]

110. Ambrogio, M.W.; Toro-González, M.; Keever, T.J.; McKnight, T.E.; Davern, S.M. Poly(lactic-co-glycolic acid) Nanoparticles as Delivery Systems for the Improved Administration of Radiotherapeutic Anticancer Agents. *ACS Appl. Nano Mater.* **2020**, *3*, 10565–10570. [CrossRef]

111. Chung, Y.-I.; Kim, J.C.; Kim, Y.H.; Tae, G.; Lee, S.-Y.; Kim, K.; Kwon, I.C. The effect of surface functionalization of PLGA nanoparticles by heparin- or chitosan-conjugated Pluronic on tumor targeting. *J. Control. Release* **2010**, *143*, 374–382. [CrossRef] [PubMed]

112. Pavot, V.; Berthet, M.; Rességuier, J.; Legaz, S.; Handké, N.; Gilbert, S.C.; Paul, S.; Verrier, B. Poly(lactic acid) and poly(lactic-*co*-glycolic acid) particles as versatile carrier platforms for vaccine delivery. *Nanomedicine* **2014**, *9*, 2703–2718. [CrossRef] [PubMed]

113. Kohnepoushi, C.; Nejati, V.; Delirezh, N.; Biparva, P. Poly Lactic-co-Glycolic Acid Nanoparticles Containing Human Gastric Tumor Lysates as Antigen Delivery Vehicles for Dendritic Cell-Based Antitumor Immunotherapy. *Immunol. Investig.* **2019**, *48*, 794–808. [CrossRef] [PubMed]

114. Kroll, A.V.; Fang, R.H.; Jiang, Y.; Zhou, J.; Wei, X.; Yu, C.L.; Gao, J.; Luk, B.T.; Dehaini, D.; Gao, W.; et al. Nanoparticulate Delivery of Cancer Cell Membrane Elicits Multiantigenic Antitumor Immunity. *Adv. Mater.* **2017**, *29*. [CrossRef]

115. Yang, R.; Xu, J.; Xu, L.; Sun, X.; Chen, Q.; Zhao, Y.; Peng, R.; Liu, Z. Cancer Cell Membrane-Coated Adjuvant Nanoparticles with Mannose Modification for Effective Anticancer Vaccination. *ACS Nano* **2018**, *12*, 5121–5129. [CrossRef]

116. Zhang, Y.; Liu, M.D.; Li, C.X.; Li, B.; Zhang, X.Z. Tumor Cell Membrane-Coated Liquid Metal Nanovaccine for Tumor Prevention. *Chin. J. Chem* **2020**, *38*, 595–600. [CrossRef]

117. Mishra, P.; Nayak, B.; Dey, R. PEGylation in anti-cancer therapy: An overview. *Asian J. Pharm. Sci.* **2016**, *11*, 337–348. [CrossRef]

118. Cruz, L.J.; Tacken, P.J.; Fokkink, R.; Figdor, C.G. The influence of PEG chain length and targeting moiety on antibody-mediated delivery of nanoparticle vaccines to human dendritic cells. *Biomaterials* **2011**, *32*, 6791–6803. [CrossRef]

119. Liu, C.M.; Chen, G.B.; Chen, H.H.; Zhang, J.B.; Li, H.Z.; Sheng, M.X.; Weng, W.B.; Guo, S.M. Cancer cell membrane-cloaked mesoporous silica nanoparticles with a pH-sensitive gatekeeper for cancer treatment. *Colloid Surf. B* **2019**, *175*, 477–486. [CrossRef]

120. Callmann, C.E.; Cole, L.E.; Kusmierz, C.D.; Huang, Z.; Horiuchi, D.; Mirkin, C.A. Tumor cell lysate-loaded immunostimulatory spherical nucleic acids as therapeutics for triple-negative breast cancer. *Proc. Natl. Acad. Sci. USA* **2020**, *117*, 17543–17550. [CrossRef]

121. Won, J.E.; Byeon, Y.; Wi, T.I.; Lee, J.M.; Kang, T.H.; Lee, J.W.; Shin, B.C.; Han, H.D.; Park, Y.-M. Enhanced Antitumor Immunity Using a Tumor Cell Lysate-Encapsulated CO$_2$-Generating Liposomal Carrier System and Photothermal Irradiation. *ACS Appl. Bio Mater.* **2019**, *2*, 2481–2489. [CrossRef] [PubMed]

122. Fang, R.H.; Hu, C.-M.J.; Luk, B.T.; Gao, W.; Copp, J.A.; Tai, Y.; O'Connor, D.E.; Zhang, L. Cancer Cell Membrane-Coated Nanoparticles for Anticancer Vaccination and Drug Delivery. *Nano Lett.* **2014**, *14*, 2181–2188. [CrossRef] [PubMed]

123. Kastl, L.; Sasse, D.; Wulf, V.; Hartmann, R.; Mircheski, J.; Ranke, C.; Carregal-Romero, S.; Martínez-López, J.A.; Fernández-Chacón, R.; Parak, W.J.; et al. Multiple Internalization Pathways of Polyelectrolyte Multilayer Capsules into Mammalian Cells. *ACS Nano* **2013**, *7*, 6605–6618. [CrossRef] [PubMed]

124. Hennessy, E.J.; Parker, A.E.; O'Neill, L.A.J. Targeting Toll-like receptors: Emerging therapeutics? *Nat. Rev. Drug Discov.* **2010**, *9*, 293–307. [CrossRef] [PubMed]

125. Antonios, D.; Ade, N.; Kerdine-Romer, S.; Assaf-Vandecasteele, H.; Larange, A.; Azouri, H.; Pallardy, M. Metallic haptens induce differential phenotype of human dendritic cells through activation of mitogen-activated protein kinase and NF-kappa B pathways. *Toxicol Vitr.* **2009**, *23*, 227–234. [CrossRef]

126. Méndez, I.Z.R.; Shi, Y.; HogenEsch, H.; Hem, S.L. Potentiation of the immune response to non-adsorbed antigens by aluminum-containing adjuvants. *Vaccine* **2007**, *25*, 825–833. [CrossRef]

127. Hem, S.L.; HogenEsch, H. Relationship between physical and chemical properties of aluminum-containing adjuvants and immunopotentiation. *Expert Rev. Vaccines* **2007**, *6*, 685–698. [CrossRef]

128. Carroll, E.C.; Jin, L.; Mori, A.; Munoz-Wolf, N.; Oleszycka, E.; Moran, H.B.T.; Mansouri, S.; McEntee, C.P.; Lambe, E.; Agger, E.M.; et al. The Vaccine Adjuvant Chitosan Promotes Cellular Immunity via DNA Sensor cGAS-STING-Dependent Induction of Type I Interferons. *Immunity* **2016**, *44*, 597–608. [CrossRef]

129. Engering, A.J.; Cella, M.; Fluitsma, D.; Brockhaus, M.; Hoefsmit, E.C.M.; Lanzavecchia, A.; Pieters, J. The mannose receptor functions as a high capacity and broad specificity antigen receptor in human dendritic cells. *Eur. J. Immunol.* **1997**, *27*, 2417–2425. [CrossRef]

130. Nolan, R.A.; Reeb, K.L.; Rong, Y.; Matt, S.M.; Johnson, H.S.; Runner, K.; Gaskill, P.J. Dopamine activates NF-kappaB and primes the NLRP3 inflammasome in primary human macrophages. *Brain Behav. Immun Health* **2020**, *2*, 100030. [CrossRef]

131. Torchilin, V.P. Recent advances with liposomes as pharmaceutical carriers. *Nat. Rev. Drug Discov.* **2005**, *4*, 145–160. [CrossRef] [PubMed]

132. Zhang, H. Thin-film hydration followed by extrusion method for liposome preparation. In *Liposomes: Methods and Protocols*; Springer: New York, NY, USA, 2017; Volume 1522, pp. 17–22. ISBN 9781461491644.

133. Song, H.; Huang, P.; Niu, J.; Shi, G.; Zhang, C.; Kong, D.; Wang, W. Injectable polypeptide hydrogel for dual-delivery of antigen and TLR3 agonist to modulate dendritic cells in vivo and enhance potent cytotoxic T-lymphocyte response against melanoma. *Biomaterials* **2018**, *159*, 119–129. [CrossRef] [PubMed]

134. Nayak, A.K.; Das, B. Introduction to polymeric gels. In *Polymeric Gels*; Woodhead Publishing: Philadelphia, PA, USA, 2018; pp. 3–27.

135. Bencherif, S.A.; Sands, R.W.; Ali, O.A.; Li, A.; Lewin, S.A.; Braschler, T.M.; Shih, T.-Y.; Verbeke, C.S.; Bhatta, D.; Dranoff, G.; et al. Injectable cryogel-based whole-cell cancer vaccines. *Nat. Commun.* **2015**, *6*, 1–13. [CrossRef] [PubMed]

136. Zariri, A.; van der Ley, P. Biosynthetically engineered lipopolysaccharide as vaccine adjuvant. *Expert Rev Vaccines* **2015**, *14*, 861–876. [CrossRef]

137. Han, J.E.; Wui, S.R.; Kim, K.S.; Cho, Y.J.; Cho, W.J.; Lee, N.G. Characterization of the Structure and Immunostimulatory Activity of a Vaccine Adjuvant, De-O-Acylated Lipooligosaccharide. *PLoS ONE* **2014**, *9*, e85838. [CrossRef]

138. Chou, Y.-J.; Lin, C.-C.; Dzhagalov, I.; Chen, N.-J.; Lin, C.-H.; Lin, C.-C.; Chen, S.-T.; Chen, K.-H.; Fu, S.-L. Vaccine adjuvant activity of a TLR4-activating synthetic glycolipid by promoting autophagy. *Sci. Rep.* **2020**, *10*, 1–15. [CrossRef]

139. Jeong, S.; Choi, Y.; Kim, K. Engineering Therapeutic Strategies in Cancer Immunotherapy via Exogenous Delivery of Toll-like Receptor Agonists. *Pharmaceutics* **2021**, *13*, 1374. [CrossRef]

140. Xie, J.; Qian, J.; Wang, S.; Freeman, M.E.; Epstein, J.; Yi, Q. Novel and Detrimental Effects of Lipopolysaccharide on In Vitro Generation of Immature Dendritic Cells: Involvement of Mitogen-Activated Protein Kinase p38. *J. Immunol.* **2003**, *171*, 4792–4800. [CrossRef]

141. Herve, C.; Laupeze, B.; Del Giudice, G.; Didierlaurent, A.M.; Da Silva, F.T. The how's and what's of vaccine reactogenicity. *NPJ Vaccines* **2019**, *4*, 39. [CrossRef]

142. Zhao, H.; Wu, L.; Yan, G.; Chen, Y.; Zhou, M.; Wu, Y.; Li, Y. Inflammation and tumor progression: Signaling pathways and targeted intervention. *Signal Transduct. Target. Ther.* **2021**, *6*, 1–46. [CrossRef]

143. Penna, G.; Roncari, A.; Amuchastegui, S.; Daniel, K.C.; Berti, E.; Colonna, M.; Adorini, L. Expression of the inhibitory receptor ILT3 on dendritic cells is dispensable for induction of CD4+Foxp3+ regulatory T cells by 1,25-dihydroxyvitamin D. *Blood* **2005**, *106*, 3490–3497. [CrossRef] [PubMed]

144. Mirza, Z.; Soto, E.R.; Dikengil, F.; Levitz, S.M.; Ostroff, G.R. Beta-Glucan Particles as Vaccine Adjuvant Carriers. *Bacteriophages* **2017**, *1625*, 143–157. [CrossRef]

145. Soto, E.R.; Caras, A.C.; Kut, L.C.; Castle, M.K.; Ostroff, G.R. Glucan Particles for Macrophage Targeted Delivery of Nanoparticles. *J. Drug Deliv.* **2011**, *2012*, 1–13. [CrossRef] [PubMed]

146. Mattner, F.; Fleitmann, J.-K.; Lingnau, K.; Schmidt, W.; Egyed, A.; Fritz, J.; Zauner, W.; Wittmann, B.; Gorny, I.; Berger, M.; et al. Vaccination with poly-L-arginine as immunostimulant for peptide vaccines: Induction of potent and long-lasting T-cell responses against cancer antigens. *Cancer Res.* **2002**, *62*.

147. Schmidt, W.; Buschle, M.; Zauner, W.; Kirlappos, H.; Mechtler, K.; Trska, B.; Birnstiel, M.L. Cell-free tumor antigen peptide-based cancer vaccines. *Proc. Natl. Acad. Sci. USA* **1997**, *94*, 3262–3267. [CrossRef]

148. Lopez-Castejon, G.; Brough, D. Understanding the mechanism of IL-1beta secretion. *Cytokine Growth Factor Rev.* **2011**, *22*, 189–195. [CrossRef]

149. Kawahara, M.; Takaku, H. A tumor lysate is an effective vaccine antigen for the stimulation of CD4+T-cell function and subsequent induction of antitumor immunity mediated by CD8+T cells. *Cancer Biol. Ther.* **2015**, *16*, 1616–1625. [CrossRef]

150. Meng, Z.; Zhang, Y.; Zhou, X.; Ji, J.; Liu, Z. Nanovaccines with cell-derived components for cancer immunotherapy. *Adv. Drug Deliv. Rev.* **2022**, *182*, 114107. [CrossRef]

151. Matsuo, K.; Yoshie, O.; Kitahata, K.; Kamei, M.; Hara, Y.; Nakayama, T. Recent Progress in Dendritic Cell-Based Cancer Immunotherapy. *Cancers* **2021**, *13*, 2495. [CrossRef]

152. Liu, J.; Fu, M.; Wang, M.; Wan, D.; Wei, Y.; Wei, X. Cancer vaccines as promising immuno-therapeutics: Platforms and current progress. *J. Hematol. Oncol.* **2022**, *15*, 1–26. [CrossRef]

 *pharmaceutics*

Review

Nanotechnology in Immunotherapy for Type 1 Diabetes: Promising Innovations and Future Advances

Saumya Nigam [1,2], Jack Owen Bishop [1,2], Hanaan Hayat [1,3], Tahnia Quadri [1,3], Hasaan Hayat [1,2] and Ping Wang [1,2,*]

[1] Precision Health Program, Michigan State University, East Lansing, MI 48824, USA; nigamsau@msu.edu (S.N.); bishop91@msu.edu (J.O.B.); hayathan@msu.edu (H.H.); quadrita@msu.edu (T.Q.); hayathas@msu.edu (H.H.)

[2] Department of Radiology, College of Human Medicine, Michigan State University, East Lansing, MI 48824, USA

[3] Lyman Briggs College, Michigan State University, East Lansing, MI 48824, USA

* Correspondence: wangpin4@msu.edu

Abstract: Diabetes is a chronic condition which affects the glucose metabolism in the body. In lieu of any clinical "cure," the condition is managed through the administration of pharmacological aids, insulin supplements, diet restrictions, exercise, and the like. The conventional clinical prescriptions are limited by their life-long dependency and diminished potency, which in turn hinder the patient's recovery. This necessitated an alteration in approach and has instigated several investigations into other strategies. As Type 1 diabetes (T1D) is known to be an autoimmune disorder, targeting the immune system in activation and/or suppression has shown promise in reducing beta cell loss and improving insulin levels in response to hyperglycemia. Another strategy currently being explored is the use of nanoparticles in the delivery of immunomodulators, insulin, or engineered vaccines to endogenous immune cells. Nanoparticle-assisted targeting of immune cells holds substantial potential for enhanced patient care within T1D clinical settings. Herein, we summarize the knowledge of etiology, clinical scenarios, and the current state of nanoparticle-based immunotherapeutic approaches for Type 1 diabetes. We also discuss the feasibility of translating this approach to clinical practice.

Keywords: autoimmunity; B cells; beta cells; cell therapy; immune checkpoint molecules; immunotherapy; microRNA; nanoparticles; stem cells; type 1 diabetes

Citation: Nigam, S.; Bishop, J.O.; Hayat, H.; Quadri, T.; Hayat, H.; Wang, P. Nanotechnology in Immunotherapy for Type 1 Diabetes: Promising Innovations and Future Advances. *Pharmaceutics* **2022**, *14*, 644. https://doi.org/10.3390/pharmaceutics14030644

Academic Editors: João Paulo Figueiró Longo and Lúis Alexandre Muehlmann

Received: 30 December 2021
Accepted: 8 March 2022
Published: 15 March 2022

Publisher's Note: MDPI stays neutral with regard to jurisdictional claims in published maps and institutional affiliations.

1. Introduction

Diabetes is a chronic health condition that affects the metabolism of glucose in the body due to decreased insulin secretion from the pancreatic islets. Diabetes is further classified into three main types: (i) Type 1, (ii) Type 2, and (iii) gestational diabetes. Out of these, Type 1 diabetes (T1D) is caused by an autoimmune disorder that inhibits the production of insulin in affected individuals [1,2]. Chronic autoimmune diseases are the consequence of the mistaken recognition of self-proteins (antigens) as foreign by the immune system, which leads to an immune response and subsequent destruction of the targeted tissues [3]. T1D is characterized by selective loss of insulin-producing beta cells of islets of Langerhans in the pancreas. Such a loss disrupts the glycemic homeostasis of the body. The loss of beta cell mass occurs when autoreactive T cells migrate to the Langerhans islets and cause local inflammation [4,5]. This autoimmune-mediated beta cell destruction can be asymptomatic for years prior to the manifestation of clinical symptoms. Currently, there is no clinical "cure" of T1D. The condition is managed using insulin and its variates which attempt to maintain the blood glucose levels within the healthy range. The condition furthers needs to be supplemented with proper diet, exercise, and weight management [6,7]. A big limitation of the conventionally prescribed insulin replacement therapy is its life-long dependence on such treatment [8–10]. Factors including lack of specificity, altered effects, and diminished potency can also hinder the patient's recovery.

In recent years, various studies have been undertaken to limit or reverse this autoimmune-mediated beta cell damage and a variety of strategies have been developed to enhance beta cell survival and/or islet regeneration [11–13]. Immunotherapy deals with the development of preclinical therapeutic strategies for disease treatment targeted towards the body's immune system. The immune system comprises a complex collaboration of cells working together to counteract the pathogens which invade the body. This intricacy makes the immunotherapeutic approach a particularly challenging task, as it involves various measures to prevent, diagnose, and treat a variety of disorders [14,15]. These approaches focus on the combination of immunity analysis and clinical chemistry to develop alternative therapeutic methodologies. It has been shown that autoantigen in combination with an appropriate immunomodulator holds potential in tackling a variety of autoimmune diseases [16]. These two compounds, when delivered simultaneously, affect the action of immune cells by modulating their autoreactivity. This immune modulation can play a key role in delaying and reducing the onset of T1D and is being explored using functional nanomaterials [17]. However, the ideal therapeutic approach for T1D should be long-lasting, render elevated antigen-specific tolerance, and integrate personalized or broad antigen specificity.

Towards this end, various nanoparticulate systems have been developed and have been transformative in the field of nanomedicine. Nanotechnology provides us with this tailorable ability to work at the atomic and molecular levels and opportunities to understand and create nanoparticulate platforms with new intrinsic properties attributed to their nanoscale size (<100 nm). For use in biomedical applications, the nanomaterials must exhibit unique properties such as aqueous stability, biocompatibility, and interactive functional groups. A variety of "smart" functional nanomaterials have made significant advances in the areas of drug delivery and drug discovery, biosensing, cell labeling and transplantation, gene therapy, immune therapy, and diagnoses and imaging [18–23]. The most attractive trait of these nanosystems is their large surface area/volume ratio, which allows their surface to carry and deliver multiple molecules to the same site of interest. Recent years have seen extensive evaluation and applications of nanosystems such as polymeric, lipid-based, liposomes, dendrimers, and inorganic nanoparticles (Metal oxides, gold, rare earth metals, graphene etc.). The range of nanoparticles used in recent preclinical studies is enormously extensive.

Various materials and compositions are being explored to both target and modulate the specific immune cell populations for the treatment of T1D. However, an absolute "cure" remains a challenge [24]. Insulin, its variates, and other therapeutic molecules which can modulate hyperglycemia have already been extensively delivered using nanoparticles to remediate the condition [25–29]. However, for the use in immunotherapy of T1D, these functional nanomaterials are required to conjugate and deliver immunomodulators to desired localized sites while maintaining its activity. Combing through a wide variety of targets, immunomodulatory agents accumulate locally and initiate specific immune responses. Development of tolerogenic vaccines for T1D and beta cell replacement therapies have shown promise by using nanoparticulate platforms [30–32]. By delivering tolerogenic agents and autoantigens, these nanomaterials can also be used to restore and enhance immune tolerance against T1D [33].

In this regard, this manuscript summarizes our knowledge of diabetic etiology, clinical scenarios, and the application of nanoparticle-based delivery approaches in targeting the immune system towards the treatment of T1D. Figure 1 illustrates various immunotherapeutic strategies currently under exploration to combat T1D discussed in the current review.

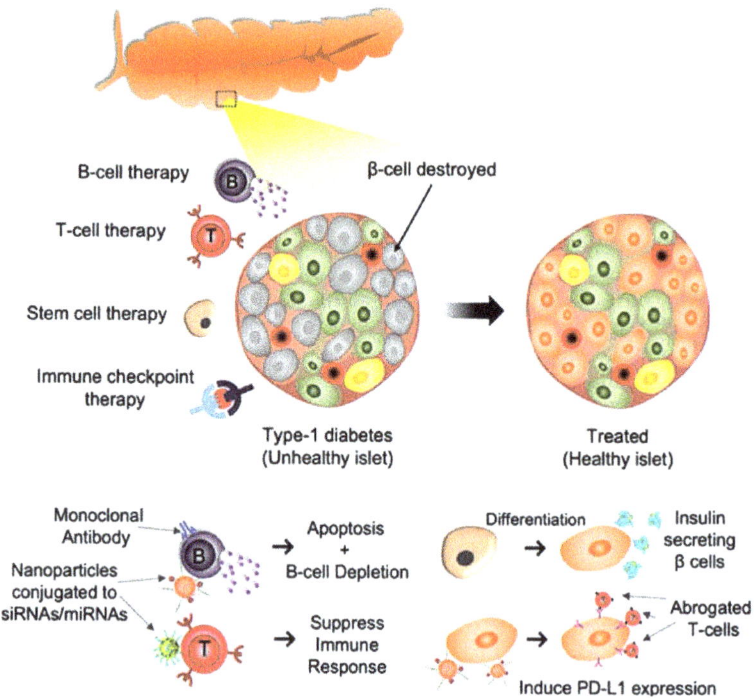

Figure 1. Schematic depiction of immunotherapies for T1D described below. T Cell Therapy: T cells modulated by nanoparticles conjugated to siRNA/miRNA leads to immunosuppression. B Cell Therapy: Modulation of B cells via monoclonal antibodies and/or nanoparticles conjugated to siRNA/miRNA leads to B cell apoptosis and depletion. Immune Checkpoint therapy: Nanoparticles conjugated to siRNA/miRNA induces PD-L1/CTLA4 expression resulting in abrogation of proximal T cells. Stem Cell Therapy: Differentiated stem cells transform into insulin producing beta cells.

1.1. Understanding T1D: Etiology and Current Clinical Scenario

T1D is a chronic autoimmune disorder characterized by the loss of insulin-producing beta cells within pancreatic islets. The immune-mediated destruction of beta cells results in insufficient levels of insulin production, creating a dependence on exogenous sources of insulin in affected individuals [2,34]. Both genetic and environmental factors contribute to the development of T1D. Typically, genetic predispositions are the result of polymorphic alleles coding for human leukocyte antigen (HLA), insulin gene promoter, and cytotoxic T lymphocyte antigen-4. In-fact these genetic disturbances account for 55–65% of T1D cases [35–37]. In the vast majority of T1D cases, pathogenesis is identified by the presence of several pancreatic autoantibodies. These autoantibodies include antibodies to insulin (IAA), islet cell cytoplasmic antibodies (ICA), insulinoma-associated 2 or protein tyrosine phosphatase antibodies (IA-2), zinc transporter8 (ZnT8), and glutamic acid decarboxylase (GAD65). The number of autoantibodies present in circulation and their abundance contribute directly to an individual's likelihood for developing T1D [38]. Autoantibodies provoke CD4$^+$ and CD8$^+$ T cells to migrate into pancreatic islets, triggering insulitis. Once these lymphocytes infiltrate the intra-islet space, they proliferate and attack endogenous beta cells, leading to the cessation of insulin production within pancreatic islets and subsequent loss of glycemic homeostatic mechanisms [39–41]. Insufficient insulin production prevents the inhibition of lipolysis leading to uncontrolled fat metabolism and the accumulation of ketone bodies in the blood. The buildup of ketone bodies such as acetoacetate and β-hydroxybutyrate causes ketoacidosis. In the absence of compensatory mechanisms, ketoacidosis results in loss of consciousness, cerebral edema, mental confusion, coma, and even death [42].

Studies investigating the molecular mechanisms behind T1D progression have determined that an individual's predisposition towards developing this autoimmune disorder is controlled by complex interactions between a multitude of genetic loci and environmental factors. As previously mentioned, polymorphic alleles at the HLA locus account for nearly half the familial clustering of T1D. HLA molecules are divided into class I (A, B, and C), and class II (DR, DQ, and DP). While HLA class I molecules are expressed on the surface of nearly all cells, HLA class II molecules are only expressed on the surface of activated T cells, B cells, and antigen-presenting cells [36,43]. Abnormalities affecting the genes which code for class II molecules contribute significantly towards T1D susceptibility, specifically those mapped to HLA-DQ/DR. Haplotypes corresponding to DR3-DQ2 and DR4-DQ8 have been observed in approximately 90% of patients diagnosed with T1D [44,45]. One study revealed that expression of human DR3 and/or DQ8 in non-diabetes-prone mice generated a loss of immune tolerance to the beta cell autoantigen glutamic acid decarboxylase, resulting in spontaneous insulitis [46]. Prior to insulitis, antigen-presenting cells (APCs) carrying beta cell peptides migrate towards lymph nodes located near the pancreas. There they present autoantigens to autoreactive CD4$^+$ T lymphocytes. This interaction triggers a signaling cascade leading to the activation of autoreactive CD8$^+$ T cells, which in turn infiltrate pancreatic islets. Once inside, CD8$^+$ T cells begin lysing beta cells, thereby eliminating the islets' ability to produce insulin [1,47,48]. Despite this, residual levels of beta cell mass have been detected long after diagnosis, however, the cause of this remains unknown [49].

Prevalence of T1D is dependent on several environmental factors including geographical location [50]. When examining the various components necessary for T1D development, the conjunction between epigenetic forces and genetic predisposition must be properly assessed. Epidemiological studies designed to assess regional T1D pervasiveness have observed elevated prevalence of the autoimmune disease in northern populations. Countries including the U.S., Canada, Finland, and Sweden suffer the highest rates of T1D; this phenomenon is believed to be the result of vitamin D deficiency [51–53]. Vitamin D serves an important role in the regulation of immune activity. Metabolites of vitamin D, such as 1,25-dihydroxyvitamin D3 suppress T cell proliferation and alter cytokine expression resulting in a more calculated immune response. Furthermore, vitamin D is known to heighten tolerance towards self-antigens and the development of autoantibodies [54,55]. Such factors should be taken into consideration when developing therapeutics or treatments for T1D. T1D continues to affect a growing population of affected individuals around the world and as a result, more therapies are needed to accommodate this growing complication.

The current clinical scenario for treating T1D is centered around intensive diet treatments and exogenous insulin administration. Clinicians currently utilize a conjunction of insulin pumps and continuous glucose monitors (CGM) to enhance the precision of glycemic control. Despite this, hypoglycemia remains one of the most consequential acute complications associated with insulin replacement therapy. Severe hypoglycemia, often occurring during nighttime hours, can result in seizures, coma, and death. In fact, nearly 6% of T1D related deaths are the result of nocturnal hypoglycemia [56]. Innovations in CGMs and insulin pumps have made significant progress in reducing nocturnal hypoglycemia. The implementation of threshold systems designed to suspend insulin delivery for up to 2 hours until CGM glucose is at a low threshold have displayed promising results in reducing nocturnal hypoglycemia [57]. Although significant progress has been made in this regard, insulin replacement therapy is only so effective at preventing prolonged periods of hyperglycemia and continues to suffer from the risks associated with hypoglycemia. Another roadblock to insulin replacement therapy stems from the complicated processes by which insulin is manufactured and distributed. More recently, insulin analogues such as Lispro, Aspart, and Delgludec insulins have been developed. Such analogues bear higher efficacy, exhibiting less latency and providing a longer duration of action [58,59]. However, in the U.S., the price of insulin and insulin analogues has increased exponentially over the past several decades. Today, monthly out-of-pocket costs for T1D patients can range between \$75–2000 depending on insurance coverage and insulin requirements, making proper treatment inaccessible for some individuals [60]. With the life-long insulin replacement therapy required to treat T1D patients, alternative forms of treatment must be developed to address the growing cohort.

T1D is more uncommon than Type 2 diabetes, resulting in a limited understanding of the immunology of T1D when compared to its counterpart. A variety of cell types are involved in the onset of T1D and involve an intricate cascade of interactions between immune cells and the islet beta cells. Immune cells responsible for this autoimmunity participate in both innate and adaptive immune responses. The initial trigger responsible for directing the autoimmune attack is still unclear, however, most researchers attribute it to an individual's genetic predisposition. Understanding the role genetically directed antigens play is necessary for identifying checkpoints which can be targeted to generate a cure or reversal of T1D phenotype. In T1D, APCs and varying lymphocytes interact to generate immune responses when presenting pathogens [61,62]. These lymphocytes include the more commonly explored T cells, as well as the more unexplored class of B cells.

1.2. T Cell Based Therapy

Elaborate studies in the past 10 years have improved our understanding regarding the adverse effects hyperactive T cells have on endogenous beta cells. These potentially pathogenic migratory cells comprise $CD4^+$ and $CD8^+$ T cell subpopulations, B cells, dendritic cells, and macrophages, which have specificity towards islets of Langerhans. The T cells are held in check by various regulatory mechanisms and by a special T cell population known as regulatory T cells (T_{regs}). An imbalance/defect in this control mechanism and/or dysfunctional Treg population might be one of the causes for the onset of T1D [63]. Various attempts have been made in the direction of understanding and identifying the T cell markers which could mediate the strengthening of immunoregulation [64]. Sorting out the molecular profile of the dysfunctional T_{reg} population and introduction of immunomodulatory agents against these markers may reveal promising targets for T cell immunotherapy [65–68]. The current and developing immunotherapies aim at either preventing the autoimmune response or re-establishing the regulatory control over the autogenic T cell population.

The $CD4^+$ T cells do not cause beta cell death through direct contact, but rather secrete cytokines to promote recruitment of other immune cells. These inflammatory cytokines, such as IFNγ, IL-1β, and TNFα, also stimulate beta cell death, thereby aggravating islet loss during T1D. On the other hand, $CD8^+$ T cells lead to beta cell death through direct contact with the beta cells [5,69]. $CD4^+$ T cells differentiate into a variety of helper T cells, which have their unique cytokine profiles that give them effector functions adapted to a variety of infections [70]. Manipulating these effector or regulatory $CD4^+$ T cells response is a promising immunotherapy strategy in various autoimmune disorders. Keeping these factors in mind, Eichmann et al. studied the effects of co-stimulation blockade using abatacept over $CD4^+$ memory T cells and the consequent decline in the beta cell function [71]. Their treatment demonstrated a substantial alteration in the population of $CD4^+$ cells and T_{reg} cells. Their results also indicated that this approach only affects conventional $CD4^+$ but not $CD8^+$ T cell populations. Similarly, Long et al. used Teplizumab to enhance the secretion of inhibitory molecules to reduce the population of $CD4^+$ and $CD8^+$ cells, which delayed the onset of T1D [72]. Autoreactive $CD8^+$ T cells have heterogenous phenotypes and their expression is seen to be affected by the rate of progression of T1D [73]. Elevated expression of activated islet-reactive $CD8^+$ memory T cells was predominant in T1D patients who demonstrated a rapid loss of C-peptide, while expression of multiple inhibitory markers, limited cytokine levels, and reduced proliferation marked a slower rate of progression of T1D [74].

Identification of markers in correcting the function of dysfunctional T_{reg} cells can also work in the direction of reversal of autoimmune response [75]. These T_{reg}-based therapeutic approaches can be helpful to restore tolerance in the T cell-mediated autoimmune responses [66,76]. T_{regs} with phenotype $CD8^+CD25^+FOXP3^+$ have been seen to effectively suppress the activity of pathogenic T cells and decrease the population of $CD8^+$ effector T cells [77]. Serr et al. identified HLA-DQ8-restricted insulin-specific $CD4^+$ T cells and demonstrated efficient human insulin-specific $Foxp3^+$ T_{reg}-induction after subimmunogenic vaccination with strong agonistic insulin mimetopes in vivo [78]. Functional chimeric antigen receptors (CARs) against insulin in conjunction with FOXP3 can be used to modify naïve effector T cells to specific T_{reg} cells in order to redirect their specificity towards T1D [79]. This approach is expected to result in high specificity, which would minimize the off-target impacts. Modulation and engineering of these T_{regs} also face drawbacks such as insufficient population, stability of modified expression, and antigen specificity.

More recently, nanomedicine has introduced novel techniques which are significantly capable of altering the immune response [80–82]. This precise control over the immunomodulation by the use of nanoparticles are proficient in inducing immune tolerance, ranging from triggering the pathogenic T cells to T_{reg} cells, and further into effector T cell populations [83]. One of the approaches where nanoparticles have found their use is by using dextran particles to administer autoantigen and immunosuppressant (rapamycin,) which selectively affect the effector T cells without global immunosuppression [84]. It also resulted in a reduction in the proliferation of $CD4^+$ T cells while an increase was observed in the ratio of $FOXP3^+$ to $IFN\gamma^+$ T cells. These microparticle-based platforms were effective in altering multiple immune cell functions by selectively inhibiting disease-associated T cell immunity and leaving the general immune responses unbroken. Bergot et al. hypothesized whether the tolerizing immunotherapy with a single peptide might be effective to control T1D, which is guided by multiple antigens [85]. They co-encapsulated an autoantigen (chromogranin A, ChgA) along with $1\alpha,25$-dihydroxyvitamin D3 in liposomal bilayer and monitored the specific autoimmune response. Liposome administration subcutaneously, but not intravenously, induced ChgA-specific $Foxp3^+$ and $Foxp3^-$ $PD1^+$ $CD73^+$ $ICOS^+$ $IL-10^+$ peripheral regulatory T cells in prediabetic mice, and liposome administration at the onset of hyperglycemia significantly delayed diabetes progression. Their work deduced that the liposomes encapsulated the single $CD4^+$ peptide, and vitamin D3 analogues induce ChgA-specific $CD4^+$ T cells that regulate $CD4^+$ and $CD8^+$ self-antigen specificities and autoimmune diabetes in NOD mice. On similar lines, Jamison et al. fabricated poly(lactide-co-glycolide) (PLG) nanoparticles and loaded with Insulin–ChgA hybrid peptide in order to monitor the balance between effector and regulatory T cells [86]. Administration of hybrid insulin peptide-coupled PLG NPs was found to prevent diabetes by impairing the ability of $CD4^+$ T cells to produce proinflammatory cytokines through induction of anergy, leading to an increase in the ratio of $Foxp3^+$ regulatory T cells to IFN-γ^+ effector T cells. It was also observed that interleukin-2 (IL-2) could enhance the T_{regs}, which in turn maintained their control over the pathogenic T cells. Aboelnazar et al. studied this relation as a therapeutic strategy and fabricated IL-2-loaded chitosan nanoparticles [87]. They found that low availability of IL-2 in the cellular microenvironment, an inverse correlation between T_{reg} and natural killer (NK) cell expression which was also related to the expression of FOXP3 on T_{reg} cells. IL-6 receptor-mediated signaling also plays a role in development of T cells, which then take part in T1D pathogenesis. Greenbaum et al. attempted to modulate the T cell phenotypes by blocking IL-6 using tocilizumab [88]. They found that while tocilizumab reduced T cell IL-6 signaling, it did not have any effect on $CD4^+$ T cell phenotypes. No significant difference in the slowing of beta cell loss was observed. Antigen-specific T cell immune tolerance can also be induced by the use oof nanoparticles. A conjugated system of carboxylated polystyrene beads (PSB) with an immunomodulating peptide, HLA-A*02:01-restricted epitopes, was seen to successfully induce tolerance and suspend the autoimmune cascade in NOD and transgenic humanized mice [89].

These works are suggestive that engineered nanomaterials can conjugate immunomodulators and target desired precise sites of both adaptive and innate immune responses. The size and surface chemistry of these nanomaterials can be tailored according to the identified target and can be tuned to respond to specific stimuli. Administration of these modified nanoparticles to the T cell family involved in the autoimmune response to T1D can successfully aim to restore immune tolerance and regulatory functions of the immune system. These studies have been tabulated in Table 1 below.

Table 1. Summary of strategies for targeting and regulating T cell population and function towards T1D.

	APPROACH	**TARGET**	**REFERENCE/S**
1.	Teplizumab	CD4$^+$ and CD8$^+$ cells	[72]
2.	Population alteration	Autoreactive CD8$^+$ T cells	[73]
3.	Functional correction	T$_{reg}$ cells	[66,75–77]
4.	Chimeric antigen receptors	T$_{reg}$ cells	[79]
5.	Rapamycin	Selective effector T cells and CD4$^+$ T cells	[84]
6.	Liposomal formulation of Autoantigen + 1α,25-dihydroxyvitamin D3	ChgA-specific Foxp3$^+$ CD4$^+$ T cells	[85]
7.	poly(lactide-co-glycolide) nanoparticles loaded Insulin–ChgA hybrid peptide	Balance population of effector and regulatory T cells	[86]
8.	interleukin-2 (IL-2)	T$_{reg}$ cells	[86,87]
9.	tocilizumab	interleukin-2 (IL-6)	[88]
10.	Carboxylated polystyrene beads with peptide HLA-A*02:01-restricted epitopes	Antigen-specific T cell immune tolerance	[89]

1.3. B Cell Based Therapy

It is known that T1D is affiliated with the loss of tolerance by autoreactive (islet-reactive) B cells [90,91]. B lymphocytes (cells), in addition to T lymphocytes (cells), work directly with the adaptive immune system to produce cellular and humoral defense mechanisms for protection against infections or tumors [90,92]. Therefore, the depletion of B cells makes an individual highly vulnerable to opportunistic infections. However, to prevent autoimmunity from occurring, the B cells must be suppressed, rehabilitated, or culled [92]. There is also a correlation between the number of CD20$^+$ B cells and a decreasing pancreatic beta-cell count [90]. In the current realm of diabetes treatment, there exists a need to discover effective immunotherapy methods, as current treatment by daily insulin injection(s) is not a sustainable way to treat diabetes and is not a cure. The role of B cells in non-obese mice models has shown to arrest the disease development around the preinsulitis stage [93]. Therefore, B cell immunotherapy is proposed as an effective treatment for T1D patients [61].

To target B cells, monoclonal antibodies are being used to identify surface antigen markers on the surface of the cells. The biomarkers targeted on the B cell surface are those involved in the processes of maturation, differentiation, and survival [61,90,93]. Through mechanisms including, but not limited to, complement-dependent cytotoxicity, antibody-dependent cellular cytotoxicity, and induction of apoptosis, the antibodies are able to induce cell death [61]. Current approaches to antibody B cell therapy include the use of rituximab (RTX), which has proven to show low efficacy in treatment results due to the following factors: detrimental side effects caused by B cell depletion, rapid reemergence of autoreactive B cells post RTX treatment, and the depletion of regulatory cells as collateral of RTX treatment [90]. Pancreas-localized B cells may also be resistant to RTX-mediated deletion, as indicated by a down-regulation of the CD20$^+$ surface marker [93]. Due to confounding effects, it is difficult to manipulate the dose and duration of RTX treatment, hindering the effectiveness and clear outcome of the treatment [61]. Combination therapy is another approach used to deplete B cell levels with aims to control glycemia and reverse T1D. Through combination therapy, antigens and antibodies are administered together to achieve a therapeutic outcome that offers better protection than administering the agents alone [62,94,95]. Oral administration of insulin in combination with anti-CD20 antibodies shows low efficacy in reversal when compared to administering proinsulin DNA with an anti-CD20 antibody [94,95]. However, the aforementioned method is moderately effective in the prevention of T1D. Thus, the combination therapy approach to an immunological treatment of diabetes shows (limited) efficacy in the prevention of T1D, and little to no efficacy in the reversal of T1D.

A treatment is needed that can deplete B cells before the onset of hyperglycemia—a point where it is too late to arrest disease progression [62,93]. B cell depletion has been shown to arrest diabetes progression at the pre-insulitis stage, prior to T cell islet infiltration and insulitis development [96]. Further, specificity for autoreactive B cells is needed, so that pan-B cell-depletion is limited, and global immunosuppression does not occur [96–98]. The use of a site-specific receptor-mediated drug delivery system, in conjunction with small interfering RNA (siRNA), to target antiapoptotic factor B cell lymphoma/leukemia 2 (BCl2), is an effective method of gene silencing in B cell immunotherapy [99]. This can be achieved through nanomedicine, by using synthetic polymer-based nanoparticles conjugated with siRNA to specifically target autoreactive B cells [100]. Nanomedicine uses nanoscale substances and materials to monitor and treat human biological systems. Nanoparticles (NP's), a major proponent of nanomedicine, are known to couple various properties with reduced toxicity, in comparison to other bio-interfaces, all at the scale of nanometers [30].

Varying polymers such as chitosan, calcium pectinate zinc oxide, alginate, casein, and other polyester or polycationic acrylic polymers have proven to be effective oral administrative carriers for immunomodulators, insulin, and other engineered vaccines to treat T1D [98]. These polymers are characterized by nanoporous structures, in which various therapeutic agents such as insulin, proinsulin DNA, and siRNAs can be conjugated, and released upon conformational change when glucose levels fluctuate [95,98]. An increase in glucose levels, for example, would induce a conformational change in the nanoparticles, allowing for conjugated materials in the nanopores to be released and directed to the immune cells through a mechanism of biodegradation.

T1D is known to be characterized by the presence of autoantibodies (AABs) and the spreading of islet autoantigens (AAGs) during a prolonged, subclinical period, prior to the detection/diagnosis of T1D, during which it's expected that seroconversion will occur early in age, with epitope spreading indicative of disease progression [101,102]. The mechanisms leading to the generation of AABs and AAGs, and thus the onset of T1D, are considered genetic, though environmental factors can play a part. Through genetic mechanisms, T cells generate blueprints for the construction of autoreactive B cells, which in turn present beta-cell antigens to autoreactive T cells, creating a vicious loop which aids in the development and progression of T1D [102,103]. Thus, it is essential that an immunotherapeutic treatment for B cell intervention is fabricated to break the detrimental loop which leads to the overall reduction of beta cells, which can be approached by nanomedicine. To induce (therapeutic) genetic expression within the B cell environment, NPs may be enveloped with CRISPR-cas9 (a gene editing system that allows for DNA repair, deletion, or modification) that expresses various cofactors to decrease expression and induce deletion of the autoreactive B cells [103]. In turn, the reduction of autoreactive B cells would result in increased B cell tolerance, and a reduction in the decrease of beta cells. Limited studies have produced results on immunotherapeutic approaches to treating T1D with NPs by targeting B cells, however there are studies that show a successful depletion of B cells in varying autoimmune diseases, leading to the prevention, and, in some cases, a reversal of the disease [101,102,104]. An immunotherapeutic approach to treating T1D through depleting autoreactive B cells poses an effective method for the actual reversal of T1D, which is largely due to individual genetic predispositions. These genetic predispositions may be combated by using CRISPR-cas9 or other immunotherapeutic agents conjugated with the prospective cofactors to induce the cellular and humoral responses, which aims to deplete the autoreactive B cells. It is important to consider the various stages of T1D during immunotherapy, as genetic and environmental conditions will impact the efficacy of the treatment.

Thus, it is indicative that a dynamic nano-delivery system must be constructed that can bind immunotherapeutic agents and release them upon environmental and physiological change. Furthermore, the ability to have controlled release of the agents in the system allows for more tunability in response to varying conditions, environments, and stimuli. Thus, nanoparticles are highly desirable in their role as a vaccine candidate, aiming to break immune tolerance and monitor glycemic activity. In comparison to methods such as surface antigen targeting through monoclonal antibodies, nanoparticle administration to specific sites on autoreactive B cells is gaining more and more attention and relevance in present-day studies, as summarized in Table 2.

Table 2. Summary of strategies for targeting and regulating B cell population and function towards T1D.

	APPROACH	**TARGET**	**REFERENCE/S**
1.	Rituximab	Autoreactive B cells	[61,90]
2.	Combination therapy (Antigens + Antibodies)	CD20+ B cells	[62,94,95]
3.	Nanoparticles + siRNA gene silencing	Autoreactive B cells	[99,100]
4.	Depletion	Autoreactive B cells	[101,102,104]
5.	Nanoparticles + CRISPR-cas9 (Gene editing)	Autoreactive B cells	[103]

1.4. Immune Checkpoint Molecules-Based Therapy

Immune checkpoint molecules comprise a group of co-stimulatory and inhibitory proteins used to regulate the body's immune response within a specific microenvironment. Checkpoints such as CD28, a receptor commonly expressed on the surface of CD4+ T cells, binds to one of two molecules (B7.1 and B7.2). This interaction promotes proliferation of CD4+ T cells and subsequent migration towards designated target cells [105,106]. Similar processes exist with CD8+ T cells, however such stimulatory action relies more heavily on the interaction between molecules such as CD70 and CD137 or CD134 [107]. Contrarily, T cell activation can be suppressed via inhibitory signaling pathways governed by checkpoint proteins including programmed cell death protein 1 (PD-1) and its cognate ligands, programmed cell death ligand 1 (PD-L1) and PD-L2 [108]. When bound to these ligands, PD-1 works to regulate the adaptive immune response by initiating immunosuppressive signals leading to the induction of apoptosis and reduced cell proliferation [109–111]. Further examples of co-inhibitory molecules include cytotoxic T lymphocytes-associated antigen 4 (CTLA-4), a CD28 homolog which binds competitively to B7.1/2. Similar to PD-1, CTLA-4 reduces T cell activation via inhibitory signaling pathways [112]. In either case, the relationships between these varied axes serve as mechanisms for acute and precise control of the body's immune system which poses potential solutions for autoimmune diseases like T1D.

In recent years, immune checkpoint therapy has revolutionized the field of oncology. Many cancer cells possess genetic and epigenetics irregularities allowing them to utilize immune checkpoints to promote survival. Studies have found that a variety of cancer cell types upregulate PD-L1 in response to interferon gamma (IFNγ) as well as other oncogenic signaling pathways [113,114]. Consequently, tumor cells express PD-L1 to abrogate T cell mediated antitumor responses. As a result, therapeutics have been developed to interrupt the PD-1/PD-L1 axis allowing T cells to function more effectively. Such therapeutics utilize monoclonal antibodies (mAb) targeted at immune checkpoints markers (PD-1, PD-L1, and CTLA-4) to inhibit T cell suppression [115,116]. However, immune checkpoint blockades have been linked to spontaneous development of autoimmune diseases. Examples of disease resulting from treatment with anti-PD-L1 antibodies include diabetes mellitus, hepatitis, myasthenia gravis, sarcoidosis, hypothyroidism, endophthalmitis, and various skin rashes [117,118]. This relationship indicates that the absence of co-inhibitory signaling may increase the likelihood that an individual develops one of the aforementioned disorders. Several case studies have been conducted analyzing individuals who have developed late onset T1D in response to immune checkpoint therapy [118,119]. Recognizing this relationship, efforts have been made to assess whether increasing co-inhibitory signaling within pancreatic islets could increase beta cell survival.

Recent studies seeking to better understand T1D progression have determined that PD-L1 is expressed by insulin producing beta cells within pancreatic islets during insulitis. Upregulation of PD-L1 coincides with islet infiltration, as well as other factors such as increased exposure to interferons (IFN) alpha and gamma. Studies analyzing this relationship have determined that the heightened presence of IFNα and IFNγ activates STAT1 and STAT2 transcription factors. This activation corresponds with increased transcription of interferon regulatory factor 1 (IRF1) and subsequent PD-L1 upregulation by pancreatic beta cells [120,121]. Unfortunately, minimal research has been conducted to appraise the potential benefit of increasing PD-L1 or CTLA-4 expression in beta cells to enhance survival during T1D progression. However, preliminary research has demonstrated the protective effect of organ-specific PD-L1 expression in transgenic NOD mice. Wang et al. found that the severity of insulitis in PD-L1 transgenic NOD mice was significantly reduced when compared to controls [122]. Furthermore, islets transplanted into diabetic recipients persisted for a significantly longer period of time when compared to non-transgenic controls. Despite this, development of T1D remained constant between experimental and control groups. Another study attempting to increase survival rates among transplanted human islet-like organoids (HILOs) within NOD mice determined that overexpression of PD-L1 contributed significantly to the HILOs' survival rate within a diabetic mouse model. Without disturbing insulin production, PD-L1$^+$ HILOs maintained glucose homeostasis for more than 50 days whereas PD-L1$^-$ HILOs were only able to maintain glucose homeostasis for approximately 10 days [123]. These data present a potential therapeutic benefit to immune checkpoint therapy in T1D.

A significant roadblock when utilizing immune checkpoint proteins in a clinical setting stems from the mechanism by which PD-L1 and CTLA-4 overexpression is induced. One potential solution presents itself in the form of iron-oxide nanoparticles (NPs). Nanoparticles conjugated to various microRNAs (miRNAs) can be used to induce overexpression of co-inhibitory molecules for the purpose of protecting endogenous beta cells [124]. While the literature pertaining to this specific topic is limited, studies attempting to enhance cancer therapeutics have determined that PD-L1 regulation can be achieved via NPs conjugated to miR-200c. Such a combination has proven to inhibit PD-L1 expression, especially when compared to naked miR-200c [125]. Further examples of miRNAs which contribute to the regulation of PD-L1 and CTLA-4 include miR-138-5p, miR-513, miR-200a, and miR-34a [126–128]. NPs serve as an ideal delivery vehicle for miRNA-based therapeutic payloads [129]. Contrarily, NPs can be used to deliver antisense oligonucleotides designed to increase expression of co-inhibitory molecules such as PD-L1 and CTLA-4 within pancreatic islets. These immune checkpoint protein molecules currently under investigation have been tabulated together in Table 3 for a better overview.

Despite the potential benefits associated with immune checkpoint therapy in T1D, minimal research has been conducted to further its clinical application. This particular subset of biomolecular research remains dormant while alternative therapeutic avenues are explored. Utilizing the innate mechanisms by which the immune system is stimulated/inhibited could prove useful in the battle against T1D. In conjunction with more traditional forms of treatment, immune checkpoint therapy has the potential to curb the progression of T1D and preserve insulin independence for a more prolonged period of time. Furthermore, such applications may provide clinicians with a more effective form of theranostic-based treatment for T1D patients.

Table 3. Summary of strategies for identifying and targeting immune checkpoint markers towards T1D.

	APPROACH	TARGET	REFERENCE/S
1.	CD8$^+$ T cell activation	CD70 and CD137 or CD134	[107]
2.	T cell suppression	programmed cell death protein 1 (PD-1) + ligand (PD-L1, PD-L2) upregulation	[108–111]
3.	T cell suppression	cytotoxic T lymphocytes-associated antigen 4 (CTLA-4) upregulation	[112]
4.	PD-L1 upregulation	Interferons: IFNα and IFNγ	[120,121]
5.	transplanted human islet-like organoids (HILOs)	PD-L1 upregulation	[123]
6.	SPIONs + miRNA	overexpression of co-inhibitory molecules	[124]
7.	Nanoparticles + miRNA (miR-200c, miR-138-5p, miR-513, miR-200a, and miR-34a)	PD-L1 and CTLA-4 regulation	[126–128]

1.5. Extracellular Vesicles and miRNA-Based Therapy

Extracellular vesicles (EVs) are membrane-bound small vesicular bodies released by the cells and are utilized in cell-to-cell communication/signaling. These vesicles are relatively small in size and fall under the nanoscale category. However, they preserve the ability to transport molecular cargo. EVs are categorized into three sub-classes, namely, microvesicles, exosomes, and apoptotic bodies. Distinctions are based on their size, type of originating cells, and formation mechanism. They can be released in response to a variety of external stimuli. Examples of this include changes in cell microenvironments (pH, temperature, irradiation), cellular stress, and chemically-induced activation.

EVs may also serve as a communication bridge between immune cells and beta cells. Pancreatic islets have also been shown to secrete EVs that behave in an autocrine manner to regulate beta cell proliferation and death. These islet mesenchymal stem cell-derived exosomes containing miRNAs, can activate the T cell response and stimulate the release of interferon gamma (IFN-γ) to induce autoimmune responses in T1D [130]. Recently, nucleic acids containing exosomes—especially miRNAs—have been shown to regulate communication networks between organs in pathological processes relating to diabetes. One such example includes influencing metabolic signals and insulin signals in target tissues, affecting cell viability, and modulating inflammatory pancreatic cells [131]. This also opens the possibility for exosomes to be developed and utilized as a tool to improve the islet transplant by modulating the immune response or as a biomarker of recurrent autoimmunity for islet transplant diagnosis.

A class of short noncoding RNAs of 19–22 nucleotides, known as microRNAs (miR-NAs), act as negative regulators of gene expression by partially pairing to the 3′ or 5′ of the untranslated regions of their target messenger RNAs (mRNAs) [132]. This new and fast rising technology using miRNAs has appealed to many researchers as a potential, minimally invasive biomarker for T1D due to three main reasons: miRNAs are exceptionally stable in cell-free body fluids such as serum, they have high resistance to RNAse digestion, and miRNA molecules have an ability to remain intact in extreme conditions (such as being in extended storage and going through repeated freeze–thaw cycles). Additionally, there is a strong possibility that miRNAs are involved in gene regulation of T1D development [133,134].

According to Scherm et al., miRNA expression differs in peripheral mononuclear cells (PMNC) and specific immune cell subsets, such as regulatory T cells, in T1D patients when compared to healthy individuals [135]. This uncharacteristic expression in miRNA leads to disrupted T cell differentiation and loss of function, subsequently resulting in immune activation and the onset of islet autoimmunity and initiation of T1D [136]. In order to provide an elaborate catalog of coding and noncoding miRNAs in human islet-derived exosomes, Krishnan et al. profiled such RNAs in human islet-derived exosomes and identified the RNAs which were aberrantly expressed under cytokine stress [137]. Wang et al. attempted a theranostic approach to deliver miRNA-targeting oligonucleotides conjugated iron oxide nanoparticles in order to modify their expression in pancreatic islets of NOD mice [124]. MiR-216a was identified as a pivotal point in regulating the beta cell proliferation and altering its expression levels significantly affected the progression of T1D (Figure 2). Similarly, modulating the levels of the miR-29 family (miR-29a, miR-29b, and miR-29c) via iron oxide nanoparticles serves to regulate the glucose homeostasis and overcome the hypoglycemic shock induced by diabetes [138]. The levels of miRNA-181a impaired immune tolerance and affect the function of T_{reg} cells. Attempts have been made to successfully block miRNA181a, increasing the T_{reg} induction and reducing the islet autoimmunity in mice [139]. These findings suggest that the identification and subsequent block of trigger markers might allow for the reversal of islet autoimmunity. MiRNAs pertaining to autoantibodies, such as insulin autoantibodies (IAA), islet cell cytoplasmic antibodies (ICA), insulinoma-associated 2, or protein tyrosine phosphatase antibodies (IA-2), zinc transporter8 (ZnT8), and glutamic acid decarboxylase (GAD65), trigger pancreatic T cells to initiate insulitis.

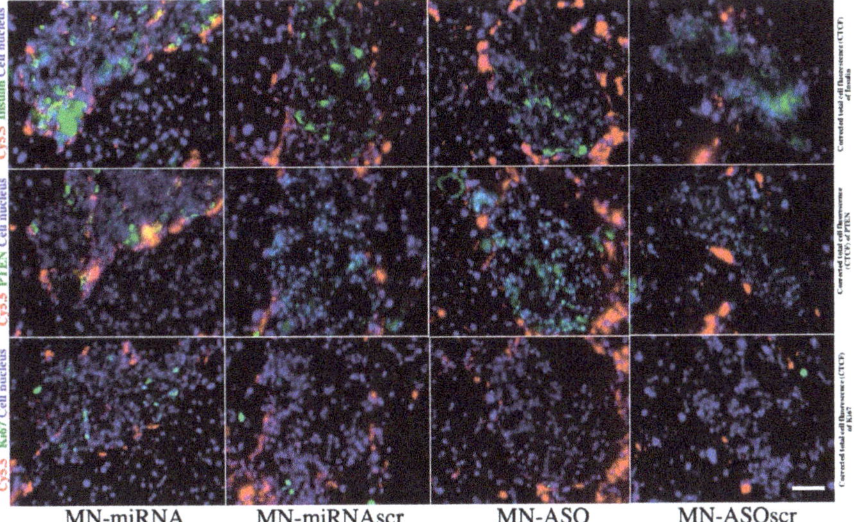

Figure 2. Fluorescence microscopy of consecutive frozen pancreatic sections from STZ-induced diabetic mice injected with MN-miRNA, MN-ASO, MN-miRNAscr, and MN-ASOscr. Animals injected with MN-miRNA showed higher insulin expression in pancreatic islets (top: green, insulin; red, Cy5.5; blue, cell nucleus) when compared to the animals injected with MN-ASO or control nanodrugs. These animals also showed downregulated PTEN expression in their islets (middle: green, PTEN; red, Cy5.5; blue, cell nucleus) when compared to the animals injected with MN-ASO or control nanodrugs. Finally, there was a notably higher cell proliferation in the islets of these animals when compared to controls (bottom green, Ki67; red, Cy5.5; blue, cell nucleus); Magnification bar = 40 μm. All experiments were performed in triplicates, reproduced with permission from Springer Nature [124].

Levels of miRNA in systemic circulation have been proposed as a new class of biomarkers for diagnosis and prognosis of T1D and this has also presented itself as a new target for modulations and therapeutics [140]. There are alterations in serum levels in newly diagnosed T1D patients, with some specific miRNAs appearing to be related to glycemic controls [141]. This newer class of potential circulating biomarkers for T1D have narrowed down their source and improved our knowledge related to the understanding of the molecular functions of these biomarkers. Akerman et al. studied the possible deviations of miRNA levels in the serum of children. They found the serum to be positive for multiple IAAs, and considered these individuals to be at high risk for T1D development [142]. They found that the serum miRNA profiles and autoantibody-positive individuals with high risk of T1D did not differ with respect to healthy, age-matched controls. Some studies have determined that beta cells initiate T1D progression through the activation of various stress pathways. This accelerates the autoimmune-mediated destruction of beta cells and the subsequent loss of insulin-producing mechanisms [130]. The aforementioned study focuses on the need to identify biomarkers in healthy beta cells, which serve as the guiding markers in identifying and monitoring dysfunctional cells. These approaches can not only help to monitor dysfunctional beta cells, but also improve the diagnostics for early detection of T1D. In this context, Bertoccini et al. focused on levels of circulating miR-375, an alleged biomarker of beta-cell death. They observed that an increase in miR-375 was indicative of later onset of T1D, suggesting residual beta-cell function [143]. MiR-375 was directly correlated to the population of viable beta cells that were under autoimmune attack. These results strongly support the potential of miR-375 as an efficient biomarker for T1D diagnosis and prognosis. Bearing this in mind, Lakhter et al. have analyzed the effects of miR-21-5p upregulation on beta cell survival and functionality [144]. Their study determined that the levels of extravesicular-associated miR-21-5p increase significantly in the T1D developing microenvironment and thus, can serve as an efficient biomarker in early T1D detection. However, they noted that utilizing miR-21-5p as an identifying biomarker has limitations due to the abundance of miR-21-5p in circulation, as well as its presence in multiple tissue types. This limits our capability to extrapolate the exact source/reason of the increased levels. Along similar lines, Santos et al. investigated the roles of circulating miR-101-3p and miR-204-5p with respect to T1D progression [134]. Their work concluded that circulating levels of miR-101-3p are higher in T1D patients and healthy individuals with autoantibodies. Based on this data they inferred that miR-101-3p plays an important role in pathways preceding the onset of T1D and can function as an important marker for diagnosis of T1D.

EVs and miRNAs serve as promising biomarker candidates with potential to assist in early T1D diagnosis and prognosis. In comparison to using naked miRNAs, the methods utilizing the conjugated complexes to nanoparticles or nanoscale vesicles have an advantage in terms of ease of administration and in vivo imaging. Although in its nascent stage, this theranostic approach is gaining increased attention and relevance in present-day studies. Table 4 presents variety of miRNA targeting strategies recently studied for T1D.

It has been reported that miRNA expression differs in peripheral blood mononuclear cells and in specific immune cell subsets, such as regulatory T cells, in T1D patients when compared to healthy individuals [135]. This uncharacteristic expression of miRNA leads to disrupted T cell differentiation and loss of function, which subsequently results in immune activation and the onset of islet autoimmunity and initiation of T1D. Researchers have also found that EVs such as islet mesenchymal stem cell-derived exosomes containing miRNAs can activate the T cell response and stimulate the release of interferon gamma (IFN-γ) to induce autoimmune responses in T1D [131].

Table 4. Summary of strategies for targeting microRNAs towards T1D.

	Approach	Target	Reference/S
1.	SPIONs + miR-216a	Expression modulation	[124]
2.	SPIONs + miR-29 family	miR-29a, miR-29b, and miR-29c levels' modulation	[138]
3.	T_{reg} induction	Block miRNA181a	[139]
4.	Diagnosis and prognosis of T1D	miRNA in systemic circulation	[140]
5.	Biomarker of beta-cell death	Circulating miR-375	[143]
6.	Beta cell survival	miR-21-5p upregulation	[144]
7.	Diagnosis of T1D progression	circulating miR-101-3p and miR-204-5p	[134]

1.6. Stem Cell Targeted Therapy

Stem cell therapy has recently gained further attention and momentum as a promising approach to curing T1D through transplantation of (differentiated) stem cell-derived beta cells that are capable of producing insulin in vivo [145]. This form of regenerative medicine influenced therapy relies on the transplantation of autologous stem cell grafts which can act in immunomodulation or assist in insulin production [146]. The aim of this form of therapy is to provide longitudinal resolve to patients suffering from T1D. This also contributes to the mission of precision medicine by providing a long-term solution to patient complications and removing the reliance on expensive, short term therapies such as insulin injections, which can be difficult to obtain and manage across the stratified sociocultural spectrum in the United States and globally [147]. The value added by clinical translation of stem cell therapy for T1D supersedes previous treatments in the temporal dimension due to its ability to provide functioning beta cells to the patient longitudinally, thus providing for a prolonged period of insulin production during which the patient may not need to rely on other drugs or therapies. Very quickly stem cell therapy can then alter the social landscape of treatment for T1D and similar autoimmune diseases through multiple dimensions by, providing relief to the patient both physiologically and financially. Due to the complexity of autoimmune diseases and their turbulent nature, it is the hope of many scientists that stem cell therapy can provide a long-term resolution to such diseases. It is evident that stem cell therapy has the potential to introduce a great deal of paradigm shifts in the current approach to treatment of autoimmune disease such as T1D.

Currently, there are various approaches to stem cell-based transplantation and therapy for T1D that are being explored in the clinic [148–150]. These approaches often involve immunosuppression of the patient to tolerable levels and subsequent transplantation of the autologous stem cells in the patient to avoid immune rejection and further patient autoreactivity [151]. One such study observed the role of autologous nonmyeloablative hematopoietic stem cell transplantation (AHST). Following this treatment, all but 1 of the 15 patients of various gender and age 14 to 31 were able to remain insulin independent for at least 6 months. The study also showed increased C-peptide levels and decreased anti-GAD antibodies, which is a clinically used biomarker for the diagnosis of T1D [148]. This clinical study is evidence of the longitudinal improvement in patient symptoms and physiological complications that result from T1D. Furthermore, it highlights the impact of a combined approach in which immunosuppression followed by AHST is considered the standard. Another study performing autologous stem cell transplantation for the treatment of T1D used mesenchymal stromal stem cells derived from patient umbilical cord, which were transplanted for treatment of T1D [150]. In this study, improvement in C-peptide levels and reduction in insulin dependence was observed for all patients who underwent treatment. This study highlights another type of stem cell, in this case mesenchymal stromal cells, that can be clinically used in transplantation models for diabetic patients. However, it is evident that these models may require imaging tools and modalities that can allow for the visualization of such parameters as transplant density, biodistribution, viability, and immunogenicity [152–154]. This can permit both short term and longitudinal monitoring of the transplant in the patient and allow for timely intervention, as is the case in many post-transplant graft loss incidents [155]. It can also allow for the visualization of certain molecular (bio)markers that can indicate the existence of specific cellular states or functions [156–158].

Presently, the main modalities used for molecular imaging of stem cells and stem cell transplants are magnetic resonance imaging (MRI) or magnetic particle imaging (MPI) [159,160]. Each of these modalities rely on the utilization of superparamagnetic iron oxide nanoparticles (SPIONs) for targeted molecular imaging of extracellular and intra-cellular markers. These SPIONs have previously been altered for targeting of immune cells, islet cells, and stem cells amongst many other cell targets [161–163]. They have also been explored for monitoring cell transplants longitudinally [164]. This is particularly useful in the context of providing tools for monitoring of stem cell transplants for the treatment of T1D because of the possibility for post-transplant rejection, due to host immune rejection, issues with cell transplant procedure, or cell viability post-transplant in the patient. The use of nanoparticles to monitoring stem cell transplants during treatment of autoimmune diseases such as T1D opens new doors to paradigms of theranostics and precision medicine, as autologous stem cells are typically used for transplantation and therapeutic purposes. The dynamic array of moieties in the domain of radionuclides, small nucleic acids, and antibodies that can conjugate to nanoparticles provide a platform for targeting specific cells from a heterogenous distribution and performing various forms of combined therapy and imaging (theranostics) [165]. In the context of guiding and improving therapeutic outcomes, nanoparticle-based tracking of stem cells has provided a far more effective and reliable method in vivo when compared to conventional methods such as labeling cells with organic dye or directly labeling them with fluorescent probes, because of their optimal magnetic and optical properties. Although there are a wide variety of structural platforms and nano-materials with which to engineer theranostic nanoparticles such as silicon and quantum dots, the most biocompatible and clinically used tool for contrast enhancement and targeted therapy are SPIONs. Several studies have highlighted the ability of SPIONs to image and track stem cells post-transplantation in both mice and humans. One such study labeled MSCs with SPIONs and encapsulated these in collagen-based microcapsules for monitoring of the cells post-transplantation [166]. Although this study focused on use of SPIONs for monitoring MSC transplantation for the treatment of myocardial infarctions, SPIONs can also be used for the labeling of differentiated beta cells derived from iPSCs for monitoring of transplant thereof in vivo [167]. Wang et al. have shown imaging of endogenous beta-cell mass through targeting of the glucagon like peptide 1 receptor (GLP-1R) [168]. This was done through conjugation of the exendin 4 to magnetic nanoparticles and subsequent injection of this probe in mice. The group was able to show specific accumulation of the probe in GLP-1R expressing endogenous beta cells and indicated the correlation between reduced signal interference with decreasing beta cell mass over time. This approach can be extrapolated to instances of imaging GLP-1R-expressing, stem cell-derived beta cell transplantations. This also provides a mechanism for the direct targeting of endogenous and transplanted beta cells, regardless of origin, to deliver interventional nanodrugs and therapeutic molecules in vivo. Additionally, these SPIONs have enabled the use of an emerging imaging modality of MPI. Prior studies have performed MPI of human islet cells labeled with dextran-coated SPIONs and transplanted under the left kidney capsule of mice [164]. However, the limitations of MPI result mainly from its inability to decipher viable vs. non-viable cell transplants, especially after a brief period of time where dead cells and their nanoparticles can undergo degradation and generate false positive signals that do not originate from live cells [169]. Despite these limitations, nanoparticles are gaining popularity in their use for monitoring such cell transplants and are continuously being explored as a platform.

2. Conclusions and Future Perspectives

In the last century, exogenous insulin therapy has transformed diabetes therapeutics in clinical settings. Since T1D has been recognized as an autoimmune disease, efforts have been made to advance our knowledge of disease mechanisms, its progression, and prevention of the associated autoimmune responses. This understanding serves as our base for designing novel therapeutic strategies in the form of targeted immunotherapeutic approaches. This review summarizes a variety of immunotherapy strategies currently being tested and utilized to cure T1D in an effort to improve the quality of clinical treatment provided to the patient. In an intricate cascade of events involving the onset of T1D, various checkpoints have been identified and have shown success in achieving targeted immunotherapy. However, they are still limited in their ability to maintain long-term glycemic homeostasis and normal insulin secretion. Since 70–90% of beta cell mass is dysfunctional or destroyed by the time clinical help is sought, identification of early-stage immunological biomarkers and intervention may be more beneficial in facilitating an early assessment of T1D. Stem cell-based beta cell regeneration approaches also need to be included in such combinatorial treatment methodologies as it pursues the ultimate objective when beta cells have been damaged. Immunotherapies, focused on a beta cell-regenerating agent and an immunomodulator, represent a promising strategy for finding a cure of T1D.

Nanoparticles have already proven their potential in a targeting a variety of disorders as they offer remarkable clinical diagnostic and therapeutic prospects. More recent works have also shown that combining these immunotherapies with nanoparticulate systems possess enhanced functionalities and have the potential to specifically target the immune check points and slow/arrest the rate of T1D progression. The studies show encouraging results in incorporating the nanoplatforms in-line with the existing immunotherapies towards combating T1D. Although in its nascent stage, it is anticipated that this combinational approach would prove to be a promising avenue to achieve a complete reversal and reset of the dysfunctional immune system in individuals with T1D.

Author Contributions: Conceptualization, P.W.; writing-original draft preparation, S.N., J.O.B., H.H. (Hanaan Hyat), T.Q. and H.H. (Hasaan Hyat); writing—review and editing, P.W.; supervision, P.W. All authors have read and agreed to the published version of the manuscript.

Funding: This research received no external funding.

Institutional Review Board Statement: Not applicable.

Informed Consent Statement: Not applicable.

Data Availability Statement: Not applicable.

Conflicts of Interest: The authors declare no conflict of interest.

References

1. DiMeglio, L.A.; Evans-Molina, C.; Oram, R.A. Type 1 diabetes. *Lancet* **2018**, *391*, 2449–2462. [CrossRef]
2. Katsarou, A.; Gudbjörnsdottir, S.; Rawshani, A.; Dabelea, D.; Bonifacio, E.; Anderson, B.J.; Jacobsen, L.M.; Schatz, D.A.; Lernmark, Å. Type 1 diabetes mellitus. *Nat. Rev. Dis. Primers* **2017**, *3*, 17016. [CrossRef] [PubMed]
3. Marrack, P.; Kappler, J.; Kotzin, B.L. Autoimmune disease: Why and where it occurs. *Nat. Med.* **2001**, *7*, 899–905. [CrossRef] [PubMed]
4. Bach, J.-F. Insulin-Dependent Diabetes Mellitus as an Autoimmune Disease. *Endocr. Rev.* **1994**, *15*, 516–542. [CrossRef] [PubMed]
5. Burrack, A.L.; Martinov, T.; Fife, B.T. T Cell-Mediated Beta Cell Destruction: Autoimmunity and Alloimmunity in the Context of Type 1 Diabetes. *Front. Endocrinol.* **2017**, *8*, 343. [CrossRef] [PubMed]
6. Riddell, M.C.; Gallen, I.W.; Smart, C.E.; Taplin, C.E.; Adolfsson, P.; Lumb, A.N.; Kowalski, A.; Rabasa-Lhoret, R.; McCrimmon, R.J.; Hume, C.; et al. Exercise management in type 1 diabetes: A consensus statement. *Lancet Diabetes Endocrinol.* **2017**, *5*, 377–390. [CrossRef]
7. Lennerz, B.S.; Barton, A.; Bernstein, R.K.; Dikeman, R.D.; Diulus, C.; Hallberg, S.; Rhodes, E.T.; Ebbeling, C.B.; Westman, E.C.; Yancy, W.S.; et al. Management of Type 1 Diabetes With a Very Low–Carbohydrate Diet. *Pediatrics* **2018**, *141*, e20173349. [CrossRef]
8. Pickup, J.C. Insulin-Pump Therapy for Type 1 Diabetes Mellitus. *N. Engl. J. Med.* **2012**, *366*, 1616–1624. [CrossRef]
9. Zinman, B. Newer insulin analogs: Advances in basal insulin replacement. *Diabetes Obes. Metab.* **2013**, *15*, 6–10. [CrossRef]

10. Vardi, M.; Jacobson, E.; Nini, A.; Bitterman, H. Intermediate acting versus long acting insulin for type 1 diabetes mellitus. *Cochrane Database Syst. Rev.* **2008**, *2008*, CD006297. [CrossRef]

11. Brawerman, G.; Thompson, P.J. Beta Cell Therapies for Preventing Type 1 Diabetes: From Bench to Bedside. *Biomolecules* **2020**, *10*, 1681. [CrossRef] [PubMed]

12. Bourgeois, S.; Sawatani, T.; Van Mulders, A.; De Leu, N.; Heremans, Y.; Heimberg, H.; Cnop, M.; Staels, W. Towards a Functional Cure for Diabetes Using Stem Cell-Derived Beta Cells: Are We There Yet? *Cells* **2021**, *10*, 191. [CrossRef] [PubMed]

13. Barra, J.M.; Kozlovskaya, V.; Kharlampieva, E.; Tse, H.M. Localized Immunosuppression With Tannic Acid Encapsulation Delays Islet Allograft and Autoimmune-Mediated Rejection. *Diabetes* **2020**, *69*, 1948. [CrossRef] [PubMed]

14. Waldmann, T.A. Immunotherapy: Past, present and future. *Nat. Med.* **2003**, *9*, 269–277. [CrossRef]

15. Till, S.J.; Francis, J.N.; Nouri-Aria, K.; Durham, S.R. Mechanisms of immunotherapy. *J. Allergy Clin. Immunol.* **2004**, *113*, 1025–1034. [CrossRef]

16. Kopan, C.; Tucker, T.; Alexander, M.; Mohammadi, M.R.; Pone, E.J.; Lakey, J.R.T. Approaches in Immunotherapy, Regenerative Medicine, and Bioengineering for Type 1 Diabetes. *Front. Immunol.* **2018**, *9*, 1354. [CrossRef]

17. Garciafigueroa, Y.; Trucco, M.; Giannoukakis, N. A brief glimpse over the horizon for type 1 diabetes nanotherapeutics. *Clin. Immunol.* **2015**, *160*, 36–45. [CrossRef]

18. Tang, Z.; Kong, N.; Ouyang, J.; Feng, C.; Kim, N.Y.; Ji, X.; Wang, C.; Farokhzad, O.C.; Zhang, H.; Tao, W. Phosphorus Science-Oriented Design and Synthesis of Multifunctional Nanomaterials for Biomedical Applications. *Matter* **2020**, *2*, 297–322. [CrossRef]

19. Makvandi, P.; Wang, C.-y.; Zare, E.N.; Borzacchiello, A.; Niu, L.-n.; Tay, F.R. Metal-Based Nanomaterials in Biomedical Applications: Antimicrobial Activity and Cytotoxicity Aspects. *Adv. Funct. Mater.* **2020**, *30*, 1910021. [CrossRef]

20. Duan, C.; Liang, L.; Li, L.; Zhang, R.; Xu, Z.P. Recent progress in upconversion luminescence nanomaterials for biomedical applications. *J. Mater. Chem. B* **2018**, *6*, 192–209. [CrossRef]

21. Liu, Y.; Shi, J. Antioxidative nanomaterials and biomedical applications. *Nano Today* **2019**, *27*, 146–177. [CrossRef]

22. Gim, S.; Zhu, Y.; Seeberger, P.H.; Delbianco, M. Carbohydrate-based nanomaterials for biomedical applications. *WIREs Nanomed. Nanobiotechnol.* **2019**, *11*, e1558. [CrossRef] [PubMed]

23. Liu, W.-L.; Zou, M.-Z.; Qin, S.-Y.; Cheng, Y.-J.; Ma, Y.-H.; Sun, Y.-X.; Zhang, X.-Z. Recent Advances of Cell Membrane-Coated Nanomaterials for Biomedical Applications. *Adv. Funct. Mater.* **2020**, *30*, 2003559. [CrossRef]

24. Stabler, C.L.; Li, Y.; Stewart, J.M.; Keselowsky, B.G. Engineering immunomodulatory biomaterials for type 1 diabetes. *Nat. Rev. Mater.* **2019**, *4*, 429–450. [CrossRef]

25. Thondawada, M.; Wadhwani, A.D.; Palanisamy, D.S.; Rathore, H.S.; Gupta, R.C.; Chintamaneni, P.K.; Samanta, M.K.; Dubala, A.; Varma, S.; Krishnamurthy, P.T.; et al. An effective treatment approach of DPP-IV inhibitor encapsulated polymeric nanoparticles conjugated with anti-CD-4 mAb for type 1 diabetes. *Drug Dev. Ind. Pharm.* **2018**, *44*, 1120–1129. [CrossRef]

26. Li, X.; Zhen, M.; Zhou, C.; Deng, R.; Yu, T.; Wu, Y.; Shu, C.; Wang, C.; Bai, C. Gadofullerene Nanoparticles Reverse Dysfunctions of Pancreas and Improve Hepatic Insulin Resistance for Type 2 Diabetes Mellitus Treatment. *ACS Nano* **2019**, *13*, 8597–8608. [CrossRef]

27. Volpatti, L.R.; Matranga, M.A.; Cortinas, A.B.; Delcassian, D.; Daniel, K.B.; Langer, R.; Anderson, D.G. Glucose-Responsive Nanoparticles for Rapid and Extended Self-Regulated Insulin Delivery. *ACS Nano* **2020**, *14*, 488–497. [CrossRef]

28. Song, M.; Wang, H.; Chen, K.; Zhang, S.; Yu, L.; Elshazly, E.H.; Ke, L.; Gong, R. Oral insulin delivery by carboxymethyl-β-cyclodextrin-grafted chitosan nanoparticles for improving diabetic treatment. *Artif. Cells Nanomed. Biotechnol.* **2018**, *46*, S774–S782. [CrossRef]

29. Wang, Y.; Fan, Y.; Zhang, M.; Zhou, W.; Chai, Z.; Wang, H.; Sun, C.; Huang, F. Glycopolypeptide Nanocarriers Based on Dynamic Covalent Bonds for Glucose Dual-Responsiveness and Self-Regulated Release of Insulin in Diabetic Rats. *Biomacromolecules* **2020**, *21*, 1507–1515. [CrossRef]

30. Neef, T.; Miller, S.D. Tolerogenic Nanoparticles to Treat Islet Autoimmunity. *Curr. Diabetes Rep.* **2017**, *17*, 84. [CrossRef]

31. Cappellano, G.; Comi, C.; Chiocchetti, A.; Dianzani, U. Exploiting PLGA-Based Biocompatible Nanoparticles for Next-Generation Tolerogenic Vaccines against Autoimmune Disease. *Int. J. Mol. Sci.* **2019**, *20*, 204. [CrossRef] [PubMed]

32. Moorman, C.D.; Sohn, S.J.; Phee, H. Emerging Therapeutics for Immune Tolerance: Tolerogenic Vaccines, T cell Therapy, and IL-2 Therapy. *Front. Immunol.* **2021**, *12*, 12. [CrossRef] [PubMed]

33. Yeste, A.; Takenaka, M.C.; Mascanfroni, I.D.; Nadeau, M.; Kenison, J.E.; Patel, B.; Tukpah, A.-M.; Babon, J.A.B.; DeNicola, M.; Kent, S.C.; et al. Tolerogenic nanoparticles inhibit T cell–mediated autoimmunity through SOCS2. *Sci. Signal.* **2016**, *9*, ra61. [CrossRef] [PubMed]

34. Rabinowe, S.L.; Eisenbarth, G.S. Type I Diabetes Mellitus: A Chronic Autoimmune Disease? *Pediatr. Clin. N. Am.* **1984**, *31*, 531–543. [CrossRef]

35. Mein, C.; Esposito, L.; Dunn, M.G.; Johnson, G.C.L.; Timms, A.E.; Goy, J.V.; Smith, A.N.; Sebag-Montefiore, L.; Merriman, M.E.; Wilson, A.J.; et al. A search for type 1 diabetes susceptibility genes in families from the United Kingdom. *Nat. Genet.* **1998**, *19*, 297–300. [CrossRef]

36. Aly, T.A.; Ide, A.; Jahromi, M.M.; Barker, J.M.; Fernando, M.S.; Babu, S.R.; Yu, L.; Miao, D.; Erlich, H.A.; Fain, P.R.; et al. Extreme genetic risk for type 1A diabetes. *Proc. Natl. Acad. Sci. USA* **2006**, *103*, 14074. [CrossRef]

37. Kavvoura, F.K.; Ioannidis, J.P.A. CTLA-4 Gene Polymorphisms and Susceptibility to Type 1 Diabetes Mellitus: A HuGE Review and Meta-Analysis. *Am. J. Epidemiol.* **2005**, *162*, 3–16. [CrossRef]

38. Redondo, M.J.; Geyer, S.; Steck, A.K.; Sharp, S.; Wentworth, J.M.; Weedon, M.N.; Antinozzi, P.; Sosenko, J.; Atkinson, M.; Pugliese, A.; et al. A Type 1 Diabetes Genetic Risk Score Predicts Progression of Islet Autoimmunity and Development of Type 1 Diabetes in Individuals at Risk. *Diabetes Care* **2018**, *41*, 1887. [CrossRef]
39. Paschou, S.A.; Papadopoulou-Marketou, N.; Chrousos, G.P.; Kanaka-Gantenbein, C. On type 1 diabetes mellitus pathogenesis. *Endocr. Connect.* **2018**, *7*, R38–R46. [CrossRef]
40. Ablamunits, V.; Elias, D.; Cohen, I.R. The pathogenicity of islet-infiltrating lymphocytes in the non-obese diabetic (NOD) mouse. *Clin. Exp. Immunol.* **1999**, *115*, 260–267. [CrossRef]
41. Graham, K.L.; Krishnamurthy, B.; Fynch, S.; Ayala-Perez, R.; Slattery, R.M.; Santamaria, P.; Thomas, H.E.; Kay, T.W.H. Intra-islet proliferation of cytotoxic T lymphocytes contributes to insulitis progression. *Eur. J. Immunol.* **2012**, *42*, 1717–1722. [CrossRef] [PubMed]
42. Usher-Smith, J.A.; Thompson, M.J.; Sharp, S.J.; Walter, F.M. Factors associated with the presence of diabetic ketoacidosis at diagnosis of diabetes in children and young adults: A systematic review. *BMJ* **2011**, *343*, d4092. [CrossRef] [PubMed]
43. Michels, A.W.; Eisenbarth, G.S. Immunologic endocrine disorders. *J. Allergy Clin. Immunol.* **2010**, *125*, S226–S237. [CrossRef] [PubMed]
44. Smigoc Schweiger, D.; Mendez, A.; Kunilo Jamnik, S.; Bratanic, N.; Bratina, N.; Battelino, T.; Brecelj, J.; Vidan-Jeras, B. High-risk genotypes HLA-DR3-DQ2/DR3-DQ2 and DR3-DQ2/DR4-DQ8 in co-occurrence of type 1 diabetes and celiac disease. *Autoimmunity* **2016**, *49*, 240–247. [CrossRef] [PubMed]
45. Devendra, D.; Eisenbarth, G.S. Immunologic endocrine disorders. *J. Allergy Clin. Immunol.* **2003**, *111*, S624–S636. [CrossRef] [PubMed]
46. Abraham, R.S.; Kudva, Y.C.; Wilson, S.B.; Strominger, J.L.; David, C.S. Co-expression of HLA DR3 and DQ8 results in the development of spontaneous insulitis and loss of tolerance to GAD65 in transgenic mice. *Diabetes* **2000**, *49*, 548. [CrossRef]
47. Campbell-Thompson, M.L.; Atkinson, M.A.; Butler, A.E.; Chapman, N.M.; Frisk, G.; Gianani, R.; Giepmans, B.N.; von Herrath, M.G.; Hyöty, H.; Kay, T.W.; et al. The diagnosis of insulitis in human type 1 diabetes. *Diabetologia* **2013**, *56*, 2541–2543. [CrossRef]
48. Culina, S.; Lalanne, A.I.; Afonso, G.; Cerosaletti, K.; Pinto, S.; Sebastiani, G.; Kuranda, K.; Nigi, L.; Eugster, A.; Østerbye, T.; et al. Islet-reactive CD8$^+$ T cell frequencies in the pancreas, but not in blood, distinguish type 1 diabetic patients from healthy donors. *Sci. Immunol.* **2018**, *3*, eaao4013. [CrossRef]
49. Keenan, H.A.; Sun, J.K.; Levine, J.; Doria, A.; Aiello, L.P.; Eisenbarth, G.; Bonner-Weir, S.; King, G.L. Residual Insulin Production and Pancreatic β-Cell Turnover After 50 Years of Diabetes: Joslin Medalist Study. *Diabetes* **2010**, *59*, 2846–2853. [CrossRef]
50. Mobasseri, M.; Shirmohammadi, M.; Amiri, T.; Vahed, N.; Hosseini Fard, H.; Ghojazadeh, M. Prevalence and incidence of type 1 diabetes in the world: A systematic review and meta-analysis. *Health Promot. Perspect.* **2020**, *10*, 98–115. [CrossRef]
51. Svoren, B.M.; Volkening, L.K.; Wood, J.R.; Laffel, L.M.B. Significant Vitamin D Deficiency in Youth with Type 1 Diabetes Mellitus. *J. Pediatr.* **2009**, *154*, 132–134. [CrossRef]
52. Cooper, J.D.; Smyth, D.J.; Walker, N.M.; Stevens, H.; Burren, O.S.; Wallace, C.; Greissl, C.; Ramos-Lopez, E.; Hyppönen, E.; Dunger, D.B.; et al. Inherited Variation in Vitamin D Genes Is Associated With Predisposition to Autoimmune Disease Type 1 Diabetes. *Diabetes* **2011**, *60*, 1624. [CrossRef] [PubMed]
53. Atkinson, M.A.; Eisenbarth, G.S.; Michels, A.W. Type 1 diabetes. *Lancet* **2014**, *383*, 69–82. [CrossRef]
54. Bhalla, A.K.; Amento, E.P.; Serog, B.; Glimcher, L.H. 1,25-Dihydroxyvitamin D3 inhibits antigen-induced T cell activation. *J. Immunol.* **1984**, *133*, 1748. [PubMed]
55. Kamen, D.; Aranow, C. Vitamin D in systemic lupus erythematosus. *Curr. Opin. Rheumatol.* **2008**, *20*, 532–537. [CrossRef] [PubMed]
56. Thordarson, H.; Søvik, O. Dead in Bed Syndrome in Young Diabetic Patients in Norway. *Diabet. Med.* **1995**, *12*, 782–787. [CrossRef]
57. Brazg, R.L.; Bailey, T.S.; Garg, S.; Buckingham, B.A.; Slover, R.H.; Klonoff, D.C.; Nguyen, X.; Shin, J.; Welsh, J.B.; Lee, S.W. The ASPIRE Study: Design and Methods of an In-Clinic Crossover Trial on the Efficacy of Automatic Insulin Pump Suspension in Exercise-Induced Hypoglycemia. *J. Diabetes Sci. Technol.* **2011**, *5*, 1466–1471. [CrossRef]
58. Garg, S.K.; Voelmle, M.K.; Beatson, C.R.; Miller, H.A.; Crew, L.B.; Freson, B.J.; Hazenfield, R.M. Use of Continuous Glucose Monitoring in Subjects with Type 1 Diabetes on Multiple Daily Injections Versus Continuous Subcutaneous Insulin Infusion Therapy. *Diabetes Care* **2011**, *34*, 574. [CrossRef]
59. Franek, E.; Haluzík, M.; Canecki Varžić, S.; Sargin, M.; Macura, S.; Zacho, J.; Christiansen, J.S. Twice-daily insulin degludec/insulin aspart provides superior fasting plasma glucose control and a reduced rate of hypoglycaemia compared with biphasic insulin aspart 30 in insulin-naïve adults with Type 2 diabetes. *Diabet. Med.* **2016**, *33*, 497–505. [CrossRef]
60. Willner, S.; Whittemore, R.; Keene, D. "Life or death": Experiences of insulin insecurity among adults with type 1 diabetes in the United States. *SSM-Popul. Health* **2020**, *11*, 100624. [CrossRef]
61. Smith, M.J.; Simmons, K.M.; Cambier, J.C. B cells in type 1 diabetes mellitus and diabetic kidney disease. *Nat. Rev. Nephrol.* **2017**, *13*, 712–720. [CrossRef] [PubMed]
62. Wong, F.S.; Wen, L.; Tang, M.; Ramanathan, M.; Visintin, I.; Daugherty, J.; Hannum, L.G.; Janeway, C.A.; Shlomchik, M.J. Investigation of the Role of B-Cells in Type 1 Diabetes in the NOD Mouse. *Diabetes* **2004**, *53*, 2581. [CrossRef] [PubMed]
63. Hull, C.M.; Peakman, M.; Tree, T.I.M. Regulatory T cell dysfunction in type 1 diabetes: What's broken and how can we fix it? *Diabetologia* **2017**, *60*, 1839–1850. [CrossRef] [PubMed]

64. Lieberman, S.M.; DiLorenzo, T.P. A comprehensive guide to antibody and T-cell responses in type 1 diabetes. *Tissue Antigens* **2003**, *62*, 359–377. [CrossRef]
65. Jacobsen, L.M.; Posgai, A.; Seay, H.R.; Haller, M.J.; Brusko, T.M. T Cell Receptor Profiling in Type 1 Diabetes. *Curr. Diabetes Rep.* **2017**, *17*, 118. [CrossRef]
66. Gitelman, S.E.; Bluestone, J.A. Regulatory T cell therapy for type 1 diabetes: May the force be with you. *J. Autoimmun.* **2016**, *71*, 78–87. [CrossRef]
67. Ahmed, S.; Cerosaletti, K.; James, E.; Long, S.A.; Mannering, S.; Speake, C.; Nakayama, M.; Tree, T.; Roep, B.O.; Herold, K.C.; et al. Standardizing T-Cell Biomarkers in Type 1 Diabetes: Challenges and Recent Advances. *Diabetes* **2019**, *68*, 1366–1379. [CrossRef]
68. Pesenacker, A.M.; Chen, V.; Gillies, J.; Speake, C.; Marwaha, A.K.; Sun, A.; Chow, S.; Tan, R.; Elliott, T.; Dutz, J.P.; et al. Treg gene signatures predict and measure type 1 diabetes trajectory. *JCI Insight* **2019**, *4*, 4. [CrossRef]
69. Pugliese, A. Autoreactive T cells in type 1 diabetes. *J. Clin. Investig.* **2017**, *127*, 2881–2891. [CrossRef]
70. Couture, A.; Garnier, A.; Docagne, F.; Boyer, O.; Vivien, D.; Le-Mauff, B.; Latouche, J.-B.; Toutirais, O. HLA-Class II Artificial Antigen Presenting Cells in CD4+ T Cell-Based Immunotherapy. *Front. Immunol.* **2019**, *10*, 10. [CrossRef]
71. Eichmann, M.; Baptista, R.; Ellis, R.J.; Heck, S.; Peakman, M.; Beam, C.A. Costimulation Blockade Disrupts CD4$^+$ T Cell Memory Pathways and Uncouples Their Link to Decline in β-Cell Function in Type 1 Diabetes. *J. Immunol.* **2020**, *204*, 3129–3138. [CrossRef] [PubMed]
72. Long, S.A.; Thorpe, J.; Herold, K.C.; Ehlers, M.; Sanda, S.; Lim, N.; Linsley, P.S.; Nepom, G.T.; Harris, K.M. Remodeling T cell compartments during anti-CD3 immunotherapy of type 1 diabetes. *Cell. Immunol.* **2017**, *319*, 3–9. [CrossRef] [PubMed]
73. Wong, F.S.; Wen, L. A predictive CD8+ T cell phenotype for T1DM progression. *Nat. Rev. Endocrinol.* **2020**, *16*, 198–199. [CrossRef] [PubMed]
74. Wiedeman, A.E.; Muir, V.S.; Rosasco, M.G.; DeBerg, H.A.; Presnell, S.; Haas, B.; Dufort, M.J.; Speake, C.; Greenbaum, C.J.; Serti, E.; et al. Autoreactive CD8+ T cell exhaustion distinguishes subjects with slow type 1 diabetes progression. *J. Clin. Investig.* **2020**, *130*, 480–490. [CrossRef]
75. Ohkura, N.; Sakaguchi, S. Transcriptional and epigenetic basis of Treg cell development and function: Its genetic anomalies or variations in autoimmune diseases. *Cell Res.* **2020**, *30*, 465–474. [CrossRef]
76. Yeh, W.-I.; Seay, H.R.; Newby, B.; Posgai, A.L.; Moniz, F.B.; Michels, A.; Mathews, C.E.; Bluestone, J.A.; Brusko, T.M. Avidity and Bystander Suppressive Capacity of Human Regulatory T Cells Expressing De Novo Autoreactive T-Cell Receptors in Type 1 Diabetes. *Front. Immunol.* **2017**, *8*, 1313. [CrossRef]
77. Pellegrino, M.; Crinò, A.; Rosado, M.M.; Fierabracci, A. Identification and functional characterization of CD8+ T regulatory cells in type 1 diabetes patients. *PLoS ONE* **2019**, *14*, e0210839. [CrossRef]
78. Serr, I.; Fürst, R.W.; Achenbach, P.; Scherm, M.G.; Gökmen, F.; Haupt, F.; Sedlmeier, E.-M.; Knopff, A.; Shultz, L.; Willis, R.A.; et al. Type 1 diabetes vaccine candidates promote human Foxp3+Treg induction in humanized mice. *Nat. Commun.* **2016**, *7*, 10991. [CrossRef]
79. Tenspolde, M.; Zimmermann, K.; Weber, L.C.; Hapke, M.; Lieber, M.; Dywicki, J.; Frenzel, A.; Hust, M.; Galla, M.; Buitrago-Molina, L.E.; et al. Regulatory T cells engineered with a novel insulin-specific chimeric antigen receptor as a candidate immunotherapy for type 1 diabetes. *J. Autoimmun.* **2019**, *103*, 102289. [CrossRef]
80. Serra, P.; Santamaria, P. Nanoparticle-based approaches to immune tolerance for the treatment of autoimmune diseases. *Eur. J. Immunol.* **2018**, *48*, 751–756. [CrossRef]
81. Baekkeskov, S.; Hubbell, J.A.; Phelps, E.A. Bioengineering strategies for inducing tolerance in autoimmune diabetes. *Adv. Drug Deliv. Rev.* **2017**, *114*, 256–265. [CrossRef] [PubMed]
82. Pan, W.; Zheng, X.; Chen, G.; Su, L.; Luo, S.; Wang, W.; Ye, S.; Weng, J.; Min, Y. Nanotechnology's application in Type 1 diabetes. *WIREs Nanomed. Nanobiotechnol.* **2020**, *12*, e1645. [CrossRef] [PubMed]
83. Du, J.; Zhang, Y.S.; Hobson, D.; Hydbring, P. Nanoparticles for immune system targeting. *Drug Discov. Today* **2017**, *22*, 1295–1301. [CrossRef] [PubMed]
84. Chen, N.; Kroger, C.J.; Tisch, R.M.; Bachelder, E.M.; Ainslie, K.M. Prevention of Type 1 Diabetes with Acetalated Dextran Microparticles Containing Rapamycin and Pancreatic Peptide P31. *Adv. Health Mater.* **2018**, *7*, 1800341. [CrossRef]
85. Bergot, A.-S.; Buckle, I.; Cikaluru, S.; Naranjo, J.L.; Wright, C.M.; Zheng, G.; Talekar, M.; Hamilton-Williams, E.E.; Thomas, R. Regulatory T Cells Induced by Single-Peptide Liposome Immunotherapy Suppress Islet-Specific T Cell Responses to Multiple Antigens and Protect from Autoimmune Diabetes. *J. Immunol.* **2020**, *204*, 1787–1797. [CrossRef]
86. Jamison, B.L.; Neef, T.; Goodspeed, A.; Bradley, B.; Baker, R.L.; Miller, S.D.; Haskins, K. Nanoparticles Containing an Insulin–ChgA Hybrid Peptide Protect from Transfer of Autoimmune Diabetes by Shifting the Balance between Effector T Cells and Regulatory T Cells. *J. Immunol.* **2019**, *203*, 48–57. [CrossRef]
87. Aboelnazar, S.; Ghoneim, H.; Shalaby, T.; Bahgat, E.; Moaaz, M. Effect of low dose IL-2 loaded chitosan nanoparticles on natural killer and regulatory T cell expression in experimentally induced autoimmune type 1 diabetes mellitus. *Cent. Eur. J. Immunol.* **2020**, *45*, 382–392. [CrossRef]
88. Greenbaum, C.J.; Serti, E.; Lambert, K.; Weiner, L.J.; Kanaparthi, S.; Lord, S.; Gitelman, S.E.; Wilson, D.M.; Gaglia, J.L.; Griffin, K.J.; et al. IL-6 receptor blockade does not slow β cell loss in new-onset type 1 diabetes. *JCI Insight* **2021**, *6*, 6. [CrossRef]

89. Xu, X.; Bian, L.; Shen, M.; Li, X.; Zhu, J.; Chen, S.; Xiao, L.; Zhang, Q.; Chen, H.; Xu, K.; et al. Multipeptide-coupled nanoparticles induce tolerance in 'humanised' HLA-transgenic mice and inhibit diabetogenic CD8+ T cell responses in type 1 diabetes. *Diabetologia* **2017**, *60*, 2418–2431. [CrossRef]

90. Hamad, A.R.A.; Ahmed, R.; Donner, T.; Fousteri, G. B cell-targeted immunotherapy for type 1 diabetes: What can make it work? *Discov. Med.* **2016**, *21*, 213–219.

91. Johnson, P.; Glennie, M. The mechanisms of action of rituximab in the elimination of tumor cells. *Semin. Oncol.* **2003**, *30*, 3–8. [CrossRef] [PubMed]

92. Huda, R. New Approaches to Targeting B Cells for Myasthenia Gravis Therapy. *Front. Immunol.* **2020**, *11*, 240. [CrossRef] [PubMed]

93. Hawker, K. B cells as a target of immune modulation. *Ann. Indian Acad. Neurol.* **2009**, *12*, 221–225. [CrossRef] [PubMed]

94. Atkinson, M.A.; Roep, B.O.; Posgai, A.; Wheeler, D.C.S.; Peakman, M. The challenge of modulating β-cell autoimmunity in type 1 diabetes. *Lancet Diabetes Endocrinol.* **2019**, *7*, 52–64. [CrossRef]

95. Sarikonda, G.; Sachithanantham, S.; Manenkova, Y.; Kupfer, T.; Posgai, A.; Wasserfall, C.; Bernstein, P.; Straub, L.; Pagni, P.P.; Schneider, D.; et al. Transient B-Cell Depletion with Anti-CD20 in Combination with Proinsulin DNA Vaccine or Oral Insulin: Immunologic Effects and Efficacy in NOD Mice. *PLoS ONE* **2013**, *8*, e54712. [CrossRef]

96. Martucci, N.M.; Migliaccio, N.; Ruggiero, I.; Albano, F.; Calì, G.; Romano, S.; Terracciano, M.; Rea, I.; Arcari, P.; Lamberti, A. Nanoparticle-based strategy for personalized B-cell lymphoma therapy. *Int. J. Nanomed.* **2016**, *11*, 6089–6101. [CrossRef]

97. Stensland, Z.C.; Cambier, J.C.; Smith, M.J. Therapeutic Targeting of Autoreactive B Cells: Why, How, and When? *Biomedicines* **2021**, *9*, 83. [CrossRef]

98. Woldu, M.A.; Lenjisa, J.L. Nanoparticles and the new era in diabetes management. *Int. J. Basic Clin. Pharmacol.* **2017**, *3*, 8. [CrossRef]

99. Temchura, V.V.; Kozlova, D.; Sokolova, V.; Überla, K.; Epple, M. Targeting and activation of antigen-specific B-cells by calcium phosphate nanoparticles loaded with protein antigen. *Biomaterials* **2014**, *35*, 6098–6105. [CrossRef]

100. Hong, S.; Zhang, Z.; Liu, H.; Tian, M.; Zhu, X.; Zhang, Z.; Wang, W.; Zhou, X.; Zhang, F.; Ge, Q.; et al. B Cells Are the Dominant Antigen-Presenting Cells that Activate Naive CD4+ T Cells upon Immunization with a Virus-Derived Nanoparticle Antigen. *Immunity* **2018**, *49*, 695–708.e4. [CrossRef]

101. Selim, M.E.; Abd-Elhakim, Y.M.; Al-Ayadhi, L.Y. Pancreatic Response to Gold Nanoparticles Includes Decrease of Oxidative Stress and Inflammation In Autistic Diabetic Model. *Cell. Physiol. Biochem.* **2015**, *35*, 586–600. [CrossRef] [PubMed]

102. Fousteri, G.; Ippolito, E.; Ahmed, R.; Rahim, A.; Hamad, A. Beta-cell Specific Autoantibodies: Are they Just an Indicator of Type 1 Diabetes? *Curr. Diabetes Rev.* **2017**, *13*, 322–329. [CrossRef] [PubMed]

103. Li, M.; Fan, Y.-N.; Chen, Z.-Y.; Luo, Y.-L.; Wang, Y.-C.; Lian, Z.-X.; Xu, C.-F.; Wang, J. Optimized nanoparticle-mediated delivery of CRISPR-Cas9 system for B cell intervention. *Nano Res.* **2018**, *11*, 6270–6282. [CrossRef]

104. Chu, V.T.; Graf, R.; Wirtz, T.; Weber, T.; Favret, J.; Li, X.; Petsch, K.; Tran, N.T.; Sieweke, M.H.; Berek, C.; et al. Efficient CRISPR-mediated mutagenesis in primary immune cells using CrispRGold and a C57BL/6 Cas9 transgenic mouse line. *Proc. Natl. Acad. Sci. USA* **2016**, *113*, 12514–12519. [CrossRef] [PubMed]

105. Walunas, T.L.; Lenschow, D.J.; Bakker, C.Y.; Linsley, P.S.; Freeman, G.J.; Green, J.M.; Thompson, C.B.; Bluestone, J.A. CTLA-4 can function as a negative regulator of T cell activation. *Immunity* **1994**, *1*, 405–413. [CrossRef]

106. Tai, X.; Van Laethem, F.; Pobezinsky, L.; Guinter, T.; Sharrow, S.O.; Adams, A.; Granger, L.; Kruhlak, M.; Lindsten, T.; Thompson, C.B.; et al. Basis of CTLA-4 function in regulatory and conventional CD4+ T cells. *Blood* **2012**, *119*, 5155–5163. [CrossRef]

107. Taraban, V.Y.; Rowley, T.F.; O'Brien, L.; Chan, H.T.C.; Haswell, L.E.; Green, M.H.A.; Tutt, A.L.; Glennie, M.J.; Al-Shamkhani, A. Expression and costimulatory effects of the TNF receptor superfamily members CD134 (OX40) and CD137 (4-1BB), and their role in the generation of anti-tumor immune responses. *Eur. J. Immunol.* **2002**, *32*, 3617–3627. [CrossRef]

108. Zhang, X.; Kang, Y.; Wang, J.; Yan, J.; Chen, Q.; Cheng, H.; Huang, P.; Gu, Z. Engineered PD-L1-Expressing Platelets Reverse New-Onset Type 1 Diabetes. *Adv. Mater.* **2020**, *32*, 1907692. [CrossRef]

109. Andersen, M.H. The Balance Players of the Adaptive Immune System. *Cancer Res.* **2018**, *78*, 1379. [CrossRef]

110. Liang, S.C.; Latchman, Y.E.; Buhlmann, J.E.; Tomczak, M.F.; Horwitz, B.H.; Freeman, G.J.; Sharpe, A.H. Regulation of PD-1, PD-L1, and PD-L2 expression during normal and autoimmune responses. *Eur. J. Immunol.* **2003**, *33*, 2706–2716. [CrossRef]

111. Latchman, Y.; Wood, C.R.; Chernova, T.; Chaudhary, D.; Borde, M.; Chernova, I.; Iwai, Y.; Long, A.J.; Brown, J.A.; Nunes, R.; et al. PD-L2 is a second ligand for PD-1 and inhibits T cell activation. *Nat. Immunol.* **2001**, *2*, 261–268. [CrossRef]

112. Egen, J.G.; Kuhns, M.S.; Allison, J.P. CTLA-4: New insights into its biological function and use in tumor immunotherapy. *Nat. Immunol.* **2002**, *3*, 611–618. [CrossRef] [PubMed]

113. Dong, H.; Strome, S.E.; Salomao, D.R.; Tamura, H.; Hirano, F.; Flies, D.B.; Roche, P.C.; Lu, J.; Zhu, G.; Tamada, K.; et al. Tumor-associated B7-H1 promotes T-cell apoptosis: A potential mechanism of immune evasion. *Nat. Med.* **2002**, *8*, 793–800. [CrossRef] [PubMed]

114. Ahmadzadeh, M.; Johnson, L.A.; Heemskerk, B.; Wunderlich, J.R.; Dudley, M.E.; White, D.E.; Rosenberg, S.A. Tumor antigen–specific CD8 T cells infiltrating the tumor express high levels of PD-1 and are functionally impaired. *Blood* **2009**, *114*, 1537–1544. [CrossRef]

115. Han, Y.; Liu, D.; Li, L. PD-1/PD-L1 pathway: Current researches in cancer. *Am. J. Cancer Res.* **2020**, *10*, 727–742. [PubMed]

116. Reiss, K.A.; Forde, P.M.; Brahmer, J.R. Harnessing the power of the immune system via blockade of PD-1 and PD-L1: A promising new anticancer strategy. *Immunotherapy* **2014**, *6*, 459–475. [CrossRef] [PubMed]

117. Brahmer, J.R.; Tykodi, S.S.; Chow, L.Q.M.; Hwu, W.-J.; Topalian, S.L.; Hwu, P.; Drake, C.G.; Camacho, L.H.; Kauh, J.; Odunsi, K.; et al. Safety and Activity of Anti–PD-L1 Antibody in Patients with Advanced Cancer. *N. Engl. J. Med.* **2012**, *366*, 2455–2465. [CrossRef]

118. Kapke, J.; Shaheen, Z.; Kilari, D.; Knudson, P.; Wong, S. Immune Checkpoint Inhibitor-Associated Type 1 Diabetes Mellitus: Case Series, Review of the Literature, and Optimal Management. *Case Rep. Oncol.* **2017**, *10*, 897–909. [CrossRef]

119. Samoa, R.A.; Lee, H.S.; Kil, S.H.; Roep, B.O. Anti–PD-1 Therapy–Associated Type 1 Diabetes in a Pediatric Patient With Relapsed Classical Hodgkin Lymphoma. *Diabetes Care* **2020**, *43*, 2293. [CrossRef]

120. Colli, M.L.; Hill, J.L.E.; Marroquí, L.; Chaffey, J.; Dos Santos, R.S.; Leete, P.; Coomans de Brachène, A.; Paula, F.M.M.; Op de Beeck, A.; Castela, A.; et al. PDL1 is expressed in the islets of people with type 1 diabetes and is up-regulated by interferons-α and-γ via IRF1 induction. *EBioMedicine* **2018**, *36*, 367–375. [CrossRef]

121. Falcone, M.; Fousteri, G. Role of the PD-1/PD-L1 Dyad in the Maintenance of Pancreatic Immune Tolerance for Prevention of Type 1 Diabetes. *Front. Endocrinol.* **2020**, *11*, 11. [CrossRef] [PubMed]

122. Wang, C.-J.; Chou, F.-C.; Chu, C.-H.; Wu, J.-C.; Lin, S.-H.; Chang, D.-M.; Sytwu, H.-K. Protective Role of Programmed Death 1 Ligand 1 (PD-L1)in Nonobese Diabetic Mice. *Diabetes* **2008**, *57*, 1861. [CrossRef] [PubMed]

123. Yoshihara, E.; O'Connor, C.; Gasser, E.; Wei, Z.; Oh, T.G.; Tseng, T.W.; Wang, D.; Cayabyab, F.; Dai, Y.; Yu, R.T.; et al. Immune-evasive human islet-like organoids ameliorate diabetes. *Nature* **2020**, *586*, 606–611. [CrossRef] [PubMed]

124. Wang, P.; Liu, Q.; Zhao, H.; Bishop, J.O.; Zhou, G.; Olson, L.K.; Moore, A. miR-216a-targeting theranostic nanoparticles promote proliferation of insulin-secreting cells in type 1 diabetes animal model. *Sci. Rep.* **2020**, *10*, 5302. [CrossRef]

125. Qian, L.; Liu, F.; Chu, Y.; Zhai, Q.; Wei, X.; Shao, J.; Li, R.; Xu, Q.; Yu, L.; Liu, B.; et al. MicroRNA-200c Nanoparticles Sensitized Gastric Cancer Cells to Radiotherapy by Regulating PD-L1 Expression and EMT. *Cancer Manag. Res.* **2020**, *12*, 12215–12223. [CrossRef]

126. Wang, X.; Li, J.; Dong, K.; Lin, F.; Long, M.; Ouyang, Y.; Wei, J.; Chen, X.; Weng, Y.; He, T.; et al. Tumor suppressor miR-34a targets PD-L1 and functions as a potential immunotherapeutic target in acute myeloid leukemia. *Cell. Signal.* **2015**, *27*, 443–452. [CrossRef]

127. Chen, L.; Gibbons, D.L.; Goswami, S.; Cortez, M.A.; Ahn, Y.-H.; Byers, L.A.; Zhang, X.; Yi, X.; Dwyer, D.; Lin, W.; et al. Metastasis is regulated via microRNA-200/ZEB1 axis control of tumour cell PD-L1 expression and intratumoral immunosuppression. *Nat. Commun.* **2014**, *5*, 5241. [CrossRef]

128. Wei, J.; Nduom, E.; Kong, L.-Y.; Wang, F.; Xu, S.; Gabrusiewicz, K.; Alum, A.; Fuller, G.; Calin, G.; Heimberger, A.B. miR-138 exerts anti-glioma efficacy by targeting immune checkpoints. *J. ImmunoTherapy Cancer* **2013**, *1*, P177. [CrossRef]

129. Lee, S.W.L.; Paoletti, C.; Campisi, M.; Osaki, T.; Adriani, G.; Kamm, R.D.; Mattu, C.; Chiono, V. MicroRNA delivery through nanoparticles. *J. Control. Release* **2019**, *313*, 80–95. [CrossRef]

130. Heninger, A.-K.; Eugster, A.; Kuehn, D.; Buettner, F.; Kuhn, M.; Lindner, A.; Dietz, S.; Jergens, S.; Wilhelm, C.; Beyerlein, A.; et al. A divergent population of autoantigen-responsive CD4$^+$ T cells in infants prior to β cell autoimmunity. *Sci. Transl. Med.* **2017**, *9*, eaaf8848. [CrossRef]

131. Chang, W.; Wang, J. Exosomes and Their Noncoding RNA Cargo Are Emerging as New Modulators for Diabetes Mellitus. *Cells* **2019**, *8*, 853. [CrossRef]

132. Lu, T.X.; Rothenberg, M.E. MicroRNA. *J. Allergy Clin. Immunol.* **2018**, *141*, 1202–1207. [CrossRef] [PubMed]

133. Wang, Z.; Xie, Z.; Lu, Q.; Chang, C.; Zhou, Z. Beyond Genetics: What Causes Type 1 Diabetes. *Clin. Rev. Allergy Immunol.* **2017**, *52*, 273–286. [CrossRef]

134. Santos, A.S.; Cunha Neto, E.; Fukui, R.T.; Ferreira, L.R.P.; Silva, M.E.R. Increased Expression of Circulating microRNA 101-3p in Type 1 Diabetes Patients: New Insights Into miRNA-Regulated Pathophysiological Pathways for Type 1 Diabetes. *Front. Immunol.* **2019**, *10*, 1637. [CrossRef] [PubMed]

135. Scherm, M.G.; Daniel, C. miRNA-Mediated Immune Regulation in Islet Autoimmunity and Type 1 Diabetes. *Front. Endocrinol.* **2020**, *11*, 11. [CrossRef] [PubMed]

136. Scherm, M.G.; Serr, I.; Zahm, A.M.; Schug, J.; Bellusci, S.; Manfredini, R.; Salb, V.K.; Gerlach, K.; Weigmann, B.; Ziegler, A.-G.; et al. miRNA142-3p targets Tet2 and impairs Treg differentiation and stability in models of type 1 diabetes. *Nat. Commun.* **2019**, *10*, 5697. [CrossRef] [PubMed]

137. Krishnan, P.; Syed, F.; Jiyun Kang, N.; Mirmira, R.G.; Evans-Molina, C. Profiling of RNAs from Human Islet-Derived Exosomes in a Model of Type 1 Diabetes. *Int. J. Mol. Sci.* **2019**, *20*, 5903. [CrossRef]

138. Dini, S.; Zakeri, M.; Ebrahimpour, S.; Dehghanian, F.; Esmaeili, A. Quercetin-conjugated superparamagnetic iron oxide nanoparticles modulate glucose metabolism-related genes and miR-29 family in the hippocampus of diabetic rats. *Sci. Rep.* **2021**, *11*, 8618. [CrossRef]

139. Serr, I.; Scherm, M.G.; Zahm, A.M.; Schug, J.; Flynn, V.K.; Hippich, M.; Kälin, S.; Becker, M.; Achenbach, P.; Nikolaev, A.; et al. A miRNA181a/NFAT5 axis links impaired T cell tolerance induction with autoimmune type 1 diabetes. *Sci. Transl. Med.* **2018**, *10*, eaag1782. [CrossRef]

140. Vasu, S.; Kumano, K.; Darden, C.M.; Rahman, I.; Lawrence, M.C.; Naziruddin, B. MicroRNA Signatures as Future Biomarkers for Diagnosis of Diabetes States. *Cells* **2019**, *8*, 1533. [CrossRef]

141. Satake, E.; Pezzolesi, M.G.; Md Dom, Z.I.; Smiles, A.M.; Niewczas, M.A.; Krolewski, A.S. Circulating miRNA Profiles Associated With Hyperglycemia in Patients With Type 1 Diabetes. *Diabetes* **2018**, *67*, 1013. [CrossRef] [PubMed]
142. Åkerman, L.; Casas, R.; Ludvigsson, J.; Tavira, B.; Skoglund, C. Serum miRNA levels are related to glucose homeostasis and islet autoantibodies in children with high risk for type 1 diabetes. *PLoS ONE* **2018**, *13*, e0191067. [CrossRef]
143. Bertoccini, L.; Sentinelli, F.; Incani, M.; Bailetti, D.; Cimini, F.A.; Barchetta, I.; Lenzi, A.; Cavallo, M.G.; Cossu, E.; Baroni, M.G. Circulating miRNA-375 levels are increased in autoantibodies-positive first-degree relatives of type 1 diabetes patients. *Acta Diabetol.* **2019**, *56*, 707–710. [CrossRef] [PubMed]
144. Lakhter, A.J.; Pratt, R.E.; Moore, R.E.; Doucette, K.K.; Maier, B.F.; DiMeglio, L.A.; Sims, E.K. Beta cell extracellular vesicle miR-21-5p cargo is increased in response to inflammatory cytokines and serves as a biomarker of type 1 diabetes. *Diabetologia* **2018**, *61*, 1124–1134. [CrossRef]
145. Chhabra, P.; Brayman, K.L. Stem Cell Therapy to Cure Type 1 Diabetes: From Hype to Hope. *Stem Cells Transl. Med.* **2013**, *2*, 328–336. [CrossRef] [PubMed]
146. Fiorina, P.; Jurewicz, M.; Augello, A.; Vergani, A.; Dada, S.; La Rosa, S.; Selig, M.; Godwin, J.; Law, K.; Placidi, C.; et al. Immunomodulatory Function of Bone Marrow-Derived Mesenchymal Stem Cells in Experimental Autoimmune Type 1 Diabetes. *J. Immunol.* **2009**, *183*, 993. [CrossRef]
147. Carlsson, P.-O.; Schwarcz, E.; Korsgren, O.; Le Blanc, K. Preserved β-Cell Function in Type 1 Diabetes by Mesenchymal Stromal Cells. *Diabetes* **2015**, *64*, 587–592. [CrossRef] [PubMed]
148. Voltarelli, J.C.; Couri, C.E.B.; Stracieri, A.B.P.L.; Oliveira, M.C.; Moraes, D.A.; Pieroni, F.; Coutinho, M.; Malmegrim, K.C.R.; Foss-Freitas, M.C.; Simões, B.P.; et al. Autologous Nonmyeloablative Hematopoietic Stem Cell Transplantation in Newly Diagnosed Type 1 Diabetes Mellitus. *JAMA* **2007**, *297*, 1568–1576. [CrossRef] [PubMed]
149. D'Addio, F.; Valderrama Vasquez, A.; Ben Nasr, M.; Franek, E.; Zhu, D.; Li, L.; Ning, G.; Snarski, E.; Fiorina, P. Autologous Nonmyeloablative Hematopoietic Stem Cell Transplantation in New-Onset Type 1 Diabetes: A Multicenter Analysis. *Diabetes* **2014**, *63*, 3041. [CrossRef]
150. Cai, J.; Wu, Z.; Xu, X.; Liao, L.; Chen, J.; Huang, L.; Wu, W.; Luo, F.; Wu, C.; Pugliese, A.; et al. Umbilical Cord Mesenchymal Stromal Cell With Autologous Bone Marrow Cell Transplantation in Established Type 1 Diabetes: A Pilot Randomized Controlled Open-Label Clinical Study to Assess Safety and Impact on Insulin Secretion. *Diabetes Care* **2016**, *39*, 149. [CrossRef]
151. Lennard, A.L. Science, medicine, and the future: Stem cell transplantation. *BMJ* **2000**, *321*, 1331. [CrossRef] [PubMed]
152. Nguyen, P.K.; Riegler, J.; Wu, J.C. Stem Cell Imaging: From Bench to Bedside. *Cell Stem Cell* **2014**, *14*, 431–444. [CrossRef] [PubMed]
153. Yang, X.; Tian, D.-C.; He, W.; Lv, W.; Fan, J.; Li, H.; Jin, W.-N.; Meng, X. Cellular and molecular imaging for stem cell tracking in neurological diseases. *Stroke Vasc. Neurol.* **2021**, *6*, 121–127. [CrossRef] [PubMed]
154. Gu, E.; Chen, W.-Y.; Gu, J.; Burridge, P.; Wu, J.C. Molecular Imaging of Stem Cells: Tracking Survival, Biodistribution, Tumorigenicity, and Immunogenicity. *Theranostics* **2012**, *2*, 335–345. [CrossRef]
155. Tabbara, I.A.; Zimmerman, K.; Morgan, C.; Nahleh, Z. Allogeneic Hematopoietic Stem Cell Transplantation: Complications and Results. *Arch. Intern. Med.* **2002**, *162*, 1558–1566. [CrossRef] [PubMed]
156. Lijkwan, M.A.; Bos, E.J.; Wu, J.C.; Robbins, R.C. Role of Molecular Imaging in Stem Cell Therapy for Myocardial Restoration. *Trends Cardiovasc. Med.* **2010**, *20*, 183–188. [CrossRef] [PubMed]
157. Ping, W.; Anna, M. Molecular Imaging of Stem Cell Transplantation for Neurodegenerative Diseases. *Curr. Pharm. Des.* **2012**, *18*, 4426–4440. [CrossRef]
158. Abbas, F.; Wu, J.C.; Gambhir, S.S.; Rodriguez-Porcel, M. Molecular Imaging of Stem Cells. *Stem J.* **2019**, *1*, 27–46. [CrossRef]
159. Pei, Z.; Zeng, J.; Song, Y.; Gao, Y.; Wu, R.; Chen, Y.; Li, F.; Li, W.; Zhou, H.; Yang, Y. In vivo imaging to monitor differentiation and therapeutic effects of transplanted mesenchymal stem cells in myocardial infarction. *Sci. Rep.* **2017**, *7*, 6296. [CrossRef]
160. Zheng, B.; von See, M.P.; Yu, E.; Gunel, B.; Lu, K.; Vazin, T.; Schaffer, D.V.; Goodwill, P.W.; Conolly, S.M. Quantitative Magnetic Particle Imaging Monitors the Transplantation, Biodistribution, and Clearance of Stem Cells In Vivo. *Theranostics* **2016**, *6*, 291–301. [CrossRef]
161. Ahrens, E.T.; Bulte, J.W.M. Tracking immune cells in vivo using magnetic resonance imaging. *Nat. Rev. Immunol.* **2013**, *13*, 755–763. [CrossRef] [PubMed]
162. Hu, S.-L.; Lu, P.-G.; Zhang, L.-J.; Li, F.; Chen, Z.; Wu, N.; Meng, H.; Lin, J.-K.; Feng, H. In vivo magnetic resonance imaging tracking of SPIO-labeled human umbilical cord mesenchymal stem cells. *J. Cell. Biochem.* **2012**, *113*, 1005–1012. [CrossRef] [PubMed]
163. Zheng, L.; Wang, Y.; Yang, B.; Zhang, B.; Wu, Y. Islet Transplantation Imaging in vivo. *Diabetes Metab. Syndr. Obes. Targets Ther.* **2020**, *13*, 3301–3311. [CrossRef] [PubMed]
164. Hayat, H.; Sun, A.; Hayat, H.; Liu, S.; Talebloo, N.; Pinger, C.; Bishop, J.O.; Gudi, M.; Dwan, B.F.; Ma, X.; et al. Artificial Intelligence Analysis of Magnetic Particle Imaging for Islet Transplantation in a Mouse Model. *Mol. Imaging Biol.* **2021**, *23*, 18–29. [CrossRef]
165. Laurent, S.; Saei, A.A.; Behzadi, S.; Panahifar, A.; Mahmoudi, M. Superparamagnetic iron oxide nanoparticles for delivery of therapeutic agents: Opportunities and challenges. *Expert Opin. Drug Deliv.* **2014**, *11*, 1449–1470. [CrossRef]

166. Blocki, A.; Beyer, S.; Dewavrin, J.-Y.; Goralczyk, A.; Wang, Y.; Peh, P.; Ng, M.; Moonshi, S.S.; Vuddagiri, S.; Raghunath, M.; et al. Microcapsules engineered to support mesenchymal stem cell (MSC) survival and proliferation enable long-term retention of MSCs in infarcted myocardium. *Biomaterials* **2015**, *53*, 12–24. [CrossRef]
167. Guldris, N.; Argibay, B.; Gallo, J.; Iglesias-Rey, R.; Carbó-Argibay, E.; Kolen'ko, Y.V.; Campos, F.; Sobrino, T.; Salonen, L.M.; Bañobre-López, M.; et al. Magnetite Nanoparticles for Stem Cell Labeling with High Efficiency and Long-Term in Vivo Tracking. *Bioconjugate Chem.* **2017**, *28*, 362–370. [CrossRef]
168. Wang, P.; Yoo, B.; Yang, J.; Zhang, X.; Ross, A.; Pantazopoulos, P.; Dai, G.; Moore, A. GLP-1R–Targeting Magnetic Nanoparticles for Pancreatic Islet Imaging. *Diabetes* **2014**, *63*, 1465–1474. [CrossRef]
169. Yao, Y.; Li, Y.; Ma, G.; Liu, N.; Ju, S.; Jin, J.; Chen, Z.; Shen, C.; Teng, G. In Vivo Magnetic Resonance Imaging of Injected Endothelial Progenitor Cells after Myocardial Infarction in Rats. *Mol. Imaging Biol.* **2011**, *13*, 303–313. [CrossRef]